Lecture Notes
in Control and Information Sciences 219

Editor: M. Thoma

Springer-Verlag London Ltd.

Marie-Odile Berger, Rachid Deriche, Isabelle Herlin,
Jérome Jaffré and Jean-Michel Morel (Eds)

ICAOS '96
12th International Conference on Analysis and Optimization of Systems
Images, Wavelets and PDEs

Paris, June 26-28, 1996

INRIA

CEREMADE

Springer

Editors

Dr. Marie-Odile Berger
INRIA Lorraine, B.P. 101, 54602 Villers les Nancy Cedex, France

Dr. Rachid Deriche
INRIA Sophia, 2004 Route des Lucioles, B.P. 93
06902 Sophia Antipolis Cedex, France

Dr. Isabelle Herlin
Dr. Jérome Jaffré
INRIA Rocquencourt, Domaine de Voluceau, B.P. 105
78153 Le Chesnay Cedex, France

Professor Jean-Michel Morel
CEREMADE, Université Paris IX-Dauphine, 75775 Paris Cedex 16, France

Honorary Editors

ISBN 978-3-540-76076-4 ISBN 978-3-540-40945-8 (eBook)
DOI 10.1007/978-3-540-40945-8

British Library Cataloguing in Publication Data
A catalogue record for this book is available from the British Library

Originally published by Springer-Verlag London Limited in 1996

Typesetting: Camera ready by author

69/3830-543210 Printed on acid-free paper

12th INTERNATIONAL CONFERENCE ON ANALYSIS AND OPTIMIZATION OF SYSTEMS

IMAGES, WAVELETS AND PDE'S

Paris, June 26-28, 1996

Organized by INRIA and CEREMADE

Foreword

ICAOS'96 is the twelfth in the series of the International Conferences on Analysis and Optimization of Systems organized by INRIA and the third in the new format, that is devoted to a specific domain.

After one conference on Infinite Dimensional Systems, one on Discrete Event Systems, this time the conference is dedicated to Images, Wavelets and PDE's. This volume will show how these three fields interact to produce exciting research.

We would like to thank the CEREMADE for coorganizing this conference and the Ministère de l'Education Nationale, de l'Enseignement Supérieur et de la Recherche for providing the conference rooms.

We also would like to extend our gratitude to:

- all authors who submitted papers to this conference,
- the reviewers who accepted the responsability of selecting papers,
- the chairpersons for chairing the sessions of the conference with efficiency,
- the members of the scientific committee for their guidance in organizing this conference,
- the members of the organizing committee for all their work in preparing the conference and in assembling these proceedings.

Finally we wish to express our appreciation to the Department of Public Relations of INRIA, especially Annick Theis-Viémont and Claudie Thenault, who worked so hard to make this conference so pleasant, and to Professor Manfred Thoma and to Springer Verlag for accepting to publish this book of proccedings, like the previous ones, in the series Lecture Notes in Control and Information Sciences.

Alain Bensoussan Jacques-Louis Lions

Presentation

This books contains the proceedings of ICAOS'96, the 12th International Conference on Analysis and Optimization of Systems. This conference was coorganized by INRIA and the CEREMADE and was dedicated to Images, Wavelets and PDE's.

The aim of the conference was to discuss the impact on image analysis of recent mathematical developments in multiscale analysis, partial differential equations, variational methods ...

ICAOS'96 provided a forum for image processing researchers and mathematicians to interact and to exchange their technical knowledge and experience, theoretical or practical, in this emerging and exciting domain.

The selected papers have been organized according to the following sessions, each session corresponding to a chapter of this book:

1. Active Contours
2. Image Enhancement and Restoration, Scale-Spaces
3. Wavelets
4. Image Segmentation
5. Image Restoration
6. Coding
7. Applications

In addition to these papers, this book includes three papers corresponding to invited lectures by L. Alvarez (Universidad de las Palmas, Spain), B.M. ter Haar Romeny (University Hospital Utrecht, The Netherlands) and L. Yaroslavsky (Tel Aviv University, Israël).

Contents

Invited lectures

Active Contours

Image Enhancement and Restoration, Scale-Space

Wavelets

Image Segmentation

Image Restoration

Coding

Applications

Author Index

Invited Lectures

Images and PDE's

Luis ALVAREZ

Departamento de Informática y Sistemas
Universidad de Las Palmas
Campus de Tafira
35017 Las Palmas, España
Email: luis@amihp710.dis.ulpgc.es

1. Introduction.

In this paper I will present an overview of some results, concerning EDP's and Image Processing, obtained by the author of this job with the collaboration of the following researchers: J.Esclarín, F.Guichard, P.L.Lions, L.Mazorra, F.Morales, J.M.Morel and F.Santana that belong to the CEREMADE laboratory of the University of Paris IX and the Computer Science Department of the University of Las Palmas (The Canary Islands). This job is the result of several research projects together with both universities and has given raise to different publications such as: [ALM92], [AGLM92c], [AGLM92b], [AGLM92a], [AGLM93], [AM94a], [AM94c], [AM94d], [AM94b], [AE94].

I am going to focus my attention in the results and experiences that I have accumulated in the last years. So I am not go to present a general overview of the field which is impossible in a single paper. In particular, I will not quote, a great number of interesting aspects and results introduced for other authors.

I think that the theoretical background of the results that I'm going to present here are very well exposed in the above references. Rather than to repeat in this paper the theoretical presentation of the mathematical results, I am going to try, in this paper, to give an unified overview of the numerical analysis aspects associated to these equations. Of course, I would like to say again, that this presentation is based in my experiences and in my particular point of view.

We will study some applications of partial differential equations to image processing. For us, an image is a function $f : \Re^n \to \Re$ where in general $n = 2$, though in some cases $n = 1$ or $n = 3$ also. The starting point in most applications of PDE's to image processing is to take the original image f as the initial datum of a PDE of the form:

$$\begin{array}{rcl} u_t(t,x) & = & F(D^2u, Du, u, t)(t,x) \\ u(0,x) & = & f(x) \end{array}$$

where $u(t,x)$ represents the output image at the scale t, u_t represents the derivative of u with respect to t, D^2u is the matrix of spatial second derivatives, and Du represents the gradient. The output image $u(t,x)$ represents a modified version of the original image. The main advantages of the above PDE's formulation is that it ensures a number of good properties as the pyramidal structure of the analysis and on the other hand it allows the design of very fast and robust algorithms using the numerical techniques associated to the theory of Numerical Analysis of PDE's.

The Shape of the function $F()$ is adapted following the intended applications. Roughly Speaking, there exist two main research domains in the applications of PDE's to image processing which are:

1. Shape Analysis
   (a) Analysis and computation of invariants
   (b) Corner and Multiple Junctions Analysis and Detection.
   (c) Etc..
2. Image Enhancement
   (a) Denoising
   (b) Deblurring.
   (c) Etc..

In the Shape Analysis Applications the attention is focussed in the invariant of the equation under geometric transformations as similarity or affinity, or morphological transformations, as an histogram transformation, etc.. In [AGLM93] the reader can find a deep mathematical analysis of this invariant approach. Moreover in the Shape Analysis Applications we extract the information of the original image following its evolution across the different scales t. So we need to follow carefully such evolution.

In the Image Enhancement Applications, we are interested in the final result of the analysis, so typically we expect that the improved version of the original image is obtained for large values of the parameter t and the evolution of the image across the scale is not really important.

From a numerical point of view there is a big difference between Shape Analysis Applications and Image Enhancement Applications. Following my own experience, I would say that in Shape Analysis Applications we need to use explicit numerical squemes with a small time step Δt, so it is very expensive from a computational point of view. In the case of Image Enhancement, implicit schemes with a rather big time step Δt work very well.

From an historical point of view, the starting point of my studies in the field of computer vision was the introduction of directional diffusion operator instead of isotropic operators in order to avoid the blurring effect produced by the convolution with linear kernels. This idea appeared in [ALM92] (in collaboration with P.L Lions and J.M. Morel). This idea has been used in Shape Analysis Applications as well as in Image Enhancement Applications and we will devote a great attention in this paper to the numerical analysis of these directional operators.

The paper is organized as follow: To fit ideas, In section 2 we present some aspect of the numerical analysis of the well known heat equation (which is equivalent to the convolution with a gaussian).We will present explicit and implicit squemes. In section 3 we present some ideas about the gradient discretization. In section 4, we present the main ideas about the discretization of the directional diffusion operators and in section 4 we present some application to a number of PDE's.

2. Heat Equation Discretization.

We will begin with heat equation in dimension 1. $u_t = u_{xx}$. Developing the function in point (t, x) in Taylor series, we have:

1. $u(t+k, x) = u(t, x) + ku_t(t, x) + O(k^2)$
2. $u(t-k, x) = u(t, x) - ku_t(t, x) + O(k^2)$
3. $u(t, x+h) = u(t, x) + hu_x(t, x) + \frac{h^2}{2}u_{xx}(t, x) + \frac{h^3}{6}u_{xxx}(t, x) + O(h^4)$
4. $u(t, x-h) = u(t, x) - hu_x(t, x) + \frac{h^2}{2}u_{xx}(t, x) - \frac{h^3}{6}u_{xxx}(t, x) + O(h^4)$

If we introduce notation $u_j^n \cong u(nk, hj)$ with $n \geq 0$ and $j \in Z$, and use first expression to approximate the derivative in relation to t we will obtain:

$$u_t - u_{xx} = \frac{u_j^{n+1} - u_j^n}{k} - \frac{u_{j+1}^n - 2u_j^n + u_{j-1}^n}{h^2} + O(k) + O(h^2)$$

Similarly, if we use second expression to approximate the derivative with respect to t we will have:

$$u_t - u_{xx} = \frac{u_j^n - u_j^{n-1}}{k} - \frac{u_{j+1}^n - 2u_j^n + u_{j-1}^n}{h^2} + O(k) + O(h^2)$$

The first approximation method determines an explicit discretization scheme, and the second one determines an implicit discretization method. In order to study the behavior of these numerical schemes, Fourier analysis techniques are used. The first scheme can be written as

$$u_j^{n+1} = (1 - 2\lambda)u_j^n + \lambda(u_{j+1}^n + u_{j-1}^n)$$

where $\lambda = \frac{k}{h^2}$. If we multiply the equality by e^{-iwhj} and add in j we will obtain

$$\sum_{j=-\infty}^{\infty} u_j^{n+1} e^{-iwhj} = (1 - 2\lambda(1 - \cos(wh))) \sum_{j=-\infty}^{\infty} u_j^n e^{-iwhj}$$

From where we deduce that at iteration n, Fourier transformation of the sign corresponds to the Fourier transformation of initial signal by $(1-2\lambda(1-cos(wh)))^n$. Thus, for the numerical scheme to be stable it has to be verified that:

$$| 1-2\lambda(1-cos(wh) | \leq 1$$

From where we obtain the well known stability condition:

$$\lambda \leq \frac{1}{2}$$

Let us realize that apparently, there are two degrees of freedom in the scheme , h and k, But actually there is just one: $\lambda=\frac{k}{h^2}$. That is to say, the only important thing is this balance. That is due to scale invariance of the equation. That is to say, equation is invariant by the transformation: $(t, x, u) \rightarrow (\gamma^2 t, \gamma x, u)$. Or in other words, if we simultaneously change spacial increment h by γh and k by $\gamma^2 k$, we will exactly obtain the same scheme. This remark is important because when we normally study a discrete image or signal, we do not have information about the real distance h between two pixels. Thus is important that our algorithms are independent of the distance.

Let us now consider the implicit discretization of the same equation. That discretization takes us to the following scheme:

$$(1+2\lambda)u_j^n - \lambda(u_{j+1}^n + u_{j-1}^n) = u_j^{n-1}$$

from which, using the same technique we have

$$\sum_{j=-\infty}^{\infty} u_j^n e^{-iwhj} = \frac{1}{1+2\lambda(1-\cos(wh))} \sum_{j=-\infty}^{\infty} u_j^{n-1} e^{-iwhj}$$

Thus, as $0 < \frac{1}{1+2\lambda(1-\cos(wh))} \leq 1$ for any value of λ, we get that the scheme is unconditionally stable. A deeper study of this scheme has been realized in [AM94a] where, particularly, the next result is obtained that:

$$\frac{1}{1+2\lambda(1-\cos(wh))} = \frac{1}{\sqrt{1+4\lambda}} \sum_{j=-\infty}^{\infty} \nu^{|j|} e^{-iwhj}$$

$$\frac{1}{1+2\lambda(1-\cos(wh))} = \frac{\lambda}{\nu(1-\nu e^{-iwh})(1-\nu e^{+iwh})}$$

where $\nu = \frac{1+2\lambda-\sqrt{1+4\lambda}}{2\lambda} < 1$ is a positive constant. From the first equality we can deduce that

$$u_j^n = \frac{1}{\sqrt{1+4\lambda}} \sum_{l=-\infty}^{\infty} \nu^{|l-j|} u_l^{n-1}$$

the second one allow us to decompose the calculus of u_j^n as the composition of a casual filter and an anti-casual filter:

1. $u_j^{n+\frac{1}{3}} = u_j^n + \nu u_{j-1}^{n+\frac{1}{3}}$ for all $j \in Z$

2. $u_j^{n+\frac{2}{3}} = u_j^{n+\frac{1}{3}} + \nu u_{j+1}^{n+\frac{1}{3}}$ for all $j \in Z$

3. $u_j^{n+1} = \frac{\nu}{\lambda} u_j^{n+\frac{2}{3}}$ for all $j \in Z$.

That determines a very fast and efficient recursive scheme of implementation.

Let us now see how we can discretisize the heat equation in dimension 2, using a neighborhood of 3×3 that indicates the number of neighbors that participate in the discretization. Using the Taylor series development we obtain next expressions: (where $u = u(t,x,y)$, $u_x = u_x(t,x,y)$, etc..).

1. $u(t+k,x,y) = u + ku_t + O(k^2)$

2. $u(t-k,x,y) = u - ku_t + O(k^2)$

3. $u(t,x+h,y) = u + hu_x + \frac{h^2}{2}u_{xx} + O(h^3)$

4. $u(t,x-h,y) = u - hu_x + \frac{h^2}{2}u_{xx} + O(h^3)$

5. $u(t,x,y+l) = u + lu_y + \frac{l^2}{2}u_{yy} + O(l^3)$

6. $u(t,x,y-l) = u - lu_y + \frac{l^2}{2}u_{yy} + O(l^3)$

7. $u(t,x+h,y+l) = u+hu_x+lu_y+\frac{1}{2}(h^2u_{xx}+2hlu_{xy}+l^2u_{yy})+O((h^2+l^2)^{\frac{3}{2}})$

8. $u(t,x-h,y-l) = u-hu_x-lu_y+\frac{1}{2}(h^2u_{xx}+2hlu_{xy}+l^2u_{yy})+O((h^2+l^2)^{\frac{3}{2}})$

9. $u(t,x+h,y-l) = u+hu_x-lu_y+\frac{1}{2}(h^2u_{xx}-2hlu_{xy}+l^2u_{yy})+O((h^2+l^2)^{\frac{3}{2}})$

10. $u(t,x-h,y+l) = u-hu_x+lu_y+\frac{1}{2}(h^2u_{xx}-2hlu_{xy}+l^2u_{yy})+O((h^2+l^2)^{\frac{3}{2}})$

If we introduce the notation $u_{i,j}^n \cong u(kn,hi,lj)$ and $\Delta u = u_{xx} + u_{yy}$ we get that in order to discretisize operator Δu in a neighborhood of 3×3 points, different schemes can be used. To simplify, we will suppose that $l = h$. So we obtain then next general scheme:

$$\Delta u = \gamma \frac{u_{i+1,j+1}^n + u_{i-1,j+1}^n + u_{i-1,j+1}^n + u_{i+1,j-1}^n - 4u_{i,j}^n}{2h^2} +$$
$$+(1-\gamma)\frac{u_{i+1,j}^n + u_{i-1,j}^n + u_{i,j+1}^n + u_{i,j-1}^n - 4u_{i,j}^n}{h^2} + O(h)$$

where γ is a free parameter to choose. The election of such parameter γ will be done based on that discretization respects rotation invariances of as Δu much as possible, for that, we will consider an image in which in a neighborhood of the point (hi_0, hj_0) it has next values:

1	1	1
0	0	0
0	0	0

If we calculate Δu in the central point by means of previous formula we get:

$$\Delta u(hi_0, hj_0) = \gamma \frac{2}{2h^2} + (1-\gamma)\frac{1}{h^2}$$

Nos then, if we rotate the initial image just $\frac{\pi}{4}$ around point (hi_0, hj_0) we will obtain

1	1	0
1	0	0
0	0	0

If we again calculate Δu at the same point:

$$\Delta u(hi_0, hj_0) = \gamma \frac{1}{2h^2} + (1-\gamma)\frac{2}{h^2}$$

Thus, if we want both values of Δu to coincide, we will have to choose $\gamma = \frac{2}{3}$. Speaking in signal theory terms, the calculus of Δu will take us to convolution the image with the next mask:

$$\frac{1}{h^2}\begin{array}{|c|c|c|}\hline \frac{1}{3} & \frac{1}{3} & \frac{1}{3} \\ \hline \frac{1}{3} & -\frac{8}{3} & \frac{1}{3} \\ \hline \frac{1}{3} & \frac{1}{3} & \frac{1}{3} \\ \hline \end{array}$$

If we denote $\Delta u^n_{i,j} \cong \Delta u(kn, hi, hj)$ we can discretisize just as we did in dimension 1 the heat equation obtaining explicit or implicit schemes, according to the value of the derivative in relation of regard to time forward or backward.

2.1. Gradient discretization.

Following the previous Taylor series development we will have the next expression for the gradient:

$$(u^n_{i,j})_x = (1-\gamma)\frac{(u^n_{i+1,j} - u^n_{i-1,j})}{2h} +$$
$$+\gamma \frac{(u^n_{i+1,j+1} - u^n_{i-1,j+1} + u^n_{i+1,j-1} - u^n_{i-1,j-1})}{4h}$$

$$(u^n_{i,j})_y = (1-\gamma)\frac{(u^n_{i,j+1} - u^n_{i,j-1})}{2h} +$$
$$+\gamma \frac{(u^n_{i+1,j+1} - u^n_{i+1,j-1} + u^n_{i-1,j+1} - u^n_{i-1,j-1})}{4h}$$

where γ is again a parameter to choose. Taking into account that the gradient norma is invariant to rotations, it is particularly, invariant to 45 degrees rotations, from where we deduce, using the same argument used for Δu, that $\gamma = 2 - \sqrt{2}$. So we are calculating u_x using the mask:

$$\frac{1}{4h}\begin{array}{|c|c|c|}\hline -(2-\sqrt{2}) & 0 & (2-\sqrt{2}) \\ \hline -2(\sqrt{2}-1) & 0 & 2(\sqrt{2}-1) \\ \hline -(2-\sqrt{2}) & 0 & (2-\sqrt{2}) \\ \hline \end{array}$$

and u_y using:

$$\frac{1}{4h}\begin{array}{|c|c|c|}\hline -(2-\sqrt{2}) & 2(\sqrt{2}-1) & -(2-\sqrt{2}) \\ \hline 0 & 0 & 0 \\ \hline (2-\sqrt{2}) & -2(\sqrt{2}-1) & (2-\sqrt{2}) \\ \hline \end{array}$$

A very interesting remark is that we can get to the same discretization of the gradient using an optimization method. Really, if we consider the linear form

$$p(x,y) = u^n_{i,j} + ax + by$$

and try to adjust a, b for that linear form to approximate the best that we can to the image in a neighborhood 3×3 of the point, that will lead as to a function of energy. If we consider as energy the function:

$$\begin{aligned} E(a,b) &= \left(u^n_{i+1,j} - u^n_{i,j} - ha\right)^2 + \left(u^n_{i-1,j} - u^n_{i,j} + ha\right)^2 + \\ &\left(u^n_{i,j+1} - u^n_{i,j} - hb\right)^2 + \left(u^n_{i,j-1} - u^n_{i,j} + hb\right)^2 + \\ &\tfrac{1}{\sqrt{2}}\left(u^n_{i+1,j+1} - u^n_{i,j} - ha - hb\right)^2 + \tfrac{1}{\sqrt{2}}\left(u^n_{i-1,j-1} - u^n_{i,j} + ha + hb\right)^2 \\ &+ \tfrac{1}{\sqrt{2}}\left(u^n_{i-1,j+1} - u^n_{i,j} - hb + ha\right)^2 + \tfrac{1}{\sqrt{2}}\left(u^n_{i+1,j-1} - u^n_{i,j} + hb - ha\right)^2 \end{aligned}$$

where diagonal elements have been modulated by the distance to the central point ($\sqrt{2}$). Then we can easily obtain that the values of a, b that optimize previous energy correspond exactly to the values of $(u^n_{i,j})_x$ and $(u^n_{i,j})_y$ previously calculated.

Of course, there are many other ways to discretisize the gradient. The way we have proposed has a very good symmetry degree. However it presents a weak point, and it is that in case of in a isolated discontinuity where the central point in the image has a value, and in other points it has another value, the gradient will be $(0,0)$ following this discretization, that can be harmful in certain applications. For example, in some models we have seen in previous section, the norm of the gradient is used to discriminate elements of low importance or noise. In these cases, for calculating the norm of the gradient it is more convenient to modify previous formulas by substituting $(u^n_{i+1,j} - u^n_{i-1,j})$ by $|u^n_{i+1,j} - u^n_{i,j}| + |u^n_{i,j} - u^n_{i-1,j}|$, $(u^n_{i+1,j+1} - u^n_{i-1,j+1})$ by $|u^n_{i+1,j+1} - u^n_{i,j+1}| + |u^n_{i,j+1} - u^n_{i-1,j+1}|$, etc.. That hence determines an approximation of $|(u^n_{i,j})_x|$ and $|(u^n_{i,j})_y|$, and to eliminate the absolute value of the gradient we will have to discriminate among 4 possibilities which are : $(|(u^n_{i,j})_x|, |(u^n_{i,j})_y|)$, $(-|(u^n_{i,j})_x|, |(u^n_{i,j})_y|)$, $(|(u^n_{i,j})_x|, -|(u^n_{i,j})_y|)$ and $(-|(u^n_{i,j})_x|, -|(u^n_{i,j})_y|)$. To do this, we can for example evaluate each option of the energy $E(a, b)$ previously described, and and keep the lower energy option.

3. Discretization of Directional Operators.

Let us firstly consider the linear operator

$$\Im(u) = a^2 u_{xx} + 2ab u_{xy} + b^2 u_{yy}$$

where

$$a^2 + b^2 = 1$$

$\Im(u)$ represents the second derivative of the function $u(t, x)$ in the direction (a, b). Normally, the discretization of a differential operator involves two stages. In first step, each partial differentiation u_{xx}, u_{xy}, u_{yy} is evaluated as an independent operator, and later on, its value is substituted in the differential operator expression that we want to discretisize, in this case $\Im(u)$. The approach we will use here is different. We are going to discretisize $\Im(u)$ globally as a unique differential operator. The advantage of doing it this way, es that we introduce specific properties of the operator $\Im(u)$ in the discretization that can not stated studying each partial differentiation separately. Using Taylor series development just as in the laplacian case we obtain:

$$\begin{aligned} \Im(u)_{i,j} &= \frac{1}{h^2}[(-4\lambda_0 u_{i,j} + \lambda_1(u_{i,j+1} + u_{i,j-1}) + \\ &\quad \lambda_2(u_{i+1,j} + u_{i,j+1}) + \lambda_3(u_{i+1,j-1} + u_{i+1,j-1}) + \\ &\quad + \lambda_4(u_{i+1,j+1} + u_{i+1,j+1}))] + E(h^2) \end{aligned} \tag{1}$$

where $\lambda_0, \lambda_1, \lambda_2, \lambda_3, \lambda_4$ must satisfy the following relationships:

$$\left\{ \begin{array}{l} \lambda_1 = 2\lambda_0 - a^2 \\ \lambda_2 = 2\lambda_0 - b^2 \\ \lambda_3 = -\lambda_0 + 0.5(-ab + a^2 + b^2) \\ \lambda_4 = -\lambda_0 + 0.5(+ab + a^2 + b^2) \end{array} \right. \tag{2}$$

resulting parameter λ_0 as a free parameter. Therefore the computation of the directional operator is equivalent to the convolution with the mask

$-\lambda_0 + 0.5(-ab + a^2 + b^2)$	$2\lambda_0 - a^2$	$-\lambda_0 + 0.5(+ab + a^2 + b^2)$
$2\lambda_0 - b^2$	$-4\lambda_0$	$2\lambda_0 - b^2$
$-\lambda_0 + 0.5(+ab + a^2 + b^2)$	$2\lambda_0 - a^2$	$-\lambda_0 + 0.5(-ab + a^2 + b^2)$

The choice of λ_0 is very important in order to obtain an accurate approximation of the second derivative $\Im(u)$. We notice that if we compute u_{xx}, u_{xy}, and u_{yy} separately and then we compute $\Im(u)$, in fact we introduce a particular choice of λ_0 given by $\lambda_0 = 0.5$. this choice does not produce accurate results. For instance in the case that $(a,b) = \left(\frac{1}{\sqrt{2}}, \frac{1}{\sqrt{2}}\right)$ we obtain that the associated mask is:

$-\frac{1}{4}$	$\frac{1}{2}$	$\frac{1}{4}$
$\frac{1}{2}$	-2	$\frac{1}{2}$
$\frac{1}{4}$	$\frac{1}{2}$	$-\frac{1}{4}$

Obviously, it is not a good representation of the second derivative in the direction of $\left(\frac{1}{\sqrt{2}}, \frac{1}{\sqrt{2}}\right)$ because we expect that in such direction λ_1, λ_2 and λ_4 be zero. To avoid this kind of behavior we have proposed, in previous papers to take λ_0 as

$$\lambda_0 = \begin{cases} \frac{2b^2+a^2-|ab|}{4} & if \quad |b| \geq |a| \\ \frac{2a^2+b^2-|ab|}{4} & if \quad |b| < |a| \end{cases}$$

However there was not a real mathematical justification of such choice. Some months ago we have developed another approach to the choice of λ_0 which yields to a surprising result. The new approach is based in the analysis of the error in the Taylor Development of the differential operator. This error, given by $E(h^2)$ in expression (1) can be expressed by:

$$E(h^2) = O(h^4) + \left(\tfrac{1}{12}a^2 u_{xxxx} + \tfrac{1}{12}b^2 u_{yyyy} + \tfrac{1}{3}ab\left(u_{xyyy} + u_{xxxy}\right) + (.5 - \lambda_0)\, u_{xxyy}\right) h^2$$

So, Initially, it seems that $\lambda_0 = 0.5$ is the best choice in order to minimize the error. However this choice does not take into account that we deal with a directional operator. To interpret the error in term of directional operators we expand the error as a combination of four order derivatives of u in the direction of $\xi = (a,b)$ and its orthogonal direction $\mu = (-b,a)$. That is:

$$\tfrac{1}{12}a^2 u_{xxxx} + \tfrac{1}{12}b^2 u_{yyyy} + \tfrac{1}{3}ab\left(u_{xyyy} + u_{xxxy}\right) + (.5 - \lambda_0)\, u_{xxyy} = c_1 u_{\xi\xi\xi\xi} + c_2 u_{\xi\xi\xi\mu} + c_3 u_{\xi\xi\mu\mu} + c_4 u_{\xi\mu\mu\mu} + c_5 u_{\mu\mu\mu\mu}$$

With the assistance of Maple we can compute the coefficients c_1, c_2, c_3, c_4 and c_5. From the point of view of directional operators the big component of the error is given by the term $u_{\mu\mu\mu\mu}$. If we choose λ_0 in such way that c_5 be equal to zero, we obtain:

$$\lambda_0 = .25$$

This result is surprising because after a rather complex computational analysis (the coefficients $c_1, \ldots, c_5$ are really big expressions) we expected a rather complex choice of λ_0 but we find out a extremely simple choice. We have tested this choice of λ_0 and we have obtained better results that with previous choices.

4. Discretization of nonlinear-equations

Once the differential operator $\Im(u)$ is discretisized, we can approach the discretization of the equations that involve this kind of operators. We will begin with the mean curvature equation

4.1. Mean Curvature Equation.

The mean curvature equation is given by:

$$u_t = \frac{u_y^2 u_{xx} - 2u_x u_y u_{xy} + u_x^2 u_{yy}}{u_x^2 + u_y^2}$$

We notice that the right hand of the above expression corresponds to the second derivative of u in the direction $(-\frac{u_y}{\sqrt{u_x^2+u_y^2}}, \frac{u_x}{\sqrt{u_x^2+u_y^2}})$. So in each point we can compute its gradient and then we can use the ideas developed in the above section to compute the second derivative. To discretize this equation we choose an explicit scheme of the form:

$$\frac{u_{i,j}^{n+1} - u_{i,j}^n}{k} = \Im(u^n)_{i,j}$$

So, to obtain $u_{i,j}^{n+1}$ we apply the following mask to the image given by $u_{i,j}^n$:

$\frac{k}{h^2}\left(0.25 + \frac{(u_{i,j}^n)_{x,y}}{\lVert\nabla u_{i,j}^n\rVert_2^2}\right)$	$\frac{k}{h^2}\left(0.5 - \frac{(u_{i,j}^n)_y^2}{\lVert\nabla u_{i,j}^n\rVert_2^2}\right)$	$\frac{k}{h^2}\left(0.25 - \frac{(u_{i,j}^n)_{x,y}}{\lVert\nabla u_{i,j}^n\rVert_2^2}\right)$
$\left(0.5 - \frac{(u_{i,j}^n)_x^2}{\lVert\nabla u_{i,j}^n\rVert_2^2}\right)$	$\left(1 - \frac{k}{h^2}\right)$	$\frac{k}{h^2}\left(0.5 - \frac{(u_{i,j}^n)_x^2}{\lVert\nabla u_{i,j}^n\rVert_2^2}\right)$
$\frac{k}{h^2}\left(0.25 - \frac{(u_{i,j}^n)_{x,y}}{\lVert\nabla u_{i,j}^n\rVert_2^2}\right)$	$\frac{k}{h^2}\left(0.5 - \frac{(u_{i,j}^n)_y^2}{\lVert\nabla u_{i,j}^n\rVert_2^2}\right)$	$\frac{k}{h^2}\left(0.25 + \frac{(u_{i,j}^n)_{x,y}}{\lVert\nabla u_{i,j}^n\rVert_2^2}\right)$

where $\lVert \nabla u_{i,j}^n \rVert_2^2 = (u_{i,j}^n)_x^2 + (u_{i,j}^n)_y^2$ and $(u_{i,j}^n)_{x,y} = (u_{i,j}^n)_x (u_{i,j}^n)_y$

4.2. Affine morphological invariant equation

Another important equation that we have studied with great details is the affine morphological invariant equation given by:

$$u_t = \left(u_y^2 u_{xx} - 2u_x u_y u_{xy} + u_x^2 u_{yy}\right)^{\frac{1}{3}}$$

To discretize this equation we use again an explicit method like in the previous equation. :

$$\frac{u_{i,j}^{n+1} - u_{i,j}^n}{k} = \left(G(u^n)_{i,j}\right)^{\frac{1}{3}}$$

where $G(u^n)_{i,j}$ is given by the convolution of $u_{i,j}^n$ with the following mask:

$\frac{1}{4h^2}$

$\lVert \nabla u_{i,j}^n \rVert_2^2 + (u_{i,j}^n)_{x,y}$	$2 \lVert \nabla u_{i,j}^n \rVert_2^2 - (u_{i,j}^n)_y^2$	$\lVert \nabla u_{i,j}^n \rVert_2^2 - (u_{i,j}^n)_{x,y}$
$2 \lVert \nabla u_{i,j}^n \rVert_2^2 - (u_{i,j}^n)_x^2$	$- \lVert \nabla u_{i,j}^n \rVert_2^2$	$2 \lVert \nabla u_{i,j}^n \rVert_2^2 - (u_{i,j}^n)_x^2$
$\lVert \nabla u_{i,j}^n \rVert_2^2 - (u_{i,j}^n)_{x,y}$	$2 \lVert \nabla u_{i,j}^n \rVert_2^2 - (u_{i,j}^n)_y^2$	$\lVert \nabla u_{i,j}^n \rVert_2^2 + (u_{i,j}^n)_{x,y}$

We notice that, from a numerical point of view, in the case of the mean curvature motion equation, numerical instabilities appears when $(u^n_{i,j})^2_x + (u^n_{i,j})^2_y$ is near to zero. However this problem does not appear in the case of the affine morphological invariant equation.

4.3. A denoising model equation

Finally, we present a discretization of the denoising model equation introduced in [ALM92]

$$u_t = g(\| \nabla G * u \|) \frac{u_y^2 u_{xx} - 2u_x u_y u_{xy} + u_x^2 u_{yy}}{u_x^2 + u_y^2}$$

where function $G(.)$ represents a smoothing kernel and $g(.)$ is a decreasing function. The intended application of this equation is to improve the image quality removing the noise. We are interested in the long time behavior of the solution $u(t,x)$. So an implicit method is better adapted (and much faster) than an explicit method in order to solve the equation. We will use the notation $u^n_{i,j} \cong u(nk, ih, jh)$, $g^n_{i,j} \cong g(\| \nabla G * u^n_{i,j} \|)$ and $\Im_{i,j}(D^2u^m, Du^l) = (u^l_{i,j})^2_x (u^m_{i,j})_{xx} - 2(u^l_{i,j})_x (u^l_{i,j})_y (u^m_{i,j})_{xy} + (u^l_{i,j})^2_x (u^m_{i,j})_{yy}$.

We propose the following implicit scheme:

$$\frac{u^{n+1}_{i,j} - u^n_{i,j}}{k} = g^n_{i,j} \frac{\Im_{i,j}(D^2u^{n+1}, Du^n)}{(u^n_{i,j})^2_x + (u^n_{i,j})^2_y}$$

To calculate $u^{n+1}_{i,j}$ is necessary to solve a linear equation system. That system can be written as follows:

$$a^n_{i,j} u^{n+1}_{i,j} = u^n_{i,j} + b^n_{i,j}(u^{n+1}_{i-1,j} + u^{n+1}_{i+1,j}) + c^n_{i,j}(u^{n+1}_{i,j-1} + u^{n+1}_{i,j+1}) + \ldots..$$

To approximate the solution of that system it results quite effective the use of a recursive algorithm based in the Gauss-Seidel method. Let $2m$ be the number of iterations that will be used to go from $u^n_{i,j}$ to $u^{n+1}_{i,j}$. We denote by $u^{n+\frac{l}{2m}}_{i,j}$ the approximation of $u^{n+1}_{i,j}$ at step l. To calculate $u^{n+\frac{l+1}{2m}}_{i,j}$ from $u^{n+\frac{l}{2m}}_{i,j}$, if l is odd we will travel through the image from left to right and from down to up so that:

$$u^{n+\frac{l+1}{2m}}_{i,j} = \frac{u^n_{i,j} + b^n_{i,j}(u^{n+\frac{l+1}{2m}}_{i-1,j} + u^{n+\frac{l}{2m}}_{i+1,j}) + \ldots..}{a^n_{i,j}}$$

if l es even, we do the travelling in the inverse sense so that:

$$u^{n+\frac{l+1}{2m}}_{i,j} = \frac{u^n_{i,j} + b^n_{i,j}(u^{n+\frac{l}{2m}}_{i-1,j} + u^{n+\frac{l+1}{2m}}_{i+1,j}) + \ldots..}{a^n_{i,j}}$$

this scheme converges rapidly to the solution and experimentally the number of iterations $2m$ necessary to obtain $u_{i,j}^{n+1}$ from $u_{i,j}^{n}$ depends of course on the values of h and k, though normally is between six and eight iterations. We have also experimentally proved that this implicit scheme is unconditionally stable, that allow us to take high values of k, and thus advance rapidly throughout scales.

References

[AE94] L. Alvarez and J. Esclarin. Image quantification using reaction-diffusion equations. to appear in SIAM J. on Applied Mathematics., 1994.

[AGLM92a] L. Alvarez, F. Guichard, P.L. Lions, and J.M. Morel. Analyse multiechelle de films. *C. R. Acad. Sci. Paris*, 315(1):1145–1148, 1992.

[AGLM92b] L. Alvarez, F. Guichard, P.L. Lions, and J.M. Morel. Axiomatization et nouveaux operateurs de la morphologie mathematique. *C. R. Acad. Sci. Paris*, 315(1):265–268, 1992.

[AGLM92c] L. Alvarez, F. Guichard, P.L. Lions, and J.M. Morel. Axiomes et equations fondamentales du traitement d'images (analyse multiechelle et e.d.p.). *C. R. Acad. Sci. Paris*, 315(1):135–138, 1992.

[AGLM93] L. Alvarez, F. Guichard, P.L. Lions, and J.M. Morel. Axioms and fundamental equations of image processing. *Arch. for Rat. Mechanics*, 123(3):199–257, 1993.

[ALM92] L. Alvarez, P.L. Lions, and J.M. Morel. Image selective smoothing and edge detection by nonlinear diffusion (ii). *SIAM J. on Numerical Analysis*, 29(3):845–866, 1992.

[AM94a] L. Alvarez and L. Mazorra. Signal and image restoration using shock filter and anisotropic diffusion. *SIAM J. on Numerical Analysis*, 31(2):590–605, 1994.

[AM94b] L. Alvarez and F. Morales. Affine morphological multiscale analysis of corners and multiple junctions. to appear in International Journal of Computer Vision, 1994.

[AM94c] L. Alvarez and J.M. Morel. Formalization and computational aspects of image analysis. *Acta Numerica*, pages 1–61, 1994.

[AM94d] L. Alvarez and J.M. Morel. *A morphological approach to multiscale analysis. From principles to equations*, pages 249–276. Kluwer Academic Publishers, 1994.

Scale-Space Research at Utrecht University

Bart M. ter HAAR ROMENY

Imaging Science Group, Utrecht University,
Heidelberglaan 100, 3584 CX Utrecht, the Netherlands.
bartcv.ruu.nl

Abstract. This paper describes the contributions to scale-space research by Utrecht University over the period 1989-1995. Starting with work on the fundamental basis of scale-space from causality and first principles, the relation between scale and differentiation of numerical data is found to be crucial. The main lines of research are discussed: (high order) differential and algebraic invariant image structure, relations between order, scale, noise and accuracy of Gaussian derivatives, deep structure of scale-space and the 'hyperstack', multiscale optic flow and nonlinear diffusion. Most applications are in medical imaging. An overview of current activities and bibliography concludes the paper.

1. Introduction

In the early eighties Witkin [1] and Koenderink and van Doorn [2] did pioneering work which laid the basis for scale-space theory in image analysis. Their acclaimed paper [2] has become a classic in this field. Although Lindeberg [3] and many others made considerable contributions to the developments of early concepts of scale-space, we focus on the work at Utrecht University. In 1989 at the Utrecht University the '3D Computer Vision Group' was assembled (now called: Imaging Science Group, Viergever director) as a collaboration of Utrecht University, the University Hospital Utrecht, five participating industries and supported by a grant of the Dutch Ministries of Economic Affairs and Education & Science. The research theme of the group was three-dimensional computer vision, with an emphasis on medical imaging. Fundamental scale-space research was chosen as one of the main research lines, next to volume visualisation, multimodality matching, segmentation and clinical applications, in order to establish a solid mathematical basis for computer vision and medical imaging applications. This paper summarizes the scale-space research of the group over the period 1989-1995. Due to space limitations this overview has to be general and selective[1].

2. Fundamental Basis of Scale-Space

First, a fundamental question had to be answered. What is image structure? Koenderink had given a firm start towards a solution to this problem in terms of a multiscale approach.

The mere process of observation, through a finite aperture of as yet undetermined size, is a fundamental reason for the multiscale nature of the visual field, or images.

[1]For a description of the group and full bibliography, see URL: **http://www.cv.ruu.nl.**

The stack of images, acquired through apertures with continuously increasing size, is termed the scale-space of the image. Koenderink showed [2], by the argument of *causality*, that the generating equation of a linear scale-space was the isotropic diffusion equation (with the Gaussian as its Green's function). Many other derivations of the Gaussian kernel have been given in literature (see a summary in [4]). Florack et al. [4] took the approach of defining *axioms* for the visual front end (as did Alvarez et al. [5]). In order to be totally uncommitted, there may be no preference for location, orientation and scale. Moreover, the system is supposed to be linear. ¿From these 'first principles' we get finally [4]:

$$\mathcal{G}(\vec{\omega};\sigma) = \exp\left\{-\frac{1}{2}\sigma^2\omega^2\right\} \quad or \quad G(\vec{x};\sigma) = \frac{1}{\sqrt{2\pi\sigma^2}}\exp\left\{-\frac{\vec{x}\cdot\vec{x}}{2\sigma^2}\right\}$$

So the (normalized) Gaussian kernel is found as a unique solution, when we demand total uncommitment. Because diferentiation and scaling (under the linear diffusion equation) commute, all partial derivatives of the Gaussian kernel are found as solutions (of the isotropic diffusion equation) as well, specifying a *complete family of scaled differential operators*:

$$\frac{\partial L}{\partial x} = \frac{\partial}{\partial x}(L_0 * G) = L_0 * \frac{\partial G}{\partial x}$$

This 'multiscale local jet' gives an answer to the measurement of image structure. Note that this procedure also solves the ill-posedness of numerical differentiation. Differentiation and observation are a single operation performed by integration. In a sense we can view the scaled Gaussian derivative kernels as the *physical* differential operators. The mathematical (scale-free) differentiation appears as the limiting case where $\sigma \downarrow 0$. Derivatives are not intrinsic properties of digital data: regularization is required to make them well defined. Normally this takes the form of a linear filtering. It was shown by Florack [6] that Schwartz' theory of regular tempered distributions essentially supplies the mathematical solution for the the ill-posedness problem: a regular tempered distribution associated with a real image $L(\mathbb{R}^d)$ is defined as

$$T_L : \left\{ \mathcal{S}(\mathbb{R}^d) \to \mathbb{R} : \phi \mapsto \int L(x)\phi(x)dx \right\}$$

where $\mathcal{S}$ is a class containing all smooth test functions ϕ, which decrease sufficiently fast at the boundaries. The (N^{th} order) derivative of a distribution is defined by $(D_{\mathbf{n}}T)(\phi) = (-)^n T(D_{\mathbf{n}}\phi)$. The particular choice for the test function follows from physical constraints, and leads to the Gaussian when the axioms of uncommittment are applied, as shown above. In a shortly appearing book, Florack expatiates on this fundamental view on the process of observation, stressing the *duality* of detector and source [7]. Interestingly, Nielsen, Florack and Deriche [8] showed the common framework that can be found between regularization by scale-space operators, functional minimization and edge detection filters.

Koenderink proposed an inspiring relation with biological front-end vision operators: the Gaussian derivatives might supply a taxonomy for the many complex sensitivity

profiles found in the cortical receptive fields [9, 10, 11]. As such, the brain can be seen as a geometry engine. See also [12].
The use of scale-space theory as a computational model for visual perception was further examplified by Florack et al. [13], who studied the covariant form of the diffusion equation. Allowing an image dependent transformation, in particular with rotational symmetry around a given 'foveal point', nicely explained the linear increase of receptive field size with excentricity. The appearing 'log-polar coordinates' are the canonical coordinates for the planar similarity group, i.e. the local coordinates in terms of which spatial scalings and rotations look like plain translations. A similar result was reached from the assumption of equal processing capacity per scale in the visual system [14].

3. Image Structure and Differential Invariants

Using the well-posed recipe to compute derivatives we are ready to study the differential structure of images, and apply the full machinery of differential geometry and tensor analysis on images (in $\mathbb{R}^D$). Interestingly, the image representation is differentiable both in the scale and spatial domain, enabling interpolation in any direction by local Taylor expansion (the 'multiscale local jet'). Extrapolation along the negative scale axis gives a method for deblurring Gaussian blur [15]. Higher order derivatives with respect to scale are calculated from higher order Laplacians due to the diffusion equation.
Single partial derivatives are of no use for image structure analysis, as they are dependent on a particular coordinate system. We need image properties *invariant* under a particular group of possible transformations of the coordinate system. Invariant theory is a classical field in mathematics, and recently there has been a strong increase of attention by the computer vision community [16].

3..1 Invariants and notation

The notation of the differential invariant expressions in Cartesian partial derivatives often leads to complex expressions. Several notations turn out to be efficient and elegant, showing the particular symmetries often encountered [17, 18]:

• Manifest invariant notation: Indexed quantities $Q_{i_1 \ldots i_n}$ can be considered tensor components, of a truly Cartesian invariant property Q. The partial derivatives of the image can now be given a tensorial interpretation. Due to the orthogonality property of the Cartesian group there is no distiction between covariant and contravariant tensors. Full contraction or alternation of indices, which may include the constant tensors δ_{ij} (Dirac-delta) and $\varepsilon_{i_1 \ldots i_n}$ (Lévy-Civita) gives an differential invariant property under the Cartesian group. Some examples are given in table 1, making the Einstein convention (contraction of pairs of indices) explicit.

• Gauge coordinates, because they 'fix the gauge', are particularly useful by employing locally the extra degree of freedom, e.g. for the group of orthogonal trans-

Name	Cartesian	Manifest	Gauge
Intensity	L	L	L
Gradient^2	$L_x^2 + L_y^2$	$L_i L_i$	L_w^2
Laplacian	$L_{xx} + L_{yy}$	L_{ii}	$L_{vv} + L_{ww}$
Isophote curv.	$\frac{2L_xL_yL_{xy}-L_x^2L_{yy}-L_y^2L_{xx}}{(L_x^2+L_y^2)^{3/2}}$	$\frac{L_iL_jL_{ij}-L_iL_iL_{jj}}{(L_kL_k)^{3/2}}$	$-\frac{L_{vv}}{L_w}$
Flowline curv.	$\frac{L_xL_y(L_{yy}-L_{xx})+L_{xy}(L_x^2-L_y^2)}{(L_x^2+L_y^2)^{3/2}}$	$\frac{-L_i\epsilon_{ij}L_{jk}L_k}{(L_lL_l)^{3/2}}$	$-\frac{L_{vw}}{L_w}$

Table 1: Some examples of 2D invariants for orthogonal transformations, expressed in Cartesian, manifest invariant and gauge coordinates. From [18].

formations a rotation: in 2D we define coordinates (v, w) such that v is everywhere tangential to the isophote and w perpendicular to it:

$$\partial_v = \frac{L_i \epsilon_{ij} \partial_j}{\sqrt{L_k L_k}} \qquad \partial_w = \frac{L_i \delta_{ij} \partial_j}{\sqrt{L_k L_k}}$$

Because $L_v \equiv -L_i \epsilon_{ij} L_j \equiv 0$ many expressions get a much simpler form (see table 1).

The way gauge coordinates are constructed automatically implies that all polynomial expressions in (v, w) are invariant under orthogonal transformations. Many examples are given in [18].
The 'ridge operator' L_{vv} and the gradient L_w have been successfully applied by van den Elsen, Maintz and Viergever for multimodality image matching, e.g. CT/MR or MR/SPECT [19, 20, 21]. The gradient of isophote curvature has been studied as T-junction detector [22].
For 3D the situation is more complicated, e.g. for the orthogonal gauge (u, v, w) where first order derivatives vanish: all tangential directions to the isophote surface are in a plane, and a natural choice is the direction of principal curvatures. This is worked out in detail by Salden et al. in [23].

3..2 Irreducible invariants

An essential observation made by Hilbert [24] states that any invariant can be represented as a polynomial function on a complete and finite set of *irreducible invariants*. This implies that these form the most fundamental and concise set of local properties (up to some spatial order) or 'primitives' that may describe all local intrinsic properties of a scalar image at a fixed level of resolution. This set is well known for second order properties [25], and is given in manifest notation for 2D in table 3.
For higher order ($\geq$ 3) there is no simple method to derive the complete irreducible set. Salden [26] presented a method that makes use of the *algebraic* structure of the local geometric description, i.e. the binary forms of certain order in the local Taylor

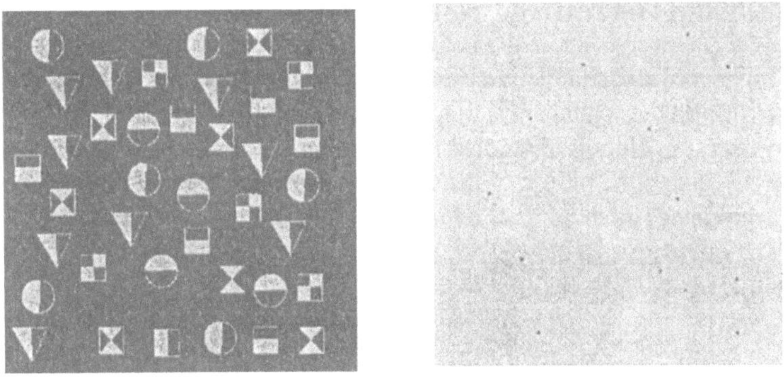

Figure 1: 4-junction detection: The pure fourth order invariant $D_4 = -I^3 + 27J^2$ (the fourth order discriminant) with $I = \frac{1}{2}L_{iijj}L_{kkll} - L_{iijk}L_{jkll} + \frac{1}{2}L_{ijkl}L_{ijkl}$ and $J = \frac{1}{8}L_{iijj}L_{kkll}L_{mmnn} - \frac{1}{8}L_{iijj}L_{klmn}L_{klmn} - \frac{1}{4}L_{iijj}L_{kklm}L_{lmnn} + \frac{1}{4}L_{iijk}L_{jklm}L_{lmnn}$, or $D = -1(L_{xxxx}L_{yyyy} - 4L_{xxxy}L_{xyyy} + 3L_{xxyy}L_{xxyy})^3 + 27(L_{xxxx}(L_{xxyy}L_{yyyy} - L_{xyyy}L_{xyyy}) - L_{xxxy}(L_{xxxy}L_{yyyy} - L_{xxyy}L_{xyyy}) + L_{xxyy}(L_{xxxy}L_{xyyy} - L_{xxyy}L_{xxyy}))^2$ calculated for a noisy set of shapes. The input image (left) is a binary test image with intensity difference of 100 units, perturbed by additive pixel uncorrelated gaussian noise with variance of 1000 units. Scale of the invariant D4: 5.5 pixel units. Note the rotation invariance and that despite the noise this fourth order property is well-represented at an appropriate scale. Resolution 512^2. (From [26]).

expansion. E.g. the fourth order binary form is given by $f(\mathbf{x}) = \frac{1}{24} L_{ijkl}\, x_i x_j x_k x_l$. The main proposition of algebra states that a function is fully described by its roots. Several theorems are available relating the *measure of coincidence of roots* to geometrical symmetries. There is also a wide algebraic literature on the study of roots, defining the *discriminant*, the *resultant* which is a measure for the coincidence of roots of polynomials of different order, and *transvectants*. The method, due to Hilbert, [24] generates complete sets of irreducible invariants for affine and orthogonal transformations. The sets for third and fourth order are increasingly more complicated, and are fully specified in [26]. Haring et al. applied these irreducibles in Kohonen neural networks for segmentation [27]. In figure 1 (from [26]) an example is given of the fourth order *discriminant* of the fourth order binary form, acting as a robust 4-junction detector.

We designed a software package (TGV: 'tools of geometry for vision'[2]) that consists of a code-generating parser frontend, to express manifest or gauge expressions of differential invariants (no matter their complexity or order, 2D, 3D, and 2D/3D/time) into cartesian Gaussian partial derivatives, performing the convolutions in the Fourier domain, and a second stage generating a full scale space with exponential scale parametrization of the invariant expression(s).

[2] Industrial proprietary rights restricts availability to labs with a formal research collaboration.

4. Gaussian Derivatives, Noise and Accuracy

Taking higher order spatial derivatives is problematic in the presence of noise. However, there is also the spatial averaging effect due to the extent (scale) of the Gaussian derivative kernel. Blom [28] studied the behaviour of noise under Gaussian derivatives: he calculated the ratio r of the variance of output noise to (additive Gaussian pixel-uncorrelated) input noise as a function of the scale of the operator and order of differentiation m_x, m_y to be:

$$r = \frac{\epsilon^2}{4\pi\sigma^2} \cdot \frac{Q_{2m_x} Q_{2m_y}}{(4\sigma^2)^{m_x+m_y}}$$

where ϵ is the interpixel distance, and the functions Q_n express the influence of the order of differentiation, and form the non-trivial part. They are defined as: $Q_0 = 1$, $Q_n = 0$ for n odd, and $Q_n = \prod_{i=1}^{n/2}(2i-1)$ for n even. For increasing order of differentiation, the influence of scale increases strongly.

A rather fundamental relation addressed by ter Haar Romeny et al. [29] is the relation between differential order, scale and accuracy of representation of Gaussian derivatives. In the Fourier domain the Gaussian kernel gets wider for decreasing scale in the spatial domain. The error can be defined as the relative energy of the aliased frequencies over the total energy:

$$err(n,\sigma) = \frac{\Delta E_n(\sigma,\omega)}{E_n(\sigma,\omega)} = \frac{\int_\pi^\infty \omega^{2n} e^{-\sigma^2\omega^2} d\omega}{\int_0^\infty \omega^{2n} e^{-\sigma^2\omega^2}} = \frac{\Gamma(\frac{1}{2}+n) - \Gamma(\frac{1}{2}+n, 0, 4\pi^2)}{\Gamma(\frac{1}{2}+n)}$$

where $\Gamma(n)$ is the Euler gamma function, and $\Gamma(n, z_0, z_1)$ is the generalized incomplete gamma function $\Gamma(n, z_0) - \Gamma(n, z_1)$. Figure 2 shows the aliasing error as a function of scale σ and order n for orders up to 10. Note that indeed for higher order a larger scale must be selected.

5. Deep Structure and the Hyperstack

It is interesting to exploit the multiscale representation of image data by studying the relation between different scales levels.

De Graaf, Vincken and Koster developed the 'hyperstack' (stack of 3D images) segmentation [30, 31, 32, 33] as a linking model based segmentation technique, originally built upon linear scale space theory. The basic idea of the hyperstack is to define relations between voxels in adjacent scale space levels, such that the levels at larger scales—containing the global information—guide the collection of voxels at the smallest scale (the original image). The entire process requires four steps: *(i)* blurring, *(ii)* linking, *(iii)* root labeling, and *(iv)* downward projection.

In the blurring phase a stack of images is created. During the linking phase voxels in two adjacent scale levels are connected by so-called child-parent linkages, with an assigned *affection* value based on heuristic and statistical features (see [32] for details). The area in which a parent in level $n+1$ is selected for a specific child

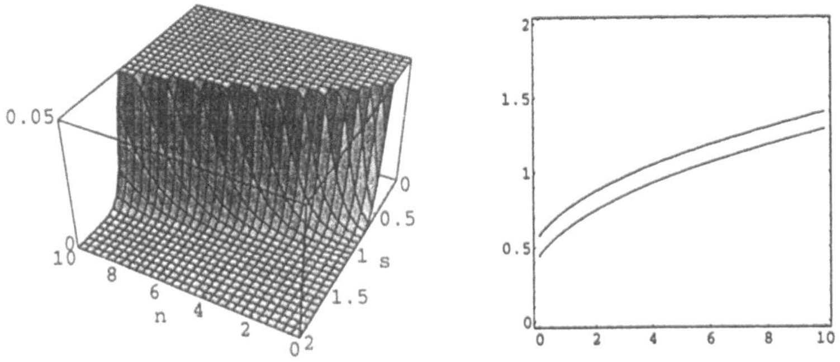

Figure 2: Maximum differential order that can be extracted with Gaussian derivative operators without aliasing as a function of scale of the Gaussian derivative kernels expressed in pixels. Left: relation between scale (s), order (n) and accuracy (vertical, in %). Right: horizontal - orders up to 10, vertical - scales up to 2 pixels. Right: same function, orders up to 10, scales up to 2 pixels. Allowed error: 5% (upper line in each plot), 1% (lower line in each plot). From [29].

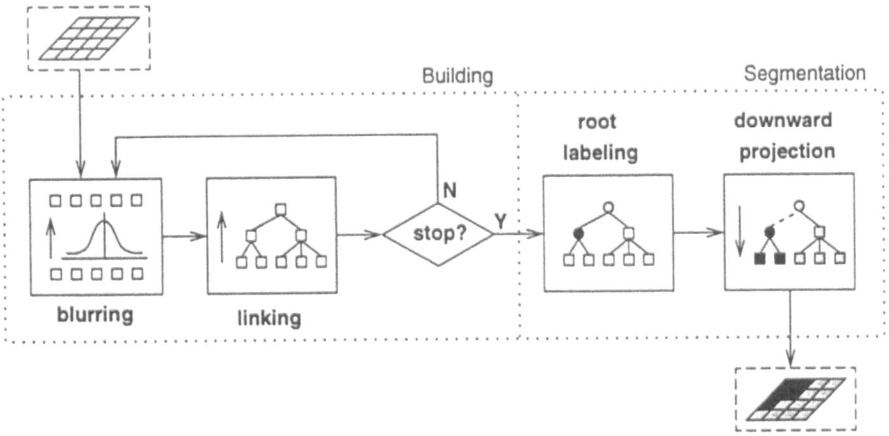

Figure 3: Schematic of the hyperstack segmentation process.

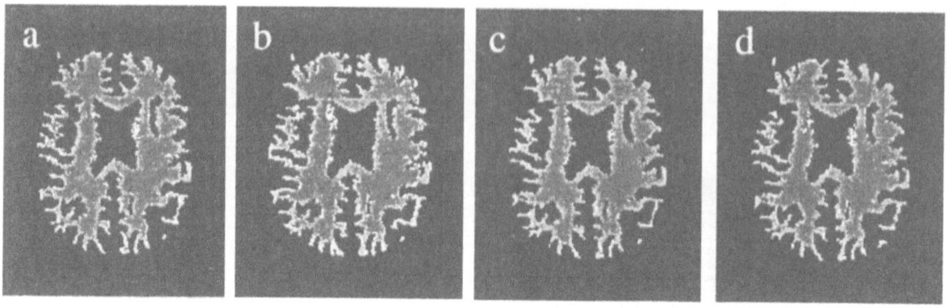

Figure 4: Segmentations of the white matter of a transversal MR brain image based on the four different scale space generators. The grey colored areas have been segmented correctly (according to the gold standard), the white colored areas correspond to erroneously segmented pixels. The figures correspond to: (a) Linear scale space implemented in the spatial domain; (b) idem, implemented in the Fourier domain; (c) Perona & Malik anisotropic diffusion; (d) Euclidean Shortening Flow. From [33].

in level n is defined by a radius $r_{n,n+1} = k \cdot \sigma_{n,n+1}$, where $\sigma_{n,n+1}$ denotes the relative σ (that corresponds with the transition from level n to level $n+1$) and k is chosen such that only parents are considered whose intensity has been influenced significantly by the child at hand. Typically, $k = 1.5$. Furthermore, the linking is a bottom-up process such that only parents that have been linked to are considered children in the next linking step. This leads to the typical tree-like structure of linkages through scale space. If the linking has converged—in the sense that only few parents are left in the top level of the hyperstack—the root labeling takes place. In this phase, the children in the tree with a relatively low affection value are labeled as *roots*, each of which represents a segment in the original image. Finally, in the down projection phase the actual segments are formed by grouping all the voxels in the ground level that are linked to a single root by following the child-parent linkages downwards.

In [33] the hyperstack—a multiscale linking model for image segmentation— is used for an in-depth comparison of four different scale space generators with respect to segmentation results. Considered are the linear (Gaussian) scale space both in the spatial and the Fourier domain, the variable conductance diffusion according to Perona & Malik, and the Euclidean shortening flow. Segmentation experiments are carried out on MR images of the brain (see figure 4), for which a gold standard is available. The hyperstack proofs to be rather insensitive to the underlying scale space generator. Currently we study the mathematical underpinning of this so far rather heuristic approach.

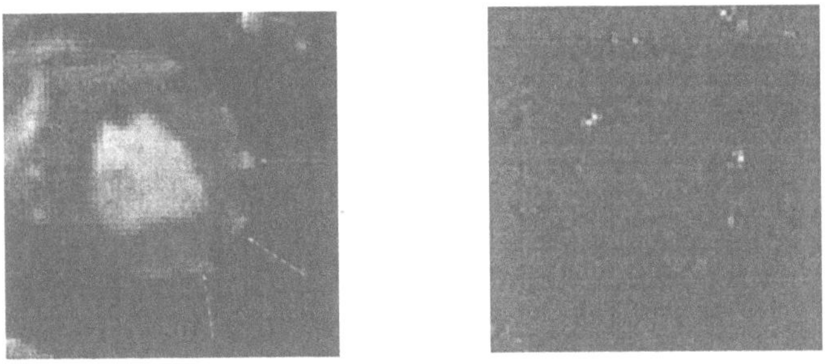

Figure 5: Left: 2D cross-section of a $3D$ MR-image of a canine heart at one instance of the cardiac cycle with epicardial markers. Right: rate of change of the corner detector $L_{xx}L_y^2 + L_{yy}L_x^2 - 2L_xL_yL_{xy}$ with $\sigma_s = \sigma_t = 1$. This detector which measures the isophote curvature κ times the gradient magnitude cubed is good at locating the implanted markers. From [36].

6. Optic Flow

We can define spatiotemporal operators that are straightforward extensions of the Gaussian kernel known from classical scale space theory[3]:

$$G(\mathbf{x}, t, \sigma_s, \sigma_t) = \frac{1}{\sqrt{2\pi\sigma_s^2}^D \sqrt{2\pi\sigma_t^2}} e^{-\frac{1}{2}(\frac{X^2}{\sigma_s^2} + \frac{t^2}{\sigma_t^2})}$$

Florack et al. generalized this idea of 'families of tuned kernels' [34] and showed how the ensemble behaviour of these co-moving derivative operators may be highly effective in the detection of transparent motion. For 'real-time' sampling Koenderink showed that causality can be guaranteed for the temporal domain when logarithmically rescaling the time-axis [35].

Florack and Nielsen [37] recently proposed a solution for optic flow 'under the aperture', i.e. as a scale-space paradigm. It essentially is a modification of the image brightness constraint in the Horn & Schunk Optic Flow Constraint Equation (OFCE). We will briefly delineate the essence. The method has been successfully implemented for the analysis of ventricular heartwall motion [38, 36]. Hereafter vectors $(\mathbf{x}, \mathbf{v})$ for place and velocity contain both the spatial and the temporal components which can be accessed by superscripts, so $\mathbf{x}^i = (t, x, y, z)$. Subscripts denote partial integration and the Einstein summation convention is in effect, i.e. a double occurrence of an index implies that we sum over the dimensions. E.g. $\mathbf{v}^i L_i = L_t + v^x L_x + v^y L_y + v^z L_z$. The traditional Horn & Schunck OFCE for scalar images states that, when co-moving with the flow, the luminance should be constant: $\frac{d\Psi}{dt} = \mathbf{v}^i \Psi_i \equiv 0$. These equations can be rewritten *under the aperture*:

$$\mathbf{v}^i \Psi_i \otimes G = 0 \quad (1)$$

which gives (using partial integration and the fact that the Gaussian decays suffi-

[3] See the intruiging RF measurements by Freeman et al.: http://totoro.berkeley.edu/.

ciently fast at the boundaries)

$$-\Psi \otimes (\mathbf{v}^i_i G + \mathbf{v}^i G_i) = 0 \quad (2)$$

To approximate the flow field we consider a polynomial approximation of the flow field to order M:

$$\mathbf{v}^i_M = \tilde{\mathbf{v}}^i + \mathbf{x}^j \tilde{\mathbf{v}}^i_j + \mathbf{x}^j \mathbf{x}^k \tilde{\mathbf{v}}^i_{jk} + \ldots . \quad (3)$$

Note that this should not be interpreted as a truncated Taylor series; all components in the expansion can change if we increase M.
Inserting equation (3) into (2) we obtain an *M*th order approximation of the flow field which can be expressed purely in the spatiotemporal operators defined above by using the fact that polynomials times (derivatives) of Gaussians can be written in other derivatives of Gaussians. Complementary equations can be obtained by differentiating the OFCE as long as no components of the velocity field arise which were not present in the expansion (3). Owing to the *aperture problem*, i.e. tangential motion is possible without affecting the image data, the solution is in general underdetermined. A number of authors differentiate the optic flow equation a number of times, *apparently* fixing the aperture problem. Such an approach may lead to a system of equations that uniquely determines (or even over-constrains!) the optic flow field, thereby violating the degrees of freedom initially present. Of course, in specific situations the implicit constraint may represent the physical situation to a good approximation in which case the resulting flow field will be in close correspondence to the true motion. In case of rigid motion this is often the case, in case of deformable motion generally not.

7. Nonlinear Diffusion

Recently many 'geometry-driven' evolution schemes have been proposed. See for a tutorial overview [39][4]. In Utrecht we have implemented a general framework to generate nonlinear multi-scale representations of image data [40, 41, 42]. The process is considered as an initial value problem with an acquired image as initial condition and a geometrical invariant as "driving force" of an evolutionary process. Geraets and Salden studied the evolution properties of Euclidean and affine curvature of space curves, defined by planar pointsets [43], while Salden et al. gave a general framework for affine and projective scale-space theories [44].
The geometrical invariants are extracted using the family of Gaussian derivative operators. These operators naturally deal with scale as a free parameter and solve the ill-posedness of differentiation. Stability requirements for numerical approximation of evolution schemes using Gaussian derivative operators are derived [41] and show that the allowed time-step is proportional to scale, and considerably larger than in nearest neighbor approximations. This approach has been used to generalize and

[4]A number of European and US based laboratories are collaborating in the 'DIFFUSION' program: http://www.cv.ruu.nl/Collaborations/node1.html. The book [39] is one of the fruits of this collaboration.

Name of flow	Image	Level set	Cons.	Flow
Linear	$\frac{\partial L}{\partial t} = \Delta L$	$\frac{\partial C}{\partial t} = \frac{-\Delta L}{L}$	L	∇L
Variable conduct.	$\frac{\partial L}{\partial t} = \nabla \cdot (c \nabla L)$	$\frac{\partial C}{\partial t} = \frac{-\nabla \cdot (c \nabla L)}{\nabla L}$	L	$c\nabla L$
Normal motion	$\frac{\partial L}{\partial t} = L_w$	$\frac{\partial C}{\partial t} = c\vec{N}$	-	-
Eucl. shortening	$\frac{\partial L}{\partial t} = L_{vv}$	$\frac{\partial C}{\partial t} = \kappa \vec{N}$	-	-
Affine shortening	$\frac{\partial L}{\partial t} = L_{vv}{}^{\frac{1}{3}} L_w{}^{\frac{2}{3}}$	$\frac{\partial C}{\partial t} = \kappa^{\frac{1}{3}} \vec{N}$	-	-
Entropy	$\frac{\partial L}{\partial t} = \alpha L_w + \beta L_{vv}$	$\frac{\partial C}{\partial t} = (\alpha + \beta\kappa)\vec{N}$	-	-

Table 2: Overview of the various evolution schemes. Cons.: Conserved property.

implement a variety of nonlinear diffusion schemes. See table 2. Weickert [45] presents a complete scale-space theory for nonlinear diffusion filtering in the continuous, semidiscrete and discrete setting. Anisotropic diffusion with a heat conduction tensor has already been studied in [46, 47, 48].

Two approaches are distinguished: evolution of the luminance function under a flow and evolution of the isoluminance curves of the image. The duality relation between curve evolution and geometric diffusion was found by Osher and Sethian, and Lopez and Morel: The level sets of an evolution of L constrained by

$$\frac{\partial L}{\partial t} = \nabla \cdot F(L_i, L_{ij}, ...)$$

evolve according to

$$\frac{\partial C}{\partial t} = \frac{-\nabla \cdot F(L_i, L_{ij}, ...)}{|\nabla L|}\vec{N}$$

and, given the curve evolution:

$$\frac{\partial C}{\partial t} = H(L_i, L_{ij}, ...)\ g(\kappa)\vec{N}$$

the luminance function evolves according to

$$\frac{\partial L}{\partial t} = -H(L_i, L_{ij}, ...)\ L\ g(\kappa)$$

Here, $\vec{N}$ is the inward unit normal, and κ isophote curvature.

We have shown how existing methods fit in and introduce some new possible choices. Among others, Niessen et al. [42] consider a combination of geometric curve evolution and gradient dependent diffusion which is of particular interest for image enhancement (see figure 6). This leads to the intuitive interpretation that the extraction of geometrical information is performed by local neighborhood operators and used

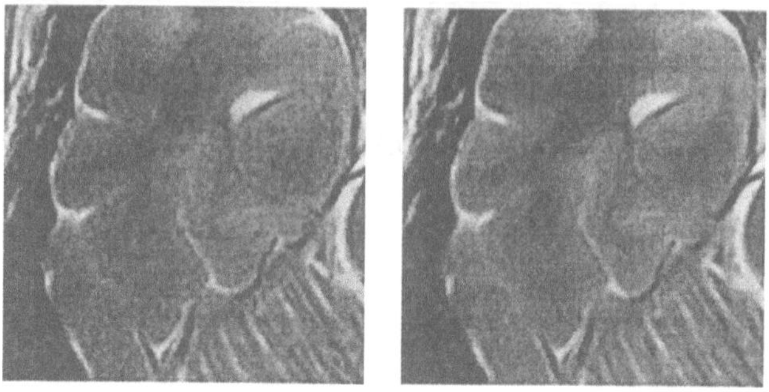

Figure 6: Left: Part of an MR-scan, used in an interactive CD-ROM based anatomical atlas [49] which shows the cerebellum and cerebrum. The data had to be resampled for optimal display (the atlas provides all three viewing directions), making the images are rather noisy. Right: after gradient dependent geometric evolution ($\frac{\partial L}{\partial t} = L_{vv}{}^{\frac{1}{3}}$) which evolves the isophotes as a function of their curvature with an extra term depending on the norm of the gradient. The process improves the visual impression of the images on the video display. From [42].

to determine the evolution of the image, while we interpret the evolution as if the isophotes move with a speed depending on an estimation of their curvature altered by a function of local image properties.

8. Current Work and Prospect

Current work is ongoing in the following areas (superv.: ter Haar Romeny, Viergever):

- multiscale differential stereo disparity operators (Maas, Nielsen)
- behaviour around singular points in the scale-space deep structure, classification, degeneracies (Kalitzin,Salden)
- multiscale optic flow (Florack, Niessen)
- affine and projective geometric scale-spaces (Salden)
- 3D nonlinear diffusion and efficient schemes (Weickert, Niessen, Salden)
- image-driven scale selection (all)
- scale-space snakes (Niessen)
- nonlinear diffusion and hyperstack segmentation (Vincken, Niessen)
- multiscale texture orientation analysis (Lopez, Niessen, Weickert)
- multiscale feature-based 3D image registration (Maintz, van den Elsen)

The scale-space concept has proven to be solid, and we can state that the study of computer vision problems 'under the aperture' has only recently taken off, with considerable result. Many problems remain to be solved. The major challenge, the exploitation of the deep structure, is now feasible. We hope to continue to contribute our part to this fascinating research area.

9. Acknowledgement

The research described is the result of teamwork. The material for this paper came from many contributions: Jan Koenderink and Andrea van Doorn (Physics Dept.), Max Viergever and Bart ter Haar Romeny (Imaging Science Group), and the PhD students / postdocs: Luc Florack, Wiro Niessen, Alfons Salden, Mads Nielsen, Robert Maas, Ruud Geraets, Andre Koster, Koen Vincken, Twan Maintz, Joachim Weickert, Stiliyan Kalitzin, Antonio Lopez, Petra van den Elsen, Hans Blom, Bas Haring, visiting professors Stephen Pizer, Jim Duncan and Jorge Llacer, and many M.Sc. students.

References

[1] A. P. Witkin, "Scale space filtering," in *Proc. International Joint Conference on Artificial Intelligence*, (Karlsruhe, Germany), pp. 1019–1023, 1983.

[2] J. J. Koenderink, "The structure of images," *Biol. Cybern.*, vol. 50, pp. 363–370, 1984.

[3] T. Lindeberg, *Scale-Space Theory in Computer Vision*. The Kluwer International Series in Engineering and Computer Science, Dordrecht, the Netherlands: Kluwer Academic Publishers, 1994.

[4] L. M. J. Florack, B. M. ter Haar Romeny, J. J. Koenderink, and M. A. Viergever, "Linear scale-space," *Journal of Mathematical Imaging and Vision*, vol. 4, no. 4, pp. 325–351, 1994.

[5] L. Alvarez, F. Guichard, P. L. Lions, and J. M. Morel, "Axiomes et equations fondamentales du traitement d'images," *C. R. Acad. Sci. Paris*, vol. 315, pp. 135–138, 1992.

[6] L. M. J. Florack, B. M. ter Haar Romeny, J. J. Koenderink, and M. A. Viergever, "Images: Regular tempered distributions," in *Proc. of the NATO Advanced Research Workshop Shape in Picture - Mathematical description of shape in greylevel images* (Y.-L. O, A. Toet, H. J. A. M. Heijmans, D. H. Foster, and P. Meer, eds.), vol. 126 of *NATO ASI Series F*, pp. 651–660, Springer Verlag, Berlin, 1994.

[7] L. M. J. Florack, *The Syntactical Structure of Scalar Images*. Dordrecht: Kluwer Academic Publishers, 1996. In preparation.

[8] M. Nielsen, L. M. J. Florack, and R. Deriche, "Regularization, scale-space, and edge detection filters," in *Proc. Fourth European Conference on Computer Vision*, (Cambridge, UK), April 14-18 1996.

[9] J. J. Koenderink and A. J. van Doorn, "Representation of local geometry in the visual system," *Biol. Cybern.*, vol. 55, pp. 367–375, 1987.

[10] J. J. Koenderink and A. J. Van Doorn, "Operational significance of receptive field assemblies," *Biol. Cybern.*, vol. 58, pp. 163–171, 1988.

[11] J. J. Koenderink and A. J. van Doorn, "Generic neighborhood operators," *IEEE Trans. Pattern Analysis and Machine Intelligence*, vol. 14, pp. 597–605, June 1992.

[12] S. M. Pizer and B. M. ter Haar Romeny, "Fundamental properties of medical image perception," *Journal of Digital Imaging*, vol. 4, pp. 1–20, Febr. 1990.

[13] L. M. J. Florack, A. H. Salden, B. M. t. Haar Romeny, J. J. Koenderink, and M. A. Viergever, "Nonlinear scale-space," *Image and Vision Computing*, vol. 13, pp. 279–294, May 1995.

[14] T. Lindeberg and B. M. ter Haar Romeny, "Linear scale-space: I. basic theory. II. early visual operations.," in *Geometry-Driven Diffusion in Computer Vision* (B. M. ter Haar Romeny, ed.), Computational Imaging and Vision, pp. 1–38,39–72, Dordrecht, the Netherlands: Kluwer Academic Publishers, 1994.

[15] B. M. ter Haar Romeny, L. M. J. Florack, M. de Swart, J. Wilting, and M. A. Viergever, "Deblurring Gaussian blur," in *Proc. Mathematical Methods in Medical Imaging II*, vol. 2299, (San Diego, CA), pp. 139–148, SPIE, July, 25-26 1994.

[16] J. L. Mundy and A. Zisserman, eds., *Geometric Invariance in Computer Vision*. Cambridge, Massachusetts: MIT Press, 1992.

[17] L. M. J. Florack, B. M. ter Haar Romeny, J. J. Koenderink, and M. A. Viergever, "Cartesian differential invariants in scale-space," *Journal of Mathematical Imaging and Vision*, vol. 3, pp. 327–348, November 1993.

[18] B. M. ter Haar Romeny, L. M. J. Florack, A. H. Salden, and M. A. Viergever, "Higher order differential structure of images," *Image and Vision Computing*, vol. 12, pp. 317–325, July/August 1994.

[19] P. A. van den Elsen, J. B. A. Maintz, E. J. D. Pol, and M. A. Viergever, "Automatic registration of CT and MR brain images using correlation of geometrical features," *IEEE Transactions on Medical Images*, vol. 14, no. 2, pp. 384–398, 1995.

[20] J. B. A. Maintz, P. A. van den Elsen, and M. A. Viergever, "Comparison of feature-based matching of CT and MR brain images," in *CVRMed, Volume 905 of Lecture Notes in Computer Science* (N. Ayache, ed.), (Berlin), pp. 219–228, Springer-Verlag, 1995.

[21] J. B. A. Maintz, P. A. van den Elsen, and M. A. Viergever, "Evaluation of ridge seeking operators for multimodality medical image matching," *IEEE Trans. Pattern Analysis and Machine Intelligence*, 1996. In press.

[22] B. M. ter Haar Romeny, L. M. J. Florack, J. J. Koenderink, and M. A. Viergever, "Invariant third order properties of isophotes: T-junction detection," in *Proc. 7th Scand. Conf. on Image Analysis* (P. Johansen and S. Olsen, eds.), (Aalborg, DK), pp. 346–353, August 1991.

[23] A. H. Salden, B. M. ter Haar Romeny, and M. A. Viergever, "Local and multilocal scale-space description," in *Proc. of the NATO Advanced Research Workshop Shape in Picture - Mathematical description of shape in greylevel images* (Y.-L. O, A. Toet, H. J. A. M. Heijmans, D. H. Foster, and P. Meer, eds.), vol. 126 of *NATO ASI Series F*, pp. 661–670, Springer Verlag, Berlin, 1994.

[24] D. Hilbert, "Ueber die vollen Invariantensystemen," *Math. Annalen*, vol. 42, pp. 313–373, 1893.

[25] L. M. J. Florack, B. M. ter Haar Romeny, J. J. Koenderink, and M. A. Viergever, "General intensity transformations and differential invariants," *Journal of Mathematical Imaging and Vision*, vol. 4, pp. 171–187, May 1994.

[26] A. H. Salden, B. M. ter Haar Romeny, L. M. J. Florack, J. J. Koenderink, and M. A. Viergever, "A complete and irreducible set of local orthogonally invariant features of 2-dimensional images," in *Proceedings 11th IAPR Internat. Conf. on Pattern Recognition* (I. T. Young, ed.), (The Hague, the Netherlands), pp. 180–184, IEEE Computer Society Press, Los Alamitos, August 30–September 3 1992.

[27] S. Haring, M. A. Viergever, and J. N. Kok, "Kohonen networks for multiscale image segmentation," *Image and Vision Computing*, vol. 12, no. 6, pp. 339–344, 1994.

[28] J. Blom, B. M. ter Haar Romeny, A. Bel, and J. J. Koenderink, "Spatial derivatives and the propagation of noise in Gaussian scale-space," *J. of Vis. Comm. and Im. Repr.*, vol. 4, pp. 1–13, March 1993.

[29] B. M. ter Haar Romeny, W. J. Niessen, J. Wilting, and L. M. J. Florack, "Differential structure of images: Accuracy of representation," in *Proc. First IEEE Internat. Conf. on Image Processing*, (Austin, TX), pp. 21–25, IEEE, November, 13-16 1994.

[30] K. L. Vincken, C. N. de Graaf, A. S. E. Koster, M. A. Viergever, F. J. R. Appelman, and G. R. Timmens, "Multiresolution segmentation of 3D images by the hyperstack," in *Proc. First Conference on Visualization in Biomedical Computing*, pp. 115–122, Los Alamitos, CA: IEEE Computer Society Press, 1990.

[31] K. L. Vincken, A. S. E. Koster, and M. A. Viergever, "Probabilistic segmentation of partial volume voxels," *Pattern Recognition Letters*, vol. 15, no. 5, pp. 477–484, 1994.

[32] A. S. E. Koster, K. L. Vincken, C. N. De Graaf, O. C. Zander, and M. A. Viergever, "Heuristic linking models in multi-scale image segmentation," *Computer Vision, Graphics and Image Processing*, 1996. In press.

[33] K. L. Vincken, W. J. Niessen, A. S. E. Koster, and M. A. Viergever, "Blurring strategies for image segmentation using a multiscale linking model," in *Proc. Computer Vision and Pattern Recognition Conference CVPR'96*, IEEE Computer Society Press, June 16-20 1996. In press.

[34] L. M. J. Florack, B. M. ter Haar Romeny, J. J. Koenderink, and M. A. Viergever, "Families of tuned scale-space kernels," in *Proceedings of the European Conference on Computer Vision* (G. Sandini, ed.), (Santa Margherita Ligure, Italy), pp. 19–23, May 19–22 1992.

[35] J. J. Koenderink, "Scale-time," *Biol. Cybern.*, vol. 58, pp. 159–162, 1988.

[36] W. J. Niessen, J. S. Duncan, B. M. ter Haar Romeny, and M. A. Viergever, "Spatiotemporal analysis of left ventricular motion," in *Medical Imaging 95: Image Processing* (M. H. Loew, ed.), pp. 250–261, SPIE Press, Bellingham, 1995.

[37] L. Florack and M. Nielsen, "The intrinsic structure of the optic flow field," Tech. Rep. 94-R033, ERCIM, 1994.

[38] W. J. Niessen, J. S. Duncan, L. M. J. Florack, B. M. ter Haar Romeny, and M. A. Viergever, "Spatiotemporal operators and optic flow," in *Physical-Based Modeling in Computer Vision* (S. T. Huang and D. N. Metaxas, eds.), pp. 78–84, IEEE Computer Society Press, 1995.

[39] B. M. ter Haar Romeny, ed., *Geometry-Driven Diffusion in Computer Vision*. Dordrecht: Kluwer Academic Publishers, 1994.

[40] W. J. Niessen, B. M. ter Haar Romeny, L. M. J. Florack, A. H. Salden, and M. A. Viergever, "Nonlinear diffusion of scalar images using well-posed differential operators," in *Conference on Computer Vision and Pattern Recognition*, (Seattle, WA), pp. 92–97, IEEE Computer Society Press, Los Alamitos, 1994.

[41] W. J. Niessen, B. M. ter Haar Romeny, and M. A. Viergever, "Numerical analysis of geometry-driven diffusion equations," in *Geometry-Driven Diffusion in Computer Vision* (B. M. ter Haar Romeny, ed.), vol. 1 of *Computational Imaging and Vision*, pp. 393–410, Dordrecht: Kluwer Academic Publishers, 1994.

[42] W. J. Niessen, B. M. ter Haar Romeny, L. M. J. Florack, and M. A. Viergever, "A general framework for geometry-driven evolution equations," *International Journal of Computer Vision*, 1996. In print.

[43] R. Geraets, A. H. Salden, B. M. ter Haar Romeny, and M. A. Viergever, "Affine scale-space for discrete pointsets," in *Proc. Soc. for Neural Networks* (C. Gielen, ed.), (Nijmegen, the Netherlands), SNN, 1995.

[44] A. H. Salden, B. M. ter Haar Romeny, and M. A. Viergever, "Affine and projective scale space theories," in *Proc. Conf. on Differential Geometry and Computer Vision: From Pure over Applicable to Applied Differential Geometry*, (Nordfjordeid, Norway), August 1–7 1995.

[45] J. Weickert, *Anisotropic Diffusion in Image Processing*. PhD thesis, Dept. of Mathematics, University of Kaiserslautern, Germany, January 1996.

[46] J. Weickert, "Anisotropic diffusion filters for image processing based quality control," in *Proc. Seventh European Conf. on Mathematics in Industry* (A. Fasano and M. Primicerio, eds.), pp. 355–362, Teubner, Stuttgart, 1994.

[47] J. Weickert, "Multiscale texture enhancement," in *Proc. Sixth Intern. Conf. on Computer Analysis of Images and Patterns (CAIP '95)* (V. Hlaváč and R. Šára, eds.), vol. 970 of *Lecture Notes in Computer Science*, pp. 230–237, Springer, Berlin, 1995.

[48] J. Weickert, "Theoretical foundations of anisotropic diffusion in image processing," in *Theoretical Foundations of Computer Vision* (W. Kropatsch, R. Klette, and F. Solina, eds.), vol. 11 of *Computing Supplement*, pp. 221–236, Springer, Wien, 1996.

[49] B. Hillen, "Interactive anatomy of craniofacial structure," tech. rep., Utrecht University, Faculty of Medicine, 1993. CD-ROM interactive.

[50] L. M. J. Florack, B. M. ter Haar Romeny, J. J. Koenderink, and M. A. Viergever, "Scale and the differential structure of images," *Image and Vision Computing*, vol. 10, pp. 376–388, July/August 1992.

[51] L. M. J. Florack, *The Syntactical Structure of Scalar Images*. PhD thesis, Utrecht University, Utrecht, The Netherlands, November 1993.

[52] B. M. ter Haar Romeny and L. M. J. Florack, "A multiscale geometric model of human vision," in *Perception of Visual Information* (W. R. Hendee and P. N. T. Wells, eds.), ch. 4, pp. 73–114, Berlin: Springer-Verlag, 1993. Second edition 1996.

[53] B. M. ter Haar Romeny, L. M. J. Florack, J. J. Koenderink, and M. A. Viergever, "Scale-space: Its natural operators and differential invariants," in *Information Processing in Medical Imaging* (A. C. F. Colchester and D. J. Hawkes, eds.), vol. 511 of *Lecture Notes in Computer Science*, (Berlin), pp. 239–255, Springer-Verlag, July 1991.

[54] J. Llacer, B. M. ter Haar Romeny, L. M. J. Florack, and M. A. Viergever, "The representation of medical images by visual response functions," *IEEE Engineering in Medicine and Biology*, vol. 3, no. 93, pp. 40–47, 1993.

[55] W. J. Niessen, K. L. Vincken, A. S. E. Koster, and M. A. Viergever, "A comparison of multiscale image representations for image segmentation," in *Proc. IEEE Workshop on Mathematical Methods in Biomedical Image Analysis*, (San Francisco), June 21-22 1996. In press.

[56] A. H. Salden, B. M. ter Haar Romeny, and M. A. Viergever, "Affine and projective invariants of space curves," in *Conference on Geometric Methods in Computer Vision II (Part of SPIE's Annual International Symposium on Optoelectronic Applied Science and Engineering)* (B. C. Vemuri, ed.), (San Diego, Cal.), pp. 64–74, SPIE, July 12-13 1993.

Local Adaptative Filters for Image Restauration and Enhancement

L. Yaroslavsky

Department of Interdisciplinary Studies, University of Tel Aviv,
Tel Aviv University, 69978 - Ramat Aviv, Tel Aviv, Israel

Summary. On a base of a new class of criteria of image processing quality, the local criteria, a unified approach to treatment and design of local adaptive linear and non-linear rank filters for image restoration and enhancement is suggested and the corresponding examples are briefly outlined. It is shown also that many of the local adaptive filters known from the literature can be regarded as special cases of the filters generated with this approach.

1. Introduction.

It is commonly accepted that the design and substantiation of optimal methods for image processing should be based on statistical representations. This approach originates from the theory of statistical inferences and communication theory where it has been shown to be adequate and very productive. According to the statistical approach, images are treated as sample functions of some statistical ensemble, and optimality of processing algorithms is understood in terms of minimization of a losses of image quality averaged over this ensemble. Many different statistical models have been investigated for this purpose. The long history of these efforts, however, shows that the "optimal" algorithms as obtained through this approach often give results that are far from always being satisfactory. Recently, a number of new local adaptive linear and nonlinear algorithms of image processing have been introduced. The classical statistical approach is especially restricted in treatment of adaptive processing which appears to be the most promising in solving practical problems. We suggest an approach to synthesis and substantiating image processing algorithms better suited to adaptive processing. This approach is based on the local criteria of processing quality introduced in Sect. 2. Examples of using this approach to synthesis of local adaptive linear and rank filters are given in Sects. 3 and 4.

2. Local criteria

Any reasonable signal processing criterion, or quality measure assumes introducing a loss function that evaluates losses due deviations of individual processed signal samples from those of an ideal signal which is supposed to be a goal of the processing and averaging the loss function values over an ensamble of the samples. In the statistical approach, the ensemble is regarded as a statistical ensemble although, for the practical purposes, ensemble average is usually substituted by averaging over an observed realisation of signal samples and an assumtion on ensemble ergodicity and stationarity is drawn in for substantiating this substitution. We modify this approach in two ways. We separate the statistical averaging from the realization averaging and introduce the realization averaging as a primary notion rather than a secondary one characteristic for the statistical approach. As for the statistical averaging of the loss function values, we will separate statistical averaging over sensor's or imaging system's random noise which is always present in the observed image signal from that over an image ensemble. Moreover we will regard the statistical averaging over an image ensemble as optional and will try to when possible avoid it in order to synthesize adaptive filters whose parameters are optimized for a particular set of image samples involved into the above realization averaging of the loss function.

For a formal definition of the local criteria, introduce the following notations. Let

$b = \{b_{m,n}\}$ be a vector of processed image samples, $\{m,n\}$ -integer coordinates of the samples;

$\hat{a} = \{\hat{a}_{m,n}\}$ - vector of the resulting image samples;

$a = \{\hat{a}_{m,n}\}$ - vector of "true" picture samples the best possible estimation of which is the processing goal ;

$LOSS(\hat{a}_{m,n}, a_{m,n})$ - loss function that measures losses due to the deviation of each (m,n)-th sample of the processed image from a true one;

$LOC(m,n;a_{k.l})$ - a locality function that is nonzero for some subset of the pixels neighboring to the pixel (k,l) on a sampling raster and zero for the rest of the pixels; the pixel (k,l) will be referred to as central pixel of the neighborhood determined by the locality function. The quality measure defining the class of the local processing criteria is then as follows:

$$\begin{aligned} AVLOSS(k,l) &= AV_{ims}AV_{bg}AV_{ob}\Big\{\sum_{m,n} LOC(m,n;a_{k,l}) \\ &\quad \times LOSS(\hat{a}_{m,n}, a_{m,n})\Big\}, \end{aligned} \tag{2.1}$$

where AV_{imsys},AV_{bg} and AV_{ob} denote statistical averaging over the intrinsic noise of the imaging system generating the picture, the background component of the picture and unknown parameters of objects under interpretation, respectively. Thus, the local criteria assign to each pixel of the processed image a value of the loss function averaged over random factors defining the given processed image as well over the neighborhood of this pixel, i.e. over a subset of the pixels defined by the locality function.

The most important features of this type of criteria are the locality function defining neighborhood of each pixel of the image and separation of the loss function averaging over this neighborhood from that over statistical parameters associated with the imaging system random noise and unknown image background and details. The notion of neighborhood is widely used in image processing as well as in signal processing in general. Most commonly, the neighborhood is understood in a spatial sense, i.e. as a set of pixels surrounding the given one spatial-wise. It is defined by its area, i.e. by the number of constituent pixels. Such neighborhood consisting of K pixels that are Spatially Nearest to the given one will be referred to as the **KSN**-neighborhood. The use of the spatial neighborhood in image processing is the way to introduce into the processing an a priori knowledge regarding the image feature that pixels geometrically close to each other belong, as a rule, to the same detail and are, therefore, highly correlated. However, the spatial nearness is not the only indicator of pixels' belonging to the same detail. Moreover, it is not the only indicator.

Therefore, we suggest to generalize the notion of the spatial neighborhood by introducing new types of the neighborhoods which unify pixels in other alternative terms. These are the neighborhoods in terms of pixel values, in terms of pixel position in the so called variational row, i.e., in the sequence of pixels ordered with respect to their increasing values, and in terms of pixel's position in the local histogram of pixel values. Neighborhoods in terms of pixel values are **EV**-neighborhood and **KNV**-neighborhood. **EV**-neighborhood is a subset of pixels $a_{m,n}$ whose values deviate from the value of the central pixel $a_{k,l}$ at most by a predefined quantities $\{\epsilon_V^+; \epsilon_V^-\}$:

$$EV\{a_{k,l}\} = \{a_{m,n} : a_{k,l} - \epsilon_V^- \leq a_{m,n} \leq a_{k,l} + \epsilon_V^+\} \tag{2.2}$$

KNV-neighborhood is defined as a subset of a given number K of pixels Nearest in Value to the central pixel:

$$\begin{aligned} KNV(a_{k,l}) &= \Big\{a(p) : a(p) \\ &= dessort\left(a_{m,n} \Big| |a_{k,l} - a_{m,n}|\right), p = 1 = 1, 2, ..., K\Big\}; \end{aligned} \tag{2.3}$$

where $dessort(a_{m,n}|d)$ is a sequence of $\{a_{m,n}\}$ sorted according to descending order of the parameter d.

The neighborhoods in terms of pixel's position in the variational row, or its rank $R(a_{k,l})$ (**R**-neighborhoods), are the **ER**-neighborhood and **KNR**-neighborhood:

$$ER\{a_{k,l}\} = \{a_{m,n} : R(a_{k,l}) - \epsilon_R^- \leq R(a_{m,n}) \leq R(a_{k,l}) + \epsilon_R^+\} \tag{2.4}$$

$$\begin{aligned} KNR(a_{k,l}) &= \Big\{a(p) : a(p) \\ &= dessort\Big(R(a_{m,n})\Big| |R(a_{k,l}) - R(a_{m,n})|\Big), \\ & p = 1 = 1, 2, ..., K\Big\}; \end{aligned} \tag{2.5}$$

Pixels belonging to the same detail of the picture often form clusters or modes in the histogram of pixel values. This fact justifies introducing a histogram neighborhood. We define the Cluster, or CL-neighborhood, as a subset of pixels whose values fall into the same cluster or mode of the histogram as the central pixel does.

The processing quality measure (2.1) can be also expressed in terms of the distribution histogram $h_{NBH}(\hat{a})$ of pixels from a neighborhood NBH defined by the locality function:

$$\begin{aligned} & AVLOSS(k,l) \\ = \; & AV_{ims} AV_{bg} AV_{ob} \Big\{ \sum_{NBH} h_{NBH}(\hat{a}) LOSS\big(\hat{a}_{m,n}, a_{m,n}\big) \Big\}, \end{aligned} \quad (2.6)$$

with

$$NBH = \big(m, n : LOC(m, n; a_{k,l}) = 1\big) \quad (2.7)$$

Here it is assumed for simplicity that weights involved in the locality function are only binary. In the case of arbitrary weights formula (2.7) is valid for the so called weighted histograms:

$$h_{NBH}(v) = \sum_{\{m,n\} \in NBH} w_{m,n} \delta(v - a_{m,n}). \quad (2.8)$$

where $\delta(.)$ is the Kronecker delta and $\{w_{m,n}\}$ are the corresponding weights. Optimal processing is assumed to be aimed at an estimate that provides minimal averaged loss function $AVLOSS(k,l)$:

$$\begin{aligned} \{\hat{a}_{k,l}\}_{opt} = \; & argmin_{M(b \to \hat{a})} AV_{ims} AV_{bg} AV_{ob} \\ & \Big\{ \sum_{m,n} LOC(m, n; a_{k,l}) LOSS\big(\hat{a}_{m,n}, a_{m,n}\big) \Big\}, \end{aligned} \quad (2.9)$$

among all possible mappings $M(b \to \hat{a})$. Consequently the corresponding algorithm will be optimal on the average over the random factors involved in averaging. However, it is often desirable to find an optimal processing algorithm for each particular picture. In these cases one must abandon in (2.1) and (2.6) averaging AV_{bg} or AV_{bg} and AV_{ob} of the loss function over the parameters of the picture under processing. The resulting algorithms, if realizable, will be adaptive to those factors which were omitted from averaging in the optimality criteria. Moreover, since the criterion defines the averaged value of the loss function within a local neighborhood and for each picture element, these algorithms will be local adaptive. This implies that image must be processed by a sliding window with the corresponding adjustment of the algorithm's parameters in each position of the window.

An important peculiarity of the criteria (2.1) and (2.6) is the fact that the locality function, by definition, is determined for each pixel by its true value which is not known. This implies that the optimal processing should be an iterative procedure:

$$\{a_{k,l}^{(i)}\}_{opt} = argmin_{M(b\to\hat{a})} AV_{ims} AV_{bg} AV_{ob} \left\{\sum_{m,n} LOC(m,n;a_{k,l}^{(i-1)}) LOSS(\hat{a}_{m,n}, a_{m,n})\right\}, \quad (2.10)$$

where i is the number of iteration.

3. Local adaptive linear filters

Linear local adaptive filters result from the optimization equation (2.9) for a quadratic loss function

$$LOSS\left(\hat{a}_{m,n}, a_{m,n}\right) = \left|\hat{a}_{m,n} - a_{m,n}\right|^2 \quad (3.1)$$

and the locality function:

$$LOC(m,n;a_{k,l}) = \begin{cases} d_{k-m,l-n}, |k-m| \le N, \ |l-n| \le N \\ 0, \text{otherwise} \end{cases} \quad (3.2)$$

that defines a **KSN** neighborhood of $(2N+1)(2N+1)$ pixels.

Linear filtering can be described as multiplication of a signal b treated as a vector by a filter matrix H

$$\hat{a} = H \times b \quad (3.3)$$

In general, the number of operations per pixel required for the computer implementation of linear filtering is proportional to the squared size of the filter matrix. Much more simple in the computer implementation are scalar linear filters that are represented by diagonal matrices. Scalar filter

$$\hat{a} = T^{-1} \times H_d \times T \times b \quad (3.4)$$

that perform filtering in the domain of an orthogonal transform T can be used as approximations to the general optimal filtering. For the scalar filter, samples $\{\hat{\alpha}_{r,s}\}$ of the filtered signal spectrum $\hat{\alpha} = T \times \hat{a}$ are obtained by multiplying the spectral samples $\{\beta_{r,s}\}$ of the input signal b by the corresponding coefficients $\{\eta_{r,s}\}$ of the filter diagonal matrix:

$$\hat{\alpha}_{r,s} = \eta_{r,s} \times \beta_{r,s} \quad (3.5)$$

Such filtering is computationally much more efficient than the general filtering if fast transforms such as FFT or transforms that can be computed via recursive algorithms are used for the orthogonal transforms. One can obtain from the Eqs.(2.1), (3.1), (3.3) and (3.4) the following general expression for the optimal scalar filter [3] :

$$\eta_{r,s}^{opt} = \frac{AV_{imsys}AV_{ob}\left(\alpha_{r,s}\beta_{r,s}^{*}\right)}{AV_{imsys}AV_{ob}\left(|\beta_{r,s}|^2\right)} \tag{3.6}$$

where * denotes complex conjugate and $\{\alpha_{r,s}\}$ and $\{\beta_{r,s}\}$ are the spectral coefficients of fragments of the "true" and the observed signals, respectively, in the window defined by the locality function (3.2). If the observed signal b can be regarded as a sum of a undistorted signal a and signal independent noise n:

$$b = a + n, \tag{3.7}$$

the optimal noise filter for suppressing the noise is as following[3] :

$$\eta_{r,s}^{opt} = \frac{AV_{n}AV_{ob}\left(|\alpha_{r,s}|^2\right)}{AV_{imsys}AV_{ob}\left(|\beta_{r,s}|^2\right)} \tag{3.8}$$

where AV_n replaces AV_{ims} in Eq.(3.6) and denotes statistical averaging the observed signal power spectrum over the noise ensemble.

If the observed signal b can be regarded as output of a certain linear system described my a matrix L mixed with additive signal independent noise n :

$$b = L \times a + n, \tag{3.9}$$

optimal restoration filter is as following [3]:

$$\eta_{r,s}^{opt} = \lambda_{r,s}^{*}\frac{AV_{n}AV_{ob}\left(|\alpha_{r,s}|^2\right)}{AV_{imsys}AV_{ob}\left(|\beta_{r,s}|^2\right)} \tag{3.10}$$

where $\lambda_{r,s}^{*}$ are complex conjugate representation coefficients of the operator L with respect to the transform T. For instance, for the Fourier transform they are samples of the system frequency response.

In the derivation of the filters (3.8) and (3.10), averaging over background component of images has been abandoned in order to make the filters adaptive to the background component of the image. Since filters (3.8) and (3.10) are defined by local power spectrum of the signal under processing, they are local adaptive. The design of these filters requires knowledge of averaged power spectra of the true and observed signals. It is assumed in our approach that they can be estimated from the observed distorted signal. For instance, for the model (3.7) of additive noise with a known spectral density $\overline{|\nu_{r,s}|^2}$ one can use the following estimation

$$|\alpha_{r,s}|^2 \approx |\beta_{r,s}|^2 - \overline{|\nu_{r,s}|^2} \tag{3.11}$$

as a zero order approximation to the "true" signal power spectrum. Experiments show [3, 4] that the filters for noise suppression designed using even such a primitive estimate of the signal local power spectrum are capable of edge preserving image smoothing. Note also that a number of known from literature local adaptive linear filters such as those described in [5 – 9] can be regarded as special cases of the filters (3.8) and (3.10).

An important issue is computer implementation of the local adaptive linear filters. Since these filters perform filtering in spectral domain, their implementation requires computation of signal local spectra for every position of the moving window. Fortunately, for such important transforms as Fourier and cosine transforms, there exist a recursive algorithm for local spectral analysis whose computational complexity can be made proportional to the linear size rather than to the area of the window or even independent on this size [4] . This makes the implementation of the described local adaptive linear filters practical even on personal computers.

4. Rank filters

Selection in the criterion (2.1) nonquadratic loss function and/or locality functions that correspond to neighborhoods other than spatial neighborhood results in nonlinear filters. For instance, choice of the loss function in a form of module of difference between the estimation and the true signal value leads to the median filters in which filtered value is obtained as median over the neighborhood defined by the locality function; choice of a quadratic loss function and the locality function defining ER- neighborhood leads to the filters based on local order statistics, in which filtered signal value is obtained as a weighted sum of the signal order statistics [1].

The median and order statistics filters are representatives of a big family of the filters that we call rank filters [2, 3]. In the rank filters, filtering is carried out in a mowing window and the filtering result is obtained as one or another parameter of the signal local histogram. Computational complexity of the rank filters is of the order of magnitude of that of the local adaptive filters for there exists a fast recursive algorithm for computation of local histograms. In order to give a compact and constructive description of the family of rank filters, we suggest to classify them by type of the operation over the selected neighborhood they perform to obtain filtered signal value.

Introduce the following basic operations:

$MEAN(NBH)$ - arithmetic mean over the neighborhood NBH;

$ROS(NBH)$ - **R**-th order statistics over the neighborhood; this includes as special case

$MED(NBH)$ - median; $MAX(NBH)$ - maximum; $MIN(NBH)$ - minimum;

$MODE(NBH)$ - the highest maximum of the histogram over the neighborhood;

$RAND(NBH)$ - a pseudo-random number taken from the same distribution as the histogram over the neighborhood;

$RANK(NBH)$ - rank of the central pixel of the window (its position in the variational row over the neighborhood);

$SIZE(NBH)$ -size (number of pixels) of the neighborhood (applicable to **EV**-, **ER**- and bf CL- neighborhoods);

Using these basic operations and selecting different types of the neighborhood, one can generate a big variety of rank filtering algorithms for image smoothing, enhancement, edge detection, segmentation., etc. Here are some examples.

Rank filters for smoothing additive and impulse noise are, respectively,:

$$\hat{a}^{(i)} = SMTH(NBH(\hat{a}^{(i-1)})) \tag{4.1}$$

and

$$\hat{a}_{(i)} = \begin{cases} \hat{a}^{(i-1)}, \text{ if } \left|\hat{a}^{(i-1)} - SMTH(NBH(\hat{a}^{(i-1)}))\right| \leq thr; \\ SMTH(NBH(\hat{a}^{(i-1)})), \text{ otherwise}; \end{cases} \tag{4.2}$$

where $SMTH$ is one of the smoothing operations $MEAN$, MED,*ROS*, $MODE$ or $RAND$, *thr* is a detection threshold of the impulse interference, and i is the number of iterations. Since image segmentation can be treated as smoothing aimed at obtaining a piece-wise constant image, the above smoothing filters can also be used for image segmentation.

Two modifications of rank order filters for local contrast enhancement are "unsharp masking" filters:

$$\hat{a} = G(a - SMTH(NBH(\hat{a}))) + SMTH(NBH(\hat{a})) \tag{4.3}$$

where G is an enhancement coefficient, and local histogram equalization:

$$\hat{a} = RANK(NBH). \tag{4.4}$$

Numerous rank filters described in literature (e.g. [1]) can be regarded as special cases of the described family.

References

[1] I. Pitas and A.N. Venetsanopoulos, Nonlinear Digital Filters. Principles and Applications. Boston: Kluwer Acad. Publ., 1990.

[2] V. Kim and L. Yaroslavskii, "Rank algorithms for picture processing", CVGIP, vol. 35, pp. 234-258, 1986.

[3] L.P. Yaroslavsky, "Linear and rank adaptive filters for picture processing", in Digital Image Processing and Computer Graphics. Theory and Applications, L.Dimitrov and E. Wenger, Eds., Wien, Muenchen: R. Oldenburg, pp. 374, 1991.

[4] L.P. Yaroslavsky, Digital Picture Processing. An Introduction. Springer Verlag, Heidelberg, 1985

[5] D.T. Kuang, A.A. Sawchuk, T.C. Strand and P. Chavel, "Adaptive noise smoothong filter for images with signal-dependent noise", IEEE Trans, vol. PAMI-7, pp. 165-177, 1985.

[6] S.S. Jiang and A. Sawchuk, "Noise apdating repeated filters and other adaptive smoothing filters using local image statistics", Applied Optics, vol. 25, pp. 2326-2336, 1986.

[7] J.S. Lim, Two dimensional image and signal processing. Englewood Cliffs, N.J.: Prentice Hall, 1990.

[8] J.S. Lee, "Digital image enhancement and noise filtering by use of local statistics", IEEE Trans., vol. PAMI-2, pp. 165-168, 1980.

[9] M. Unser, "Improved restoration of noisy images by adaptive least-squares post-filtering", Signal Processing, vol. 20, pp. 3-14, May 1990.

Active Contours

3D Active Contours

V. Caselles[1], R. Kimmel[2], G. Sapiro[3], and C. Sbert[1]

[1] Dpt. of Mathematics, Univ. Illes Balears. Ctra. Valldemossa km 7.5, Palma de Mallorca. Spain, dmivca0@ps.uib.es, dmicsj0@ps.uib.es
[2] Dpt. Electrical Engineering, Technion, I.I.T., Haifa 32000, Israel
[3] Hewlett-Packard Labs, 1501 Page Mill Road, Palo Alto, CA 95304

Summary. A novel geometric approach for three dimensional object segmentation is presented. The scheme is based on geometric deformable surfaces moving towards the object to be detected. We show that this model is equivalent to the computation of surfaces of minimal area in a Riemannian space. The new approach is stable, robust and automatically handles changes in the surface topology during the deformation.
Key words: 3D object segmentation, dynamic surface, Riemannian geometry, minimal surfaces.

1. Introduction

One of the basic problems in image analysis is object detection, associated with the problem of boundary detecion. One solution to this problem was given firstly by Kass et al. [6]. They proposed the model of *snakes* or *active contours* which is based on deforming an initial contour or surface towards the boundary of the object to be detected. The deformation is obtained by trying to minimize an energy functional such that its minima is obtained at the boundary of the object.

Let $C : [0,1] \to \mathcal{R}^2$ be a parametrized curve and I a given image where we want to detect the objects boundaries. The snake model associated to C an energy given by

$$E(C) = \alpha \int_0^1 |C'(t)|^2 dt + \beta \int_0^1 |C''(t)|^2 dt - \lambda \int_0^1 |\nabla I(C(t))| dt \qquad (1.1)$$

where $\alpha, \beta, \lambda \geq 0$. The first two terms basically control the smoothness of the contours to be detected and the third term is responsible for attracting the contour towards the object.

This energy model is not capable of changing its topology. The topology of the final curve will be as that of the initial one. This may be a problem when an un-known number of objects must be simultaneously detected. It is clear that classical snake model can be generalized to 3D images, where the boundaries of the objects are surfaces.

Recently, Caselles et al. [2] have proposed a geometrical model of deformable contours, later a similar one has been proposed by Malladi et al. [7]. These models are based on the theory of surfaces evolution and geometric flows. Assume that the deforming curve C is given by a level set of a function

$u : \mathcal{R}^2 \to \mathcal{R}$, then, the deformation of C is given via the deformation of u by an evolution equation based on mean curvature motion given by

$$u_t = |\nabla u| g(I) div\left(\frac{\nabla u}{|\nabla u|}\right) + g(I)\nu|\nabla u| \tag{1.2}$$

$$u(0,x) = u_0(x) \tag{1.3}$$

where $g(I) = \frac{1}{1+(DG_\sigma * I)^2}$ is the stopping factor, ν is a positive real constant and u_0 is a smoothed version of $1-\chi_C$, where χ_C is the characteristic function of a set C containing the object of interest in the image I.

This model allows automatic changes in topology. Thereby, several objects can be detected simultaneously. The model can easily extended to 3D object detection.

In [3], they have shown the relation between these two approaches for two dimensional object. They proposed a model *Geodesic Snakes* that unifies both in the sense that, the classical snakes are equivalent to geodesic computation and assuming a level set representation geodesic curve can be found by a geometric flow similar to the above.

Here we extend the results in [3] to three dimensional object detection. We show that the desired boundary is given by a minimal surface in a Riemannian space defined by the image. The plan of the paper is as follows. In Section 2 we describe the 2D geodesic active contours. In Section 3 we present the three dimensional deformable model as minimal surface. Correctness of the model, existence and uniqueness results are presented in Section 4. Experimental results are given in Section 5.

2. Geodesic Active Contours

Let us consider a particular case of 1.1, where $\beta = 0$ and replacing the edge detector $|\nabla I|$ by a general function $g(|\nabla I|)^2$ of the gradient such that $g(r) \to 0$ as $r \to \infty$, we obtain

$$E(C) = \alpha \int_0^1 |C'(t)|^2 dt + \lambda \int_0^1 g(|\nabla I(C(t))|)^2 dt = E_{int} + E_{ext} \tag{2.1}$$

This functional is not intrinsic. In order to solve this problem and with the help of Maupertuis Principle, they proved ([3]) that the minimum of 2.1 with the restriction of a fixed energy level E_0 (motivated by the discussion on ideal edge they choose $E_0 = 0$), is given by a geodesic curve in a Riemannian space with the metric defined by $g_{ij} = (E_0 + \lambda g(|\nabla I|)^2)\delta_{ij}$. In other word to minimize 2.1 is equivalent to solve

$$Min_C \int_0^1 g(|\nabla I(C(\tau))|)|C'(\tau)| d\tau \tag{2.2}$$

Assuming C is a level set of a function u, then the level set formulation of the steepest descent method says that solving the above geodesic problem is equivalent to search for the steady state ($\frac{\partial u}{\partial t} = 0$) of the following evolution equation

$$\frac{\partial u}{\partial t} = |\nabla u| div \left(g(I) \frac{\nabla u}{|\nabla u|} \right) = g(I) |\nabla u| div \left(\frac{\nabla u}{|\nabla u|} \right) + \nabla g \cdot \nabla u \qquad (2.3)$$

Comparing the equation 2.3 with 1.2, we see that the term $\nabla g \cdot \nabla u$ is missing in the old model. This new term directs the curve towards the boundary of the objects ($-\nabla g$ points toward the center of the boundary) and eventually force it to stay there.

In the old model the curve stops when $g = 0$, this happens only along an ideal edge, this makes the geometrical model 1.2 inappropiate for the detection of boundaries in real images.

3. Three Dimensional Deformable Models as Minimal Surfaces

We extend the model in the previous section to 3D surfaces by computing surfaces of minimal area, where area is defined in a given Riemannian space as well. In the case of surface we consider the "weighted" area

$$A_R = \int \int g(I) da \qquad (3.1)$$

where da is the Euclidean element of area. In A_R the area element is given by $g(I)da$. This is the 3D analogue of the metric used in [3] to construct the geodesic active contour model. The basic element of our deformable model will be given by minimizing 3.1 by means of an evolution equation obtained from its Euler-Lagrange.

Computing now the Euler-Lagrange of A_R, we get

$$S_t = (g\mathbf{H} - \nabla g \cdot \mathbf{N})\mathbf{N} \qquad (3.2)$$

where S is the 3D surface, $\mathbf{H}$ is its mean curvature and $\mathbf{N}$ its inner unit normal. Taking a level set representation, in analogy with 2.3, the steepest descent method to minimize 3.1 gives

$$\frac{\partial u}{\partial t} = |\nabla u| div \left(g(I) \frac{\nabla u}{|\nabla u|} \right) = g(I) |\nabla u| div \left(\frac{\nabla u}{|\nabla u|} \right) + \nabla g \cdot \nabla u \qquad (3.3)$$

We can add a constant force as a constraint obtaining the general *minimal surfaces model* for object detection

$$\frac{\partial u}{\partial t} = |\nabla u| div \left(g(I) \frac{\nabla u}{|\nabla u|} \right) + \nu g(I) |\nabla u| \qquad (3.4)$$

This is the flow we will further analyze and use for 3D object detection. It has the same properties and geometric characteristics as the geodesic active contours, allowing automatic changes in topology, several objects can be detected simultaneously, without previous knowledge of their exact number.

4. Main Results

4.1 Existence and Uniqueness Results

As shown in the previous section, our 3D object detection model is given by

$$\frac{\partial u}{\partial t} = |\nabla u| div \left(g(I) \frac{\nabla u}{|\nabla u|} \right) + \nu g(I) |\nabla u| \quad (t,x) \in [0,\infty[\times \mathcal{R}^3 \tag{4.1}$$

$$u(0,x) = u_0(x) \qquad x \in \mathcal{R}^3 \tag{4.2}$$

This model should be solved in $R = [0,1]$ with Neumann boundary conditions. To simplify, and as is usual done in the literature we extend the images by reflection to $\mathcal{R}^3$ and we look for solutions of 4.1 which are periodic.

Existence and uniqueness results for equation 4.1 can be proved using the theory of viscosity solutions [5]. We have

Theorem 4.1. *Let $W^{1,\infty}$ denote the space of bounded Lipschitz functions in $\mathcal{R}^3$. assume that $g \geq 0$ is such that $\sup\{|\nabla g^{\frac{1}{2}}(x)| : x \in \mathcal{R}^3\} < \infty$ and $\sup\{|\partial_{ij} g(x)| : x \in \mathcal{R}^3, i,j = \{1,2,3\}\} < \infty$. Let $u_0, v_0 \in C(\mathcal{R}^3) \cap W^{1,\infty}(\mathcal{R}^3)$. Then*

1. *The equation 4.1 admits a unique viscosity solution $u \in C([0,\infty[\times \mathcal{R}^3) \cap L^\infty(0,T; W^{1,\infty}(\mathcal{R}^3))$ for all $T < \infty$. Moreover, it satisfies*

$$\inf_{\mathcal{R}^3} u_0 \leq u(t,x) \leq \sup_{\mathcal{R}^3} u_0 .$$

2. *Let $v \in C([0,\infty[\times \mathcal{R}^3)$ be the viscosity solution of 4.1 with initial condition v_0. Then for all $T \in [0,\infty[$ we have*

$$\sup_{0 \leq t \leq T} \| u(t,x) - v(t,x) \|_{L^\infty(\mathcal{R}^3)} \leq \| u_0(x) - v_0(x) \|_{L^\infty(\mathcal{R}^3)},$$

which means that the solution is stable.

The proof of this Theorem follows the same steps of the corresponding proof for the model 1.2 (see [2], Theorem 3.1).

4.2 Correctness of the Geometric Minimal Surfaces Model

By correctness we mean that a smooth contour in an ideal image with no noise should be recovered by the model.

To study the asymptotic behavior of the equation

$$\frac{\partial u}{\partial t} = |\nabla u| div \left(g(I) \frac{\nabla u}{|\nabla u|} \right) + \nu g(I) |\nabla u| \quad (t, x) \in [0, \infty[\times \mathcal{R}^3 \tag{4.3}$$

with initial condition

$$u(0, x) = u_0(x) \quad \forall x \in \mathcal{R}^3, \tag{4.4}$$

we assume that $S = \{x \in [0,1]^3 : g(x) = 0\}$ is a compact surface of class C^2. S divides the cube $[0,1]^3$ in two regions: the interior region $I(S)$ and the exterior region $E(S)$. The initial datum u_0 will be taken in $C^2(\mathcal{R}^3)$ periodic with fundamental domain $[0,1]^3$ and vanishing in an open neighborhood of $S \cup I(S)$. Let $u(t,x)$ be the unique viscosity solution of 4.3 given by Theorem 4.1 above. We follow the evolution of the set $G(t) = \{x \in [0,1]^3 : u(t,x) = 0\}$ whose boundary $S(t)$ we are interested in.

The main result we have proved is the following (see [4])

Theorem 4.2. *Suppose that $S = \{x \in [0,1]^3 : g(x) = 0\}$ is a smooth compact surface. If the constant ν is sufficiently large, then $S(t)$ converges to S in the Hausdorff distance as $t \longrightarrow \infty$.*

Remark 4.1. If we consider the model 4.3 with $\nu = 0$, then the above result holds if we take our initial surface sufficiently close to S.

Remark 4.2. If $u_0(x)$ is taken as

$$u_0(x) = \begin{cases} > 0 & x \in \overline{B(y,r)} \subset I(S) \\ 0 & x \quad \text{in a neighborhood of} \quad S \cup E(S) \end{cases}$$

i.e. a function vanishing in a neighborhood of $S \cup E(S)$, then the above result holds. We may recover a surface S starting from its inner region.

5. Experimental Results

We now present some examples of our minimal surfaces deformable model. The numerical implementation is based on the algorithm for surface evolution via level sets developed by Osher and Sethian [8] and recently used by many authors for different problems in computer vision and image processing. The algorithm allows the evolving surface to change topology without monitoring the deformation.

Figure 5.1 presents the evolution of an initial sphere (left) surrounding a torus (right). Note that the model detects the different topology.

Figure 5.2 presents an example of the detection of a knotted surface. This object is composed by two linked torus. The initial condition is an ellipsoid surrounding the two torus (top left). Note how the model manages to split and detect the knotted surface (bottom right).

A medical image is given in Figure 5.3 and 5.4. Figure 5.3 presents the 3D detection of a tumor in an MRI image. The initial 3D shape is presented in the left (small sphere inside the tumour) and the final shape, the weighted minimal surface, is presented in the right. Figure 5.4 shows slices of this 3D detection, together with the corresponding MRI data.

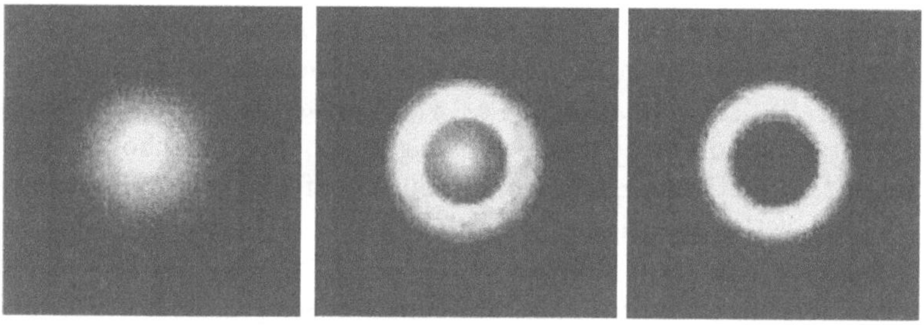

Fig. 5.1. Evolution of a sphere until it reproduces the torus

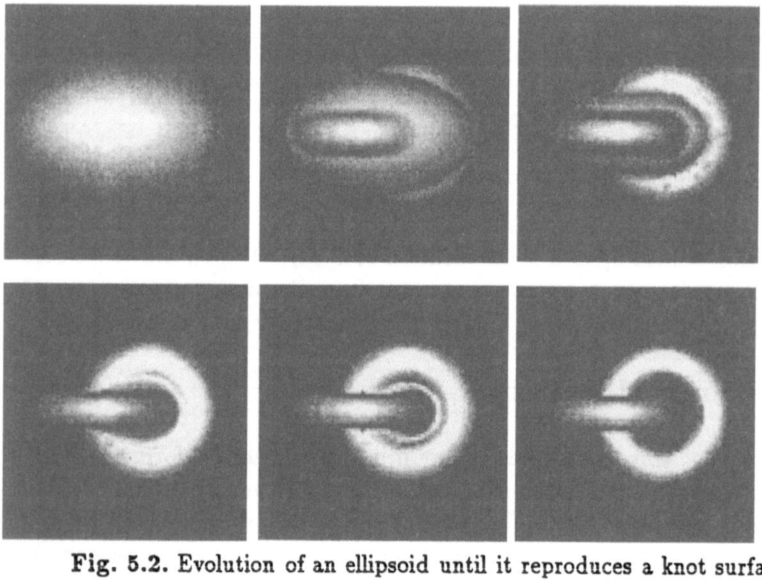

Fig. 5.2. Evolution of an ellipsoid until it reproduces a knot surface

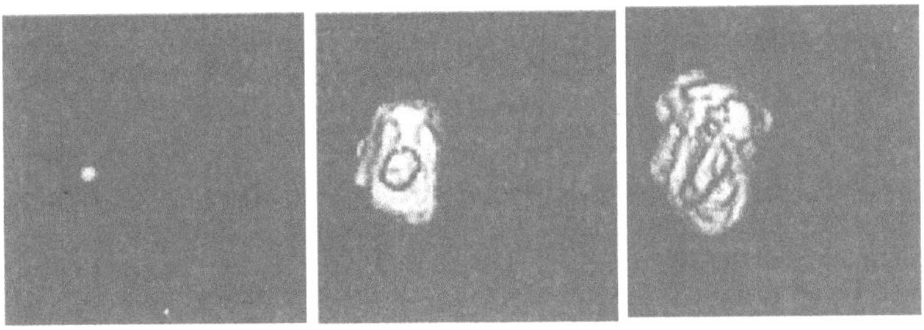

Fig. 5.3. Evolution of a sphere inside a brain tumour until it recovers the surface of the tumour

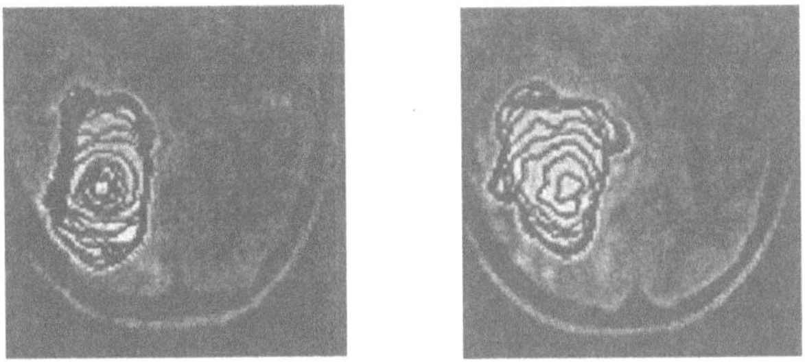

Fig. 5.4. Two slices of the above tumour with the corresponding sections of the deformable surface

References

1. Adalsteinsson D. and Sethian J.A. (1993): A fast level set method for propoagating interfaces. LBL TR-University of Berkeley.
2. Caselles V., Catté F., Coll T. and Dibos F. (1993): A geometric model for active contours. Numerische Mathematik **66**, pp 1-31.
3. Caselles V., Kimmel R. and Sapiro G. (1994): Geodesic active contours. HP Labs Technical Report.
4. Caselles V., Kimmel R., Sapiro G. and Sbert C. (1995): Minimal surfaces: A three dimensional segmentation approach. Technical Report, Department of electrical engineering, Technion-Israel Institute of Technology.
5. Crandall M.G., Ishii H. and Lions P.L. (1992): User's guide to viscosity solutions of second order partial linear differential equations. Bulletin of American Math. Society **27**, pp. 1-67.
6. Kass M., Witkin A. and Terzopoulos D. (1988): Snakes: Active contour models. International Journal of Computer Vision **1**, pp. 321-331.
7. Malladi R., Sethian J.A. and Vemuri B.C. (1995): Shape modelling with front propagation: A level set approach. IEEE Trans. on PAMI.
8. Osher S.J. and Sethian J.A. (1988): Fronts propagation with curvature dependent speed: Algorithms based on Hamilton-Jacobi formulations. Journal of Computational Physics **79**, pp. 12-49.

Regularization properties for Minimal Geodesics of a Potential Energy

Laurent COHEN[1] and Ron KIMMEL[2]

[1] CEREMADE, U.R.A. CNRS 749, Université Paris IX-Dauphine, Place du Marechal de Lattre de Tassigny, 75775 Paris CEDEX 16, France
cohen@ceremade.dauphine.fr

[2] Lawrence Berkeley Laboratory, University of California, Berkeley, Mailstop 50A-2129 LBL UC Berkeley, California 94720, USA
ron@csr.lbl.gov

Abstract. Some new results on our approach [2] of edge integration for shape modeling are presented. It enables to find the global minimum of active contour models' energy between two points. Initialization is made easier and the curve cannot be trapped at a local minimum by spurious edges. We modified the "snake" energy by including the internal regularization term in the external potential term. Our method is based on the interpretation of the snake as a path of minimal length on a surface or minimal cost. We then make use of level sets propagation to find the shortest path which is the global minimum of the energy among all paths joining two endpoints.
We show that our energy, though only based on a potential integration along the curve, has a regularization effect like snakes. We show a relation between the maximum curvature along the resulting contour and the potential generated from the image.
Keywords: Shape modeling, Deformable Models, Weighted distance transform, Shape Segmentation, Feature Extraction, Energy Minimization, P.D.E.'s, Curve Evolution.

1. Introduction

An active contour model for boundary integration and features extraction, introduced in [7], has been considerably used and studied during the last years.

Although the smoothing effect of the snakes may overcome small defaults in the data, spurious edges generated by noise or in a complex image may stop the evolution of the curve so that it might be trapped by an insignificant local minimum of the energy. The inflation or expansion force [3] helps to prevent the contour from being trapped by isolated edges into a local minimum.

In this paper we present some results on a new approach, introduced in [2], for finding the global minimum for energy minimizing curves. Only endpoints are needed as an easy initialization and we are guaranteed that the global minimum is found between these points and spurious edges cannot lead to a local minimum. The deformable contour model is a mapping $\mathcal{C}(s) = (x(s), y(s))$ where $s \in \Omega = [0,1]$ with an energy of the following form:

$$E(\mathcal{C}) = \int_{\Omega} \frac{w_1}{2}\|\mathcal{C}_s(s)\|^2 + \frac{w_2}{2}\|\mathcal{C}_{ss}(s)\|^2 + P(\mathcal{C}(s))ds \qquad (1.1)$$

where P is the potential associated to the external forces.

2. Paths of Minimal Action

The minimization problem we are trying to solve is slightly different from the deformable models, though there is much in common. The reason we modified the energy is that we now have an expression where the internal regularization energy is included in the potential term. We can then solve the energy minimization in a similar way to that of finding the shortest path on a surface using the method developed in [8]. The energy of the new model has the following form:

$$E(\mathcal{C}) = \int_{\Omega} w\|\mathcal{C}_s(s)\|^2 + P(\mathcal{C}(s))ds = \int_{\Omega} \tilde{P}(\mathcal{C}(s))ds = wL + \int_{\Omega} P(\mathcal{C}(s))ds \quad (2.1)$$

Here $\mathcal{C}$ is in the space of all curves connecting two given points (restricted by boundary conditions): $\mathcal{C}(0) = p_0$ and $\mathcal{C}(L) = p_1$, where L is the length of the curve. Contrary to the classical snake energy, here s represents the arc-length parameter, which means that $\|\mathcal{C}_s(s)\|^2 = 1$. This makes the energy depend only on the geometric curve and not on the parameterization (see [1]). The regularization term with w, now exactly measures the length of the curve. Having the above minimization problem in mind, we first search for the *surface of minimal action* U starting at $p_0 = \mathcal{C}(0)$. At each point p of the image plane, the value of this surface U corresponds to the minimal energy integrated along a path starting at p_0 and ending at p.

$$U(p) = \inf_{\mathcal{C}(L)=p} \left\{ \int_{\mathcal{C}} \tilde{P} ds \right\} \quad (2.2)$$

In [8], a method to determine the shortest path on a surface between a start point p_0 and a destination p_1 was presented. Applying these ideas to minimize our energy (2.1), it is possible to formulate a partial differential evolution equation describing the set of equal energy contours $\mathcal{L}$ in 'time'. These are the level set curves of U defined by equation (2.2). The evolution equation is of the form:

$$\frac{\partial \mathcal{L}(s,t)}{\partial t} = \frac{1}{\tilde{P}} \boldsymbol{n}(s,t), \quad (2.3)$$

where $\tilde{P} = P + w$ and $\boldsymbol{n}(s,t)$ is the normal to the closed curve $\mathcal{L}(.,t) : S^1 \to \mathbb{R}^2$. This evolution equation is initialized by a curve $\mathcal{L}(s,0)$ which is a small circle surrounding the point p_0. It corresponds to a null energy. This evolution equation (2.3) is similar to a balloon evolution [3] with an inflation force depending on the potential.

This equation is solved using the Eulerian formulation for curve evolution introduced in [10] to overcome numerical difficulties and handle topological changes.

Data: given by $\tilde{P}$ and the two endpoints p_0 and p_1.
Step 1: Minimal Action U_0 from p_0 using front propagation which finds level set curves $\mathcal{L}$ of U_0 starting from an infinitesimal circle centered at p_0 (Osher-Sethian).
Step 2: Backpropagation: tracking the minimal path by gradient descent on U_0 starting from p_1 ending at p_0.

We have just presented a sketch of the algorithm. A synthetic example is presented in Figure 2.1. Observe the way the level curves propagate faster along the road.

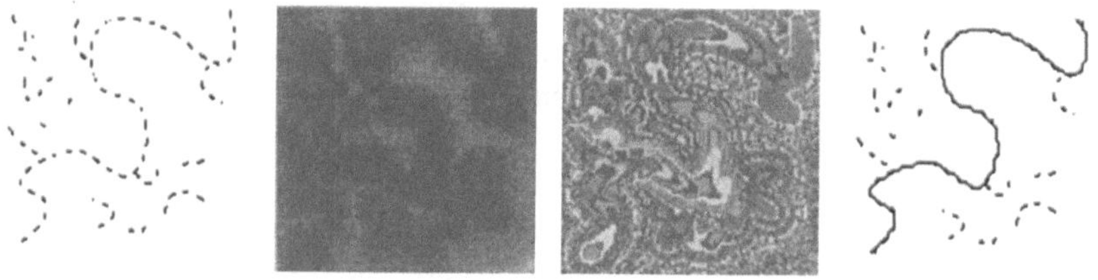

Fig. 2.1. Line image. From left to right: original, potential, minimal action (random look up table to show the level set propagation starting from the bottom left), minimal path between bottom left and top right.

3. Regularization properties

3.1 Curvature Bound

We now show how the constant w and the potential P in the energy of (2.1) influence the smoothness of the solution minimizing the energy E and make it behave like a regular snake.

We shall make use of the following lemmas to introduce an upper bound on the curvature along the resulting contour $C(s)$ by controlling the potential P.

Lemma 3.1. *The curvature magnitude $|\kappa| = \|C_{ss}\|$ along the geodesics minimizing*

$$\int_{\Omega} P(C(s))ds, \tag{3.1}$$

where s is the arclength parameter, is bounded by

$$|\kappa| \leq \sup_{\Omega} \left\{ \frac{\|\nabla P\|}{P} \right\}. \tag{3.2}$$

Proof. Following [1], the Euler-Lagrange equation of (3.1) is given by

$$PC_{ss} - \langle \nabla P, \frac{C_{ss}}{||C_{ss}||} \rangle \frac{C_{ss}}{||C_{ss}||} = 0.$$

Using the geometrical relation $C_{ss} = \kappa \boldsymbol{n}$ we can rewrite the above expression, that indicates the curve's behavior at the minima of (3.1), as

$$P\kappa \boldsymbol{n} - \langle \nabla P, \boldsymbol{n} \rangle \boldsymbol{n} = 0.$$

This yields the following expression for the curvature along the geodesics of P:

$$\kappa = \frac{\langle \nabla P, \boldsymbol{n} \rangle}{P}.$$

Since $\boldsymbol{n}$ is a unit vector, the numerator is a projection on a unit vector operation. Thus, we can conclude that along any geodesic path minimizing (3.1) the curvature magnitude is bounded by Equation (3.2). □

Using Lemma 3.1, an *a priori* bound of the curvature magnitude may be obtained by evaluation of sup and inf of P over the image domain $\mathcal{D}$ instead of the curve domain Ω in (3.2). We readily have the following result which applies to our case with the energy of (2.1):

Lemma 3.2. *Given a potential $P \geq 0$, and let $\tilde{P} = w + P$, the curvature magnitude $|\kappa| = ||C_{ss}||$ along the geodesics minimizing the energy of (2.1) is bounded by*

$$|\kappa| \leq \frac{\sup_{\mathcal{D}}\{||\nabla P||\}}{w}. \tag{3.3}$$

Proof. Since $P \geq 0$ we have that $\inf_{\mathcal{D}}\{\tilde{P}\} \geq w$. Using this relation and Equation (3.2), we have:

$$\begin{aligned} |\kappa| \leq \sup_{\Omega} \left\{ \frac{||\nabla \tilde{P}||}{\tilde{P}} \right\} = \sup_{\Omega} \left\{ \frac{||\nabla P||}{P + w} \right\} &\leq \sup_{\mathcal{D}} \left\{ \frac{||\nabla P||}{P + w} \right\} \\ &\leq \frac{\sup_{\mathcal{D}}\{||\nabla P||\}}{w} \end{aligned}$$

□

Equation (3.3) enables us to control the behavior of any geodesic minimizing (2.1), and especially the minimal geodesics that interest us. Lemma 3.1 also gives a nice interpretation of the connection between the curvature of the resulting contour, and the ratio between the gradient magnitude and the value of the potential P. When the curve's normal is orthogonal to the slope of P, so that the curve is directed towards the valley, then the curvature is zero implying a straight line. While if the curve travels along a contour of equal height in P, then the normal $\boldsymbol{n}$ coincides with the slope of P and the curvature increases causing the curve to bend and direct the curve to flow into the valley, where the potential is lower.

The conclusion is that to decrease the limit of the curvature magnitude of the geodesics in equation (3.3), and thereby lead to a smoothing effect on the resulting contour, we have two different ways:

- Smoothing the potential (or the image) to decrease $\sup_{\mathcal{D}}\{\|\nabla P\|\}$.
- Increasing the constant w added to P increases the denominator w without affecting $\sup_{\mathcal{D}}\{\|\nabla P\|\}$. This gives a justification for calling w a regularization parameter in Section 2..

Figure 3.1 shows the effect of changing w on the solution. The potential is based on the image gradient like in [7].

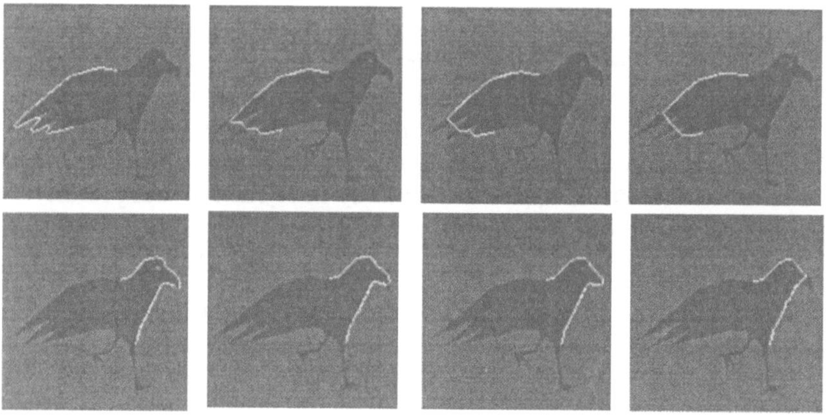

Fig. 3.1. Regularization effect by increasing the coefficient w from left to right.

3.2 Case of Attraction Potential

As introduced in [3], previous local edge detection might be taken into account as data for defining the potential. Indeed, since the gradient norm usually changes its values along a boundary contour, this operation assigns an equal attraction weight along the boundary. Edge points are scattered over the image domain and serve as the key points in generating a single boundary contour. The difficulty here is that there is no order in the set of points and that it is unknown in advance which points belong to the boundary. This is defined as implicit constraints in [4]. One possible way of defining a potential P is as a function of the distance map [5], where each point p is assigned with a value representing the shortest Euclidean distance to an edge point:

$$d(p) = \inf_{q \in \text{ edge}} \{dist(p,q)\}, \quad \text{and} \quad P(p) = f(d(p)) \tag{3.4}$$

where $dist(p,q)$ is the Euclidean distance between the two points p and q and f is an increasing function. An example of distance map is shown in Figure 2.1. Consistent numerical approximations of (3.4) for the computation of $d_{\mathcal{E}}$ on a sequential computer involves in high complexity. Quick sequential algorithms [6] were used for defining the attraction potential in [5]. Sub-pixel estimation of the distance using a parallel algorithm was presented in [9]. It gives a high *sub-pixel* precision of the distance. This is one possible application of shortest path estimation [8]. Note also that the distance potential selection P may be also considered as the normalized force introduced in [3] for stabilizing the results (*i.e.* for $P = d_{\mathcal{E}}$ we have $||\nabla P|| = \frac{\nabla P}{||\nabla P||}$) since $||\nabla d_{\mathcal{E}}|| = 1$ almost everywhere. This last equality is useful in the context of the previous section to get an estimation of the curvature's bound when $\tilde{P} = w + d_{\mathcal{E}}$. From equation (3.3), we have:

$$|\kappa| \leq \frac{1}{w}, \tag{3.5}$$

i.e. w is the minimum curvature radius along the final contour. In the case $\tilde{P} = w + f(d_{\mathcal{E}})$, the upper bound becomes

$$|\kappa| \leq \sup_d \frac{f'(d)}{w+d}. \tag{3.6}$$

This bound can be easily found for the usual functions $f(d) = \alpha d^2$ or $f(d) = 1 - e^{-\alpha d^2}$ which corresponds to robust statistics (see [4]).

4. Examples and Results

We demonstrate the performance of the proposed algorithm by applying it to several real images. The images were scaled to 128×128 pixels. In the first example, we are interested in a road detection between two points in the image of Figure 4.1. Road areas are lighter and correspond to higher gray levels. The potential function P was thus selected to be the opposite of the gray level image itself: $P = -I$. Minimizing this potential along a curve yields a path that follows the middle of the road. Our approach can be used for the minimization of many paths emerging from the same point in one single calculation of the minimal action. Given a start point in the upper left area, the path achieving the global minimum of the energy is found between this point and four other given points to determine the roads graph in our previous image.

In the second example, we want to extract the left ventricle in an MR image of the heart area. The potential is a function of the distance to the closest edge in a Canny edge detection image (see Figure 4.2). Since it is a closed contour, given a single point, saddle point classification [2] is used to find the second end point. The closed contour is formed of the two minimal paths joining the end points.

Fig. 4.1. The initial data is shown on the left. In the middle, our path of minimal action connecting the two black points as start and end points. On the right, many paths are obtained simultaneously connecting the start point on the upper left to 4 other points.

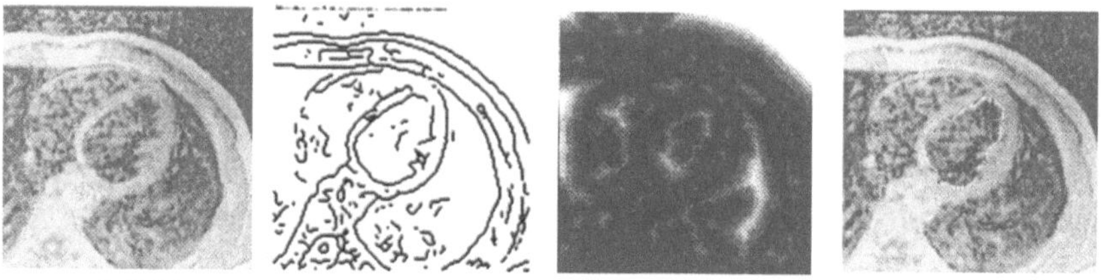

Fig. 4.2. MRI heart image, from left to right: Original image, edge image, distance map and Heart ventricle detection. The start point is on the lower left and the other one is the detected saddle.

5. Concluding Remarks

In this paper we presented some regularization properties of a method for integrating objects boundaries by searching for the path of minimal action connecting two points. The search for the global minimum makes sense only after the two end points are determined, and the 'action' or 'potential' is generated from the image data. The proposed approach makes snake initialization an easier task that requires only one or two end points and overcomes one of the fundamental problems of the active contour model, that is being trapped by a local minimum.

An upper bound over the curvature magnitude of the final contour was obtained by the ratio of gradient magnitude and the value of the potential. It was shown that controlling the smoothness of the final contour is possible by adding a regularization term to the potential function, thereby decreasing this bound.

The result of the proposed procedure may be considered either as the solution or as initial condition for classical snake models for further smoothing. Convergence to the proper smoothed version should now be almost immediate, since the global minimum should be close to its smoothed version obtained by a classical snake.

References

1. V. Caselles, R. Kimmel, and G. Sapiro. Geodesic active contours. In *Proc. Fifth IEEE ICCV*, pages 694–699, Cambridge, USA, June 1995.
2. L. Cohen and R. Kimmel. Edge integration using minimal geodesics. Technical report 9504, Ceremade, Université Paris Dauphine, January 1995.
3. Laurent D. Cohen. On active contour models and balloons. *Computer Vision and Image Understanding*, 53(2):211–218, March 1991.
4. Laurent D. Cohen. Auxiliary variables for deformable models. In *Proc. Fifth IEEE ICCV* , pages 975–980, Cambridge, USA, June 1995. Long version in *Journal of Mathematical Imaging and Vision*, 6(1), January 1996.
5. Laurent D. Cohen and Isaac Cohen. Finite element methods for active contour models and balloons for 2D and 3D images. *IEEE* , PAMI-15(11), November 1993.
6. P. E. Danielsson. Euclidean distance mapping. *Computer Vision, Graphics, and Image Processing*, 14:227–248, 1980.
7. Michael Kass, Andrew Witkin, and Demetri Terzopoulos. Snakes: Active contour models. *International Journal of Computer Vision*, 1(4):321–331, 1988.
8. R. Kimmel, A. Amir, and A. Bruckstein. Finding shortest paths on surfaces using level sets propagation. *IEEE* , PAMI-17(6):635–640, June 1995.
9. R. Kimmel, N. Kiryati, and A. M. Bruckstein. Distance maps and weighted distance transforms. *Journal of Mathematical Imaging and Vision*, 1995. Special Issue on Topology and Geometry in Computer Vision, to appear.
10. S. J. Osher and J. A. Sethian. Fronts propagation with curvature dependent speed: Algorithms based on Hamilton-Jacobi formulations. *Journal of Computational Physics*, 79:12–49, 1988.

Multi-Resolution algorithms for Active Contour Models

Bertrand Leroy[1], Isabelle L. Herlin[1], and Laurent D. Cohen[2]

[1] INRIA - AIR project, Rocquencourt, B.P. 105, 78153 Le Chesnay cedex, France. email: Bertrand.Leroy@inria.fr, Isabelle.Herlin@inria.fr

[2] CEREMADE, U.R.A. CNRS 749, Université Paris IX- Dauphine, Place du Maréchal de Lattre de Tassigny 75775 Paris cedex 516, France
email: Laurent.Cohen@ceremade.dauphine.fr

Summary. Multi-resolution methods applied to active contour models can speed up processes and improve results. In order to estimate those improvements, we describe and compare in this paper two models using such algorithms. First we propose a multi-resolution algorithm of an improved snake model, the balloon model. Convergence is achieved on an image pyramid and parameters are automatically modified so that, at each scale, the maximal length of the curve is proportional to the image size. This algorithm leads to an important saving in computational time without decreasing the accuracy of the result at the full scale. Then we present a multi-resolution parametrically deformable model using Fourier descriptors in which the curve is first described by a single harmonic; then harmonics of higher frequencies are used so that precision increases with the resolution. We show that boundary finding using this multi-resolution algorithm leads to more stability. These models illustrate two different ways of using multi-resolution methods: the first one uses multi-resolution data, the second one applies multi-resolution to the model itself.

1. Introduction

We propose two applications of multi-resolution methods for active contour models. In section 2 a first model which consists in a multi-resolution approach of the balloon model, introduced by Cohen [2], is described. Its aim is to speed-up the process while allowing a constant accuracy of the result. Then section 3 presents a parametric model based on Fourier descriptors showing that a multi-resolution algorithm can increase the model stability. These algorithms are applied for facial features extraction. Their respective advantages are discussed in the conclusion.

2. A multi-resolution balloon model

2.1 The balloon model

The use of energy–minimizing curves, known as "snakes", to extract features of interest in an image has been introduced by Kass, Witkin and Terzopoulos [6]. Further improvements to this model were successively developed by many other authors [1, 2, 3, 8].

The contour model, as introduced in [6], is a curve $v(s) = (x(s), y(s))$ that minimizes an energy functional of the following form:

$$E(v) = \int_{\Omega} \alpha\|v'(s)\|^2 + \beta\|v''(s)\|^2 + P(v(s))ds \tag{2.1}$$

where P is the potential associated to the image I. Usually P is equal to the opposite of the square of the image gradient norm:

$$P = -|\nabla I|^2$$

If v is a local minimum for E, it satisfies the associated Euler-Lagrange equation:

$$\left\{ \begin{array}{l} -(\alpha v')' + (\beta v'')'' = F(v) = -\nabla P(v) \\ + \text{ Cyclic Boundary conditions} \end{array} \right. \tag{2.2}$$

After discretizing equation (2.2) by finite differences we obtain a linear system:

$$AV = F,$$

where A is a pentadiagonal matrix, $V = (v_i)_i$ is the vector of positions $v_i = v(ih)$, F represents the forces at these points and h is the space discretization step.

The associated evolution equation (see [2]) may be defined, after temporal discretization, by:

$$V^t = (Id + \tau A)^{-1}(V^{t-1} + \tau F(V^{t-1})) \tag{2.3}$$

where τ is the time step, Id denotes the identity matrix, t is the time parameterization index and V^t describes the curve position at step time t.

The balloon model [2] is an improvement of this classical snake model that modifies the force F by normalizing the associated potential force and by adding an internal pressure force:

$$F = -k\frac{\nabla P}{\|\nabla P\|} + \gamma\vec{n}(s) \tag{2.4}$$

where $\vec{n}(s)$ is the vector normal to the curve. The first term of equation (2.4) corresponds to the normalized image force. The second term is the internal pressure with amplitude γ.

2.2 Multi-resolution algorithm

A multi-resolution approach for the balloon model consists in solving iteratively the problem at successive scales. First, the active contour solution is searched and found at a coarse scale, needing few discretization nodes and solved by a small linear system on a small image. Then the solution curve at this coarse scale is used as initialization at a finer one. This process is iterative. A similar algorithm is presented in [4] for different scales of blurring of the potential image.

Giving α, β and γ as parameters, the initial curve is projected on a coarse image of size $2^{N-S} \times 2^{N-S}$ and discretized with 2^{M-S} nodes; where $2^N \times 2^N$ is the size of the original image, 2^M is the number of discretization nodes at the finest scale and S is the coarsest scale. When the convergence at scale S is achieved, the same process is applied at a finer scale $(S-1)$ using the solution curve obtained at scale S as initializing curve. By propagating this result from the coarser to the finer scale, we obtain a result on the initial full scale image without loss of precision. since the convergence process ends at scale 0 which corresponds to the original image.

Therefore the multi-resolution algorithm may be summarized in the following way : Given the initial guess at scale S, denoted V^*_{S+1}, the iterative scheme has the following form at scale p, decreasing from S to 0, while the image size increases from 2^{N-S} to 2^N:

$$
\begin{aligned}
1. V_p^0 &= \Pi_p(V^*_{p+1}) \\
2. V_p^t &= (Id + \tau A_p)^{-1}(V_p^{t-1} + \tau F_p(V_p^{t-1})) \qquad (2.5) \\
3. V_p^* & \quad \text{is the solution of equation (2.3) at convergence at scale p}
\end{aligned}
$$

where:

- V_p is the vector of size 2^{M-p} representing the discrete curve at scale p;
- Π_p is the projection from scale $p+1$ to scale p;
- F_p is the force vector at scale p;
- A_p the stiffness matrix of size $2^{M-p} \times 2^{M-p}$ at scale p.

Since the size of the shape to be detected decreases when scale p increase, the parameter affected to the expansion is calculated at each scale so that the limit size of the curve is constant among scales. This is achieved by using a force defined as:

$$F = \frac{\gamma}{2^p} \vec{n}(s) - k \frac{\nabla P}{\|\nabla P\|}. \qquad (2.6)$$

When there is no image force ($k = 0$), it can be shown that the limit curve is a circle whose size is chosen as a characteristic to be invariant with respect to scale. To obtain a limit perimeter equal to L, the expansion parameter must be set to:

$$\gamma = 4\pi L(\alpha + 4\pi^2 \beta) \qquad (2.7)$$

In order to have a limit length at scale p equal to $\frac{L_0}{2^p}$ the expansion value is equal must be set to $\frac{\gamma}{2^p}$.

Therefore the multi-resolution algorithm may be summarized in the following way:

1. Build the pyramid of images from scale 0 (the original image) to scale S.
2. Given an initial curve V_{init} at scale 0, construct the curve V_{S+1}^* by projecting V_{init} in the image at scale S and reducing the number of discretization points to 2^{M-S}.
3. Calculate the solution V_S^* of the iterative scheme (2.5) at scale S using V_{S+1}^* as initialization.
4. For p decreasing from $(S-1)$ to 0, calculate, at scale p, the curve V_p^* (discretized by 2^{M-p} points) using the projection of V_{p+1}^* as initialization.

The main advantage of this method, as expected with multi-resolution algorithms [10], is to reduce computing costs:

- Initial convergence leads to a rough estimation of the boundary and is achieved at a coarse scale by solving a small linear system. At a finer scale the initial curve is already close to the boundary. The number of iterations to achieve convergence when dealing a large system to solve is thus smaller than the standard method.
- Since the number of discretization points decreases at the coarser scales, computation cost at each step of convergence is also smaller than with the standard method.

Figure 2.1 shows the results of mouth extraction using three consecutive scales (for representation images have been normalized to the same size). The computation time needed for convergence has been 55% shorter with the multi-resolution balloon model than with the standard balloon model.

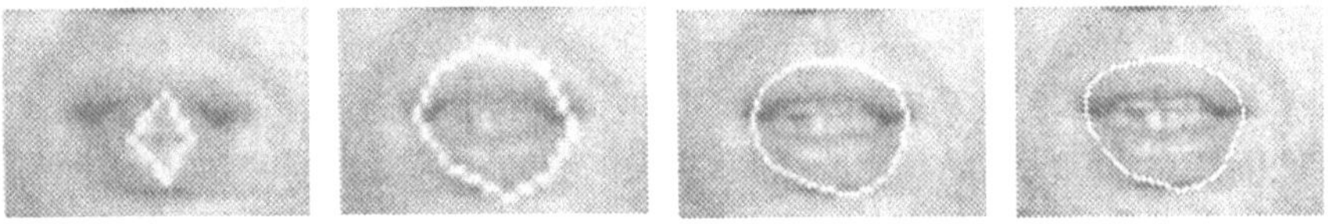

Fig. 2.1. Extraction of the mouth. From left to right: initialization at scale 2, results obtained at scales 2 then 1 and final result at scale 0.

3. A multi-resolution parametrically deformable model

3.1 Fourier descriptors active contour model

In this section, we present a parametric model based on the elliptic Fourier description. While in the snake model, the constraints on the global regularity of the contour are included in the internal energy function, these constraints may now be included in the parametrical model itself. It is possible to apply a multi-resolution algorithm to such a model by defining the scale as the number of harmonics used to describe the curve.

The use of Fourier descriptors for active contour model has been introduced by Staib and Duncan [9] in order to extract an object boundary. Their method is based on the use of probability distributions on the parameters.

In order to be less sensitive to the initial parameter value, we propose a variational approach similar to the method used in the snake model [7].

An elliptic Fourier representation of a closed curve is a parametrical curve v defined by:

$$v(\theta) = \left(\begin{array}{c} x(\theta) \\ y(\theta) \end{array} \right) = \sum_{k=0}^{N} A_k \left(\begin{array}{c} \cos(\theta k) \\ \sin(\theta k) \end{array} \right), \tag{3.1}$$

where A_k is a 2×2 matrix, N the number of harmonics used to describe the curve and θ the angular parameterization index.

The curve modeling the boundary of the object is obtained by minimizing an energy functional similar to the snake energy (2.1):

$$E(v) = \int_0^{2\pi} P(v(\theta)) + \lambda \frac{\partial v(\theta)}{\partial \theta}^2 d\theta, \tag{3.2}$$

where $\lambda \in \Re^+$.

The first term P is an image potential equal to the opposite of the square of the image gradient ($P = -|\nabla I|^2$) and the second one is an elasticity term associated to the curve tension. The energy gradient with respect to the parameters of the model is a vector of size $4N$, whose components are the partial derivatives $\frac{\partial E}{\partial a}$ with regard to each of the four elements of the N matrices A_k:

$$\frac{\partial E}{\partial a} = \int_0^{2\pi} \nabla P. \frac{\partial v(\theta)}{\partial a} + 2\lambda \frac{\partial v(\theta)}{\partial \theta \partial a} \frac{\partial v(\theta)}{\partial \theta} d\theta, \tag{3.3}$$

Given an initial set of parameters, the curve v^* that is the closest local minimum of E is obtained by applying a Newton minimization. This leads to several instability problems when high frequency harmonics are used because the curve has weak regularity constraints and can be attracted by noise points. It is thus interesting to apply this model with a multi-resolution algorithm in order to reduce the possibilities that the optimization process tends to a weak local minimum that is not meaningful.

3.2 Multi-resolution algorithm

On the contrary to the snake model, there is no geometrical constraint term in the energy functional because it is included within the model. Thus we restrict the space of admissible curves by defining the number of harmonics used to describe the curve. The aim of this multi-resolution algorithm is to obtain a better stability by increasing progressively the number of harmonics. Using only the first term of the Fourier decomposition defines an ellipse ; this was used in [5]. The iterative algorithm is as follows:

1. Describe the object shape with an ellipse curve v_1.
2. Find a curve v_1^* described by a single harmonic which is a minimum of E using v_1 as initialization.
3. For p increasing from 2 to N, using the curve v_{p-1}^* as initialization, find the curve v_p^* described by p harmonics.

Figure 3.1 shows the convergence process of the multi-resolution algorithm applied to an image of the mouth. The curve is first described by a single harmonic and evolves from an initial position to a coarse approximation of the mouth. As the number harmonics increases, the regularity constraints are smaller and the mouth boundary can be described more precisely. In order to illustrate the stability gain due to multi-resolution algorithm, figure 3.2 shows several stages of the convergence process when using standard convergence method.

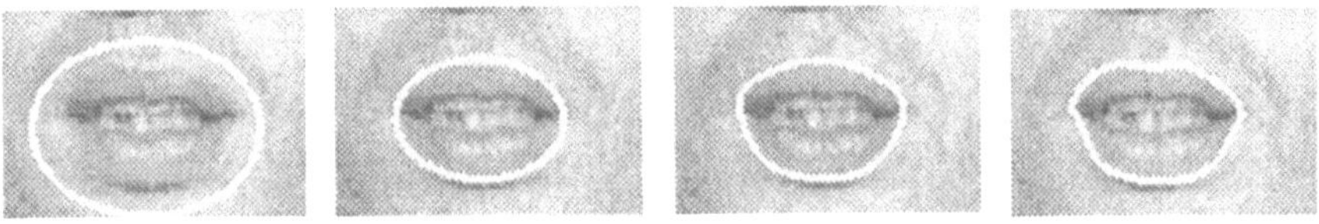

Fig. 3.1. Extraction of the mouth using the multi-resolution algorithm. From left to right: initialization and results obtain with 1, 4 then 9 harmonics.

4. Conclusion

Two different applications of multi-resolution algorithms for active contour models have been presented in this paper. For the balloon model, the multi-resolution method is applied to data. Since a large part of the convergence process is achieved at coarse scales on small images, this algorithm leads to saving in computational time. Although the expansion parameter is modified

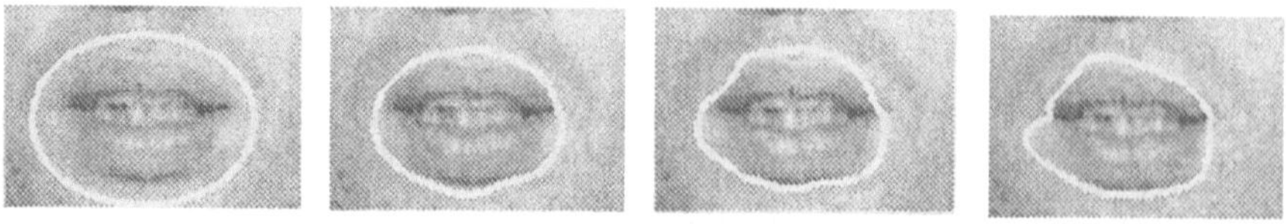

Fig. 3.2. Extraction of the mouth without the multi-resolution algorithm using 9 harmonics . From left to right: initialization and results obtained at several stages of the convergence. The right image presents the final result.

so that the maximal length of the curve is proportional to the size of the image, the other parameters bound to the regularity constraints are not modified when scale changes. With parametrically deformable model using Fourier descriptors the constraints on the global regularity of the contour are included in the parametrical model itself. Thus, by applying multi-resolution to the parameters and defining the scale as the size of the Fourier decomposition, the regularity constraints are determined by the scale. Such an algorithm can improve the model stability and allows the use of higher frequency harmonics to extract irregular object boundary. However, when extracting highly irregular boundary, a large number of harmonics are needed and it may be preferable to use the balloon model which can handle a large spectrum of shapes with limited number of parameters.

References

1. M. O. Berger and R. Mohr. Towards autonomy in active contour models. In *Proceedings of the International Conference of Pattern Recognition*, pages 847–851, Atlantic City, NJ, June 1990.
2. L. D. Cohen. On active contour models and balloons. *CVGIP: Graphical models and Image Processing*, 53(2):211–218, March 1991.
3. L. D. Cohen and I. Cohen. Finite element methods for active contour models and balloons for 2-D and 3-D images. *IEEE Transactions on Pattern Analysis and Machine Intelligence*, 15(11):1131–1147, November 1993.
4. L. D. Cohen and A. Gorre. On the convexity of the active contour energy. In *Proceedings of GRETSI*, Juan-les-Pins, September 1995.
5. K. Deng and J. N. Wilson. Contour estimation using global shape constraints and local forces. In *Proceedings of SPIE, Geometric Methods in Computer Vision*, volume 1570, pages 1–7, San Diego, California, U.S.A., July 1991.
6. M. Kass, A. Witkin, and D. Terzopoulos. Snakes: Active contour models. In *IEEE Proceedings of the International Conference on Computer Vision*, pages 259–268, London, June 1987.
7. B. Leroy, A. Chouakria, I. L. Herlin, and E. Diday. Approche géométrique et classification pour la reconnaissance de visage. In *Congrès Reconnaissance des Formes et Intelligence Artificielle*, Rennes, January 1996.

8. F. Leymarie and M. Levine. Tracking deformable objects in the plane using an active contour model. *IEEE Transactions on Pattern Analysis and Machine Intelligence*, 15(6):635–646, 1993.
9. L. H. Staib and J. S. Duncan. Boundary finding with parametrically deformable models. *IEEE Transactions on Pattern Analysis and Machine Intelligence*, 14(11):1061–1075, November 1992.
10. Demetri Terzopoulos. Multiresolution algorithms in computational vision. In *Image Understanding*, pages 225–262. S.Ullman, W.Richards, 1986.

On projective plane curve evolution

Olivier FAUGERAS[1] and Renaud KERIVEN[2]

[1] I.N.R.I.A. Sophia-Antipolis, 06561 Valbonne, France, faugeras@sophia.inria.fr
[2] E.N.P.C. CERMICS, 93167 Noisy Le Grand, France, keriven@cermics.enpc.fr

Abstract. In this paper, we investigate the evolution of curves of the projective plane according to a family of projective invariant intrinsic equations. This is motivated by previous work for the Euclidean [11, 12, 14] and the affine cases [21, 22, 3, 2] as well as by applications in the perception of two-dimensional shapes. We establish the evolution laws for the projective arclength and curvature. Among this family of equations, we define a "projective heat equation" [7] and establish the link with the projective evolution of curves in $\mathbf{R}^2$.
Keywords: multi-scale analysis, partial differential equations, projective geometry

1. Introduction

The use of partial differential equations and curve or surface evolution theory in image analysis became a major research topic in the past years (see [18]) leading to applications in image de-noising and de-blurring [19], in selective smoothing and edge detection [1, 17], in contrast enhancement [20], in shape segmentation [5]. Recently, applications were found in problems usually addressed by the computer vision community: intrinsic flows [14, 21] hold very good geometric smoothing properties and allow the computation of local differential invariants [9]. Motivated by the importance of projective geometry in computer vision, we found it natural to extend the Euclidean [14] and affine [21] cases to the projective one.

2. Geometric flows

Let $\mathcal{L}$ be a Lie group operating on some objects. A quantity q depending on these objects is called an *invariant* of $\mathcal{L}$ if, whenever a transformation $L \in \mathcal{L}$ changes q into q', we have $q' = \alpha(L)q$, where α is a function of L alone, i.e. does not depend on the object which is transformed. If $\alpha \equiv 1$, then q is called an *absolute invariant.*

Differential invariants are special invariants based on local transformations (see [13]).

Let $C : \mathbf{R} \to \mathbf{R}^2$ be a plane curve of parameter p. The first and the second differential invariants for the Euclidean group $\{m \mapsto Rm + T \mid R \text{ rotation}, T \text{ translation}\}$ are the well known Euclidean arclength v and curvature κ defined by:

$$\begin{cases} \frac{\partial v}{\partial p} & = & \|\frac{\partial C}{\partial p}\| \\ \kappa & = & \|\frac{\partial^2 C}{\partial v^2}\| \end{cases} \tag{1}$$

which are preserved by rotations and translations.

The corresponding invariants for the group of proper affine motions $\{m \mapsto Am + B \mid [A] > 0, B \in \mathbf{R}^2\}$ are the affine arclength s and curvature μ defined by:

$$\begin{cases} \frac{\partial s}{\partial p} & = & [\frac{\partial C}{\partial p}, \frac{\partial^2 C}{\partial p^2}]^{1/3} \\ \mu & = & [\frac{\partial^2 C}{\partial s^2}, \frac{\partial^3 C}{\partial s^3}] \end{cases} \tag{2}$$

which are invariants for affine proper motions, and absolute invariants for special affine motions ($\{m \mapsto Am + B \mid [A] = 1, B \in \mathbf{R}^2\}$) .

Circles (and straight lines) are the only curves with constant Euclidean curvature. In the affine case, constant affine curvature is obtained for the conics ($\mu = 0$ for a parabola, $\mu > 0$ for an ellipse and $\mu < 0$ for an hyperbola).

Given an initial plane curve $\mathcal{C}_0(p) : \mathbf{R} \to \mathbf{R}^2$, the associated geometric flow (see [16]) is the family of curves $\mathcal{C}(p,t) : \mathbf{R} \times [0, \tau) \to \mathbf{R}^2$ evolving according to the following law:

$$\begin{cases} \frac{\partial \mathcal{C}(p,t)}{\partial t} & = & \frac{\partial^2 \mathcal{C}(p,t)}{\partial r^2} \\ \mathcal{C}(p,0) & = & \mathcal{C}_0(p) \end{cases} \tag{3}$$

where r is the group arclength (v for the Euclidean geometric flow, s for the affine one). Contrary to the classical heat flow $\mathcal{C}_t = \mathcal{C}_{pp}$, these flows are intrinsic (i.e. don't depend on the parameterization p of the initial curve). They are invariant for the considered Lie group. Their "smoothing" properties may be summarized as follow ([14, 21]): closed curves evolve toward a convex one and then disappear shrinking toward a circle point (Euclidean case) or an ellipse point (affine case).

For a given group, a plane curve is defined up to a group transformation by its group arclength and curvature. Hence, it is natural to study these flows through the evolution of the arclength and curvature. With $g_e = \frac{dv}{dp}$ and $g_a = \frac{ds}{dp}$, we have:

$$\begin{cases} \frac{\partial g_e}{\partial t} & = & -g_e \kappa^2 \\ \frac{\partial \kappa}{\partial t} & = & -\kappa^3 - \frac{\partial^2 \kappa}{\partial v^2} \end{cases} \text{ and } \begin{cases} \frac{\partial g_a}{\partial t} & = & -2 g_a \mu / 3 \\ \frac{\partial \mu}{\partial t} & = & \frac{4}{3}\mu^2 + \frac{1}{3}\frac{\partial^2 \mu}{\partial s^2} \end{cases} \tag{4}$$

3. Projective geometry

Like in equations (1) and (2), it is possible to define the projective arclength and curvature of a plane curve in $\mathbf{R}^2$. However, this leads to too complex expressions. The idea is to embed such a curve in the real projective plane $\mathcal{P}^2$. One can see $\mathcal{P}^2$ as the set of the lines of $\mathbf{R}^3$ going through the origin. An element of $\mathcal{P}^2$ is represented by its homogeneous coordinates (x, y, z) where (x, y, z) and $(\lambda x, \lambda y, \lambda z), (\lambda \neq 0)$ are different coordinate vectors of the same projective point.

Let $\mathbf{B}(p) : \mathbf{R} \rightarrow \mathcal{P}^2$ be a smooth curve of the projective plane. Using standard results of projective differential geometry [4], we change $\mathbf{B}(p)$ by a scale factor $\lambda(p)$ and characterize its projective arclength σ and curvature k introducing the Cartan point $\mathbf{A} = \lambda\mathbf{B}$, and the Cartan frame $(\mathbf{A}, \mathbf{A}^{(1)}, \mathbf{A}^{(2)})$ which satisfy the projective Frenet equations:

$$\begin{aligned} \frac{d\mathbf{A}}{d\sigma} &= \mathbf{A}^{(1)} \\ \frac{d\mathbf{A}^{(1)}}{d\sigma} &= -k\mathbf{A} + \mathbf{A}^{(2)} \\ \frac{d\mathbf{A}^{(2)}}{d\sigma} &= -\mathbf{A} - k\mathbf{A}^{(1)} \end{aligned} \tag{5}$$

and the condition:

$$|\mathbf{A}\mathbf{A}^{(1)}\mathbf{A}^{(2)}| = 1 \tag{6}$$

Note that $\mathbf{B}$ and $\mathbf{A}$ are different coordinate vectors of the same projective point. The point $\mathbf{A}^{(1)}$ is on the tangent to the curve in $\mathbf{A}$ and the line $\langle \mathbf{A}, \mathbf{A}^{(2)} \rangle$ is the projective normal. Functions k and σ are invariant under the action of the projective group and characterize the curve up to a projective transformation.

The plane curves with a constant projective curvature are (see [10]):

- If $k = k_0 = -3/32^{1/3}$: the exponential $(y = e^x)$
- If $k < k_0$: the general parabola $(y = x^m, m \notin \{2, \frac{1}{2}, -1\})$
- If $k > k_0$: the logarithmic spiral $(\rho = e^{m\theta}, m \neq 0)$

4. Projective invariant intrinsic flows

The law $\mathbf{A}_t = \mathbf{A}_{\sigma\sigma}$ investigated in [7] could be thought as a natural extension of the Euclidean and affine cases. Yet, this law raises some contradictions. For instance, according to the expression of k_t in [7], curves with a constant initial curvature should evolve keeping a constant curvature. Actually, it's not the case (see [10]).

The reason why it is so is that the Cartan point $\mathbf{A}(p, t)$ is some particular representant for the projective point $\mathbf{B}(p, t)$ and depends on the curve and its spatial derivatives at (p, t). As a result, one can't expect an arbitrary differential equation $\{\mathbf{A}(p, 0) = \mathbf{A}_0(p); \mathbf{A}_t = f(p, t)\}$ to be such that $\mathbf{A}(p, t)$ will still be the Cartan point of the curve at time $t > 0$.

This leads us to consider the evolution law

$$\left\{ \begin{aligned} \mathbf{A}(p, 0) &= \mathbf{A}_0(p) \ \ (\mathbf{A}_0 \text{ Cartan point of the initial curve}) \\ \mathbf{A}_t(p, t) &= \alpha\mathbf{A} + \beta\mathbf{A}^{(1)} + \gamma\mathbf{A}^{(2)} \end{aligned} \right. \tag{7}$$

where $f(p,t)$ has been decomposed on the Cartan frame, and to find out which conditions on (α, β, γ) will assure that $\mathbf{A}(p,t)$ remains the Cartan point.

Another way to see this is to consider the surface $\mathcal{S} = \{\mathbf{A}(p,t)\}$ of $\mathbf{R}^3$. The reason why this is a well-defined surface is because there is no scale factor on $\mathbf{A}$ even though it represents a projective point of $\mathcal{P}^2$. Now, in order for (7) to be a well-defined PDE on $\mathcal{S}$,the vector $\mathbf{A}_t$ has to belong to the tangent plane $T_{\mathcal{S}}$ at (p,t). The right hand side contains the vector $\mathbf{A}^{(1)}$ which belongs to $T_{\mathcal{S}}$ but the vector $\alpha\mathbf{A} + \gamma\mathbf{A}^{(2)}$ does not in general belong to $T_{\mathcal{S}}$ unless α and γ are dependent In fact the condition is even stronger since not only $\mathbf{A}_t$ must belong to $T_{\mathcal{S}}$ but also, as stated above, $\mathbf{A}$ must remain a Cartan point.

We get the following result (see [8] for the proof):

Proposition 1 *The differential equation (7) has a meaning (i.e. $\mathbf{A}(p,t)$ is the Cartan point of the curve at time t) if and only if:*

$$\alpha = \frac{1}{3+k_\sigma} \quad [\quad -\frac{1}{3}k_{\sigma^3} - \frac{3}{2}k_{\sigma^2}\gamma_\sigma - k_\sigma(\frac{7}{3}k\gamma + \frac{17}{6}\gamma_{\sigma^2} + \beta_\sigma) - \frac{8}{3}k^2\gamma_\sigma$$
$$+k(\gamma - \frac{5}{3}\gamma_{\sigma^3}) + \gamma_{\sigma^2}/2 - \gamma_{\sigma^5}/6 \ ] \tag{8}$$

In this case, the projective arclength and curvature evolve according to:

$$\frac{g_t}{g} = \alpha + \beta_\sigma - \frac{1}{3}(k\gamma - \gamma_{\sigma^2}) \tag{9}$$
$$k_t = -\alpha_{\sigma^2} + \frac{3}{2}\gamma_\sigma + \frac{\gamma_{\sigma^4}}{6} + k(\frac{2}{3}\gamma_{\sigma^2} - 2\alpha)$$
$$+k_\sigma(\beta + \frac{7}{6}\gamma_\sigma) + \frac{\gamma}{3}(k_{\sigma^2} + 2k^2) \tag{10}$$

where $g = \frac{d\sigma}{dp}$.

Note that $\mathbf{A}_t = \mathbf{A}_{\sigma\sigma}$ is the case $(\alpha, \beta, \gamma) = (-k, 0, 1)$, thus doesn't satisfy condition (8), hence the previous contradictions.

Moreover, if β and γ are projective invariant intrinsic quantities, then α defined by equation (8) is a projective invariant intrinsic quantity too. Therefore, we get:

Corollary 1 *Let β and γ be some projective invariant intrinsic quantities, let α be defined by equation (8). The differential equation (7) defines a projective invariant intrinsic flow. The evolution of the projective arclength and curvature of the curves is given by equations (9, 10).*

5. The projective "heat flow"

Among all the possible choices for (β, γ), it turns out that the simplest one $(0,1)$ is also the right one for a projective "heat flow" extending the Euclidean and affine cases. Some intuitive justification could be:

- $\beta\mathbf{A}^{(1)}$ is on the tangent in $\mathbf{A}$. Thus, the choice of β has no importance: changing β doesn't modify the family of curves obtained but only their parameterization p (see [21]).

- $(\beta, \gamma) = (0, 1)$ are the components of $\mathbf{A}_{\sigma\sigma}$ on $(\mathbf{A}^{(1)}, \mathbf{A}^{(2)})$. The induced α could be considered as a corrected component on $\mathbf{A}$.

However, the deep reason for this choice is that it gives the same flow as $C_t = C_{\sigma\sigma}$ in $\mathbf{R}^2$ (see next section). Consequently, we have from proposition 1 the following statement:

Proposition 2 *Let α be:*

$$\alpha = \frac{1}{9 + 3k_\sigma}(3k - 7kk_\sigma - k_{\sigma^3})$$

Let $\mathbf{B}_0(p)$ be a curve of $\mathcal{P}^2$ and $\mathbf{A}_0(p)$ its Cartan points. We define its projective heat flow as the solution of:

$$\begin{cases} \mathbf{A}(p, 0) & = & \mathbf{A}_0(p) \\ \mathbf{A}_t(p, t) & = & \alpha\mathbf{A} + \mathbf{A}^{(2)} \end{cases} \tag{11}$$

Let $g = \frac{d\sigma}{dp}$. The projective arclength and curvature evolve according to:

$$\frac{g_t}{g} = \frac{-1}{9 + 3k_\sigma}(8kk_\sigma + k_{\sigma^3}) \tag{12}$$

$$k_t = \frac{2}{3}k^2 + \frac{1}{3}k_{\sigma^2} - 2\alpha k - \alpha_{\sigma^2} \tag{13}$$

6. Going back to $\mathbf{R}^2$

We prove in [8] that:

Proposition 3 *Given an initial curve in $\mathcal{P}^2$, let $\mathbf{B}_0(p)$ be any coordinate vector of it.*

1. *The flow defined by*

$$\begin{cases} \mathbf{B}(p, 0) & = & \mathbf{B}_0(p) \\ \mathbf{B}_t(p, t) & = & \mathbf{B}_{\sigma\sigma} \end{cases} \tag{14}$$

   is intrinsic and doesn't depend on the choice of $\mathbf{B}_0$ (i.e. $\mathbf{B}_0(p)$ and $\phi(p)\mathbf{B}_0(p)$ give the same family of curves).

2. *This flow is the projective heat flow defined by equation (11) up to a parameterization of the curves.*

3. Let λ be the Cartan scale factor ($\mathbf{A} = \lambda\mathbf{B}$). (σ, k, λ) define $\mathbf{B}$ up to a projective transformation. Their evolution is given by:

$$\begin{aligned}
\frac{g_t}{g} &= \frac{-1}{9+3k_\sigma}(8kk_\sigma + k_{\sigma^3} + 18\Lambda_{\sigma^2}) \\
k_t &= \frac{2}{3}k^2 + \frac{1}{3}k_{\sigma^2} - 2Pk - P_{\sigma^2} - 2k_\sigma\Lambda_\sigma \qquad (15) \\
\Lambda_t &= \frac{-1}{9+3k_\sigma}[k_{\sigma^3} + 3k_\sigma(\Lambda_\sigma^2 - 3\Lambda_{\sigma^2}) + 4k(k_\sigma - 3) + 9(\Lambda_\sigma^2 - \Lambda_{\sigma^2})] \\
&\text{where } g = \frac{d\sigma}{dp}, \ \Lambda = \log|\lambda|, \ P = \Lambda_\sigma^2 - \Lambda_{\sigma^2} - k + \Lambda_t
\end{aligned}$$

Let $C_0(p) = (x_0, y_0)$ be a real plane curve, it is then easy to prove that:

Corollary 2 *The flow defined by $\{C(p,0) = C_0 \ ; \ C_t = C_{\sigma\sigma}\}$ is a projective invariant flow. It gives the same family of curves through the map $(\frac{x}{z}, \frac{y}{z})$ as the projective heat flow (11) with initial curve $(x_0, y_0, 1)$. Let $C(p,t) = (x, y)$ and λ be the Cartan scale of $(x, y, 1)$, the evolution of the projective arclength and curvature of C is given by equations (15).*

This was already proved in [15], even though the argument in [16] about the relationship between different coordinate vectors is incorrect (see proposition 3 above)

7. Conclusion

We have established a link between the invariant projective flow defined in $\mathbf{R}^2$ [16, 15] and the one defined in $\mathcal{P}^2$ [7]. We have defined the projective heat equation in three equivalent ways: $\mathbf{A}_t = \alpha\mathbf{A} + \mathbf{A}^{(2)}$ (α given by equation (8)) or $\mathbf{B}_t = \mathbf{B}_{\sigma\sigma}$ in $\mathcal{P}^2$, and $C_t = C_{\sigma\sigma}$ in $\mathbf{R}^2$. As expected, the connection is not trivial but simple enough. The advantage of the definition in $\mathcal{P}^2$ [7] which we have modified here to make it entirely correct is that: a) it does not depend on the particular coordinates used to represent $\mathcal{P}^2$ and b) it has allowed us to establish the evolution of the projective arclength and curvature. There remains to see if it is possible to define a projective scale-space as in the Euclidean and affine cases. Of particular interest would be to compare our approach with the one developed by Dibos [6].

References

[1] L. Alvarez, P-L. Lions, and J-M. Morel. Image selective smoothing and edge detection by nonlinear diffusion (II). *SIAM Journal of numerical analysis*, 29:845–866, 1992.

[2] Luis Alvarez, Frédéric Guichard, Pierre-Louis Lions, and Jean-Michel Morel. Axiomatisation et nouveaux opérateurs de la morphologie mathématique. *C.R. Acad. Sci. Paris*, pages 265–268, 1992. t. 315, Série I.

[3] Luis Alvarez, Frédéric Guichard, Pierre-Louis Lions, and Jean-Michel Morel. Axioms and Fundamental Equations of Image Processing. Technical Report 9231, CEREMADE, 1992.

[4] Elie Cartan. *La Théorie des Groupes Finis et Continus et la Géométrie Différentielle traitée par la Méthode du Repère Mobile.* Jacques Gabay, 1992. Original edition, Gauthiers-Villars, 1937.

[5] V. Caselles, R. Kimmel, and G. Sapiro. Geodesic active contours. In *ICCV*, Cambridge, June 1995.

[6] Françoise Dibos. Projective multiscale analysis. Technical Report 9533, CEREMADE, 1995.

[7] Olivier Faugeras. On the evolution of simple curves of the real projective plane. *Comptes rendus de l'Académie des Sciences de Paris, Tome 317, Série I*, 0(6):565–570, September 1993. Also INRIA Technical report number 1998.

[8] Olivier Faugeras and Renaud Keriven. On the projective evolution of 2d curves. Technical Report 95-45, ENPC-CERMICS, November 1995.

[9] Olivier Faugeras and Renaud Keriven. Scale-spaces and affine curvature. In R. Mohr and C. Wu, editors, *Proc. Europe-China Workshop on Geometrical modelling and Invariants for Computer Vision*, pages 17–24, Xi'an, China, April 1995.

[10] Olivier Faugeras and Renaud Keriven. Some recent results on the projective evolution of 2D curves. In *ICIP*, volume 3, pages 13–16, Washington, October 1995.

[11] M. Gage and R.S. Hamilton. The heat equation shrinking convex plane curves. *J. of Differential Geometry*, 23:69–96, 1986.

[12] M. Grayson. The heat equation shrinks embedded plane curves to round points. *J. of Differential Geometry*, 26:285–314, 1987.

[13] W. Guggenheimer, Heinrich. *Differential Geometry.* Dover Publications, New York, 1977.

[14] Benjamin B. Kimia, Allen Tannenbaum, and Steven W. Zucker. On the Evolution of Curves via a Function of Curvature. I. The Classical Case. *Journal of Mathematical Analysis and Applications*, 163(2):438–458, 1992.

[15] P. J. Olver, Guillermo Sapiro, and Allen Tannenbaum. Classification and uniqueness of invariant geometric flows. *Comptes rendus de l'Académie des Sciences de Paris, Tome 319, Série I*, pages 339–344, 1994.

[16] P. J. Olver, Guillermo Sapiro, and Allen Tannenbaum. *Differential invariant signatures and flows in computer vision: A symmetry group approch*, pages 205–306. In Romeny [18], 1994.

[17] Pietro Perona and Jitendra Malik. Scale-space and edge detection using anisotropic diffusion. *IEEE Transactions on Pattern Analysis and Machine Intelligence*, 12(7):629–639, July 1990.

[18] B. Ter Haar Romeny, editor. *Geometry driven diffusion in Computer Vision*. Kluwer, 1994.

[19] L. Rudin, S. Osher, and E. Fatemi. Nonlinear total variation based noise removal algorihtms. *Physica D*, 60:259–268, 1992.

[20] Guillermo Sapiro and Vicent Caselles. Histogram modification via pde's. Technical report, HPL-TR, December 1994.

[21] Guillermo Sapiro and Allen Tannenbaum. Affine Invariant Scale Space. *IJCV*, 11(1):25–44, August 1993.

[22] Guillermo Sapiro and Allen Tannenbaum. On affine plane curve evolution. *Journal of Functional Analysis*, 119:79–120, 1994.

Image Processing of Meteorological Images With Implicit Functions

Hussein M. YAHIA, Isabelle L. HERLIN

INRIA, Rocquencourt
B.P. 105, 78153 Le Chesnay CEDEX, France.
Email: Hussein.Yahia@inria.fr, Isabelle.Herlin@inria.fr

Summary. Level sets of implicit functions are used for segmentation and tracking of deformable structures encountered in remote sensed imagery. Iso-contours are proposed as a basic tool to model motion of natural structures undergoing visco-elastic deformation and topological change. In this study we present an energy minimization process for the segmentation of deformable structures. The energy functional carries information about position, tangency and curvature. By using implicit functions, it is possible to satisfy in a simple way complex geometrical requirements about motion which would be rather difficult to ensure using a parametrized interpolating surface. The method is applied over a temporal sequence of satellite images displaying a vortex motion.

1. Introduction

As satellite acquisition allows increasingly finer temporal and spatial resolution images of complex moving structures such as sea or atmospheric streams evolution, the problem of modelling and temporal tracking of these natural structures with data compression is becoming pervasive in remote sensed imagery.

In this study we are interested in the problem of modelling and motion tracking of complex moving structures undergoing arbitrary deformations. Moreover, our work is strongly oriented towards the goal of data compression. To achieve these overall requirements, we find level sets of implicit functions quite useful both in modelling, motion tracking and data compression.

Although there already exists powerful probabilistic and parametric methods to segment an image sequence [3, 8, 5], implicit contours are particularily interesting to model natural structures displaying some kind of visco-elastic behaviour. For such structures, the use of parametric and probabilistic methodologies can be quite difficult, mostly when topological changes appear, or if a large deformation is applied to the structure.

The paper is organized as follows: in section 2 are defined the implicit functions used in this paper. Section 3 is devoted to describing the energy minimzation process used to approximate deforming structures by implicit templates. A surface interpolation model for a motion tracking problem is presented in section 4 and we show how implicit surfaces can be quite useful to solve it. In section 5 we apply these methods on a temporal sequence displaying an atmospheric vortex motion and the paper is ended by concluding remarks and perspectives in section 6.

2. Implicit Functions, Soft Objects

There are many ways of defining and using implicit functions [1, 2, 7, 10, 14]. In this work we use a specific kind of implicit functions called Soft Objects [14, 15, 16], defined by local functions which vanish outside a bounded region. Let $U \subset \mathbb{R}^2$ be an open set. We consider a finite set:

$$S = \{k_1, k_2, \cdots, k_n\}$$

of **control points** in U. To each control point k_i we associate a *radius of influence* r_i. Local functions ϕ_i are introduced:

$$\phi_i(M) = -0.44\frac{\|\overrightarrow{k_iM}\|^6}{r_i{}^6} + 1.88\frac{\|\overrightarrow{k_iM}\|^4}{r_i{}^4} - 2.44\frac{\|\overrightarrow{k_iM}\|^2}{r_i{}^2} + 1\ , \tag{2.1}$$

if M is such that $\|\overrightarrow{k_iM}\| \leq r_i$, and 0 otherwise.

To control the concave parts of an iso-contour, we also introduce **negative** control points: these are just ordinary control points, they define a force field of the same intensity but in the opposite direction. An implicit function ϕ is introduced:

$$\phi(M) = \sum_{k_i \in \mathcal{S}^+} \phi_i(M) - \sum_{k_i \in \mathcal{S}^-} \phi_i(M)\ ,$$

where $\mathcal{S}^+$ and $\mathcal{S}^-$ are respectively the set of positive and negative control points ($\mathcal{S} = \mathcal{S}^+ \cup \mathcal{S}^-$). Iso-contours are just level sets of ϕ.

3. An approximation problem

Let Γ be a polygonal contour, defined by its set of vertices:

$$V = \{P_1, P_2, \cdots, P_m\}$$

Let us consider the following problem:

Find a good approximation of Γ *as an iso-contour* $\phi^{-1}(c)$.

The subject of this section is to express this approximation problem by an energy minimization process.

Since the iso-contour must be very close to each point of Γ, each value $\phi(P_i)$ must be close to c. Hence a first energy term is defined (see [6]):

$$E_1 = \sum_{i=1}^{m} (c - \phi(P_i))^2 \tag{3.1}$$

To accurately track moving structures, it is often necessary to put constraints on tangent and curvature information.

Let us discuss the tangent first. Since Γ is polygonal the vector $\overrightarrow{P_i P_{i+1}}$ may be used as the tangent direction (denoted $\overrightarrow{T_i^1}$) at P_i on the original contour Γ. On the iso-contour we recall that tangent vectors (denoted $\overrightarrow{T_i^2}$) are given by the gradient rotated clockwise by a right angle. The energy functional tries to keep these tangent vectors $\overrightarrow{T_i^1}$ and $\overrightarrow{T_i^2}$ in the same direction. A second energy term is introduced:

$$E_2 = \sum_{i=1}^{m-1} [\ det(\overrightarrow{T_i^1}, \overrightarrow{T_i^2})\ ]^2 = \sum_{i=1}^{m-1} \begin{vmatrix} t_x^i & \frac{\partial \phi}{\partial y}(P_i) \\ t_y^i & -\frac{\partial \phi}{\partial x}(P_i) \end{vmatrix}^2 \tag{3.2}$$

where (t_x^i, t_y^i) are the components of $\overrightarrow{P_i P_{i+1}}$.

Let us now consider the curvature information. Since curvature involves second derivatives, it is necessary to approximate the polygonal contour Γ by a C^2 curve in order to compute curvature on the original contour Γ. Now, let γ be the parametrized cubic B-spline approximating [9] Γ and let $(t_i)_{1 \leq i \leq q}$ be a sequence of parameter values. The curvature C_i^1 of the parametrized curve γ at $p = \gamma(t_i)$ is

$$\frac{< \ddot{\gamma}(t_i) \mid \overrightarrow{N(\gamma(t_i))} >}{\| \frac{d\gamma}{dt}(t_i) \|^2},$$

$\overrightarrow{N(\gamma(t_i))}$ being the unit normal vector at $\gamma(t_i)$, and $< \cdot \mid \cdot >$ denoting the dot product of vectors. From standard differential geometry [12], the curvature C_i^2 of the iso-contour is obtained from the second fondamental form, evaluated at a tangent vector and divided by the square norm of that tangent vector. Hence we introduce a third energy term:

$$E_3 = \sum_{i=1}^{q} (C_i^1 - C_i^2)^2 = \sum_{i=1}^{q} [\ \frac{< \ddot{\gamma}(t_i) \mid \overrightarrow{N(\gamma(t_i))} >}{\| \frac{d\gamma}{dt}(t_i) \|^2} - \frac{S_p(v)}{\|v\|^2}\ ]^2 \tag{3.3}$$

where v is a tangent vector at the iso-contour passing through $p = \gamma(t_i)$, and S_p the second fundamental form of the iso-contour. If energy E_3 is minimized, then at each point $p = \gamma(t_i)$, the curvature of the iso-contour will be close to that of the parametrized spline.

The final form of energy is

$$E = \alpha E_1 + \beta E_2 + \gamma E_3 \tag{3.4}$$

with $\alpha + \beta + \gamma = 1$.

Since energy E is a function of the coordinates (x_i, y_i) of control points k_i and their radii of influence r_i:

$$E(x_1, y_1, x_2, y_2, \cdots, x_n, y_n, r_1, r_2, \cdots, r_n) ,$$

E is minimized on this set of variables. Since E is a positive definite sum of squares, a conjugate gradient method is used to perform the minimization of E (see [4]). After having done some experiments, we found that a good starting initialization can be generated in the following way. Positive control points are placed inside the convex parts of Γ, and negative control points outside the concave parts of Γ. For each control point, either positive or negative, the distance d between that control point and Γ is computed. Then the radius of influence of that control point is $d + K$, where K is a fixed constant. (Typically, K is chosen in order to ensure that Γ lies inside the circles defined by the control points and their radius of influence). In this paper, the value of K is 15 pixels, and the iso-value is kept fixed. For the particular type of initialization just described, an iso-value $c = 0.9$ gives very good results for a wide variety of contours (recall that the maximum value of each ϕ_i is 1).

Moreover, taking $\alpha = 0.7$, $\beta = 0.2$ and $\gamma = 0.1$ gives quite satisfactory results, although one could obviously find some examples where better results may be obtained by changing these tuning coefficients.

4. Motion tracking and surface interpolation

The previous section described a process for segmenting a still image. We are now interested in the motion of the structure. Consider an image sequence of a moving structure and suppose that each image is segmented using the method described in the previous section.

From a geometrical point of view, one may view the tracking problem in the following way: let be given two temporal occurrences of the structure, namely, two 2D boundaries. The evolution of the structure between these occurrences generates a continuous 3D surface, the third dimension being time. The main objective is to compute the best possible surface. To perform this task, one may require an evolution model, expressed as constraints applied to this surface. The problem is then to find the surface interpolating the two contours and respecting these constraints. A good interpolating surface should not cross itself, and the surface should not display too much variations in curvature.

More formally, two 2D contours C_1 and C_2 are placed in two parallel planes $t = 0$ and $t = 1$. In order to focus this study on deformation, one can first apply the rigid motion that performs the best matching between the two contours C_1 and C_2. This enables to manage the global translation motion that is often observed on meteorological and oceanographic image sequences. One then looks for an interpolating surface ϕ, with an elastic behavior, containing C_1 and C_2. The surface ϕ is described by two parameters $(s,r) \in \Omega = [0,1]^2$. The parameter r represents in fact the time.

Elasticity may be modelled by control of the first fundamental form. This form, $\mathcal{F}$ (see equation 4.1, where $< \mid >$ stands for the dot product of two vectors) expresses the distortion produced by ϕ on an elementary square in parameter's space.

$$\mathcal{F} = \begin{pmatrix} < \frac{\partial \phi}{\partial s} \mid \frac{\partial \phi}{\partial s} > & < \frac{\partial \phi}{\partial s} \mid \frac{\partial \phi}{\partial r} > \\ < \frac{\partial \phi}{\partial s} \mid \frac{\partial \phi}{\partial r} > & < \frac{\partial \phi}{\partial r} \mid \frac{\partial \phi}{\partial r} > \end{pmatrix} \tag{4.1}$$

If $\mathcal{F}$ is equal to identity, the local basis $(\frac{\partial \phi}{\partial s}, \frac{\partial \phi}{\partial r})$ of the tangent plane of ϕ is orthonormal, and the distortion is minimal. Moreover, the closer to identity is $\mathcal{F}$, the smaller is the distortion. The local distortion of the surface is then measured by the expression (4.2), where Id stands for identity matrix:

$$||\mathcal{F} - Id||^2 \, . \tag{4.2}$$

The elastic behavior is obtained by minimizing the integral of this criterion over the surface. This corresponds to the following energy functional:

$$E(\phi) = \int\int_{\Omega} \left((< \phi_s \mid \phi_s > -1)^2 + 2 < \phi_s \mid \phi_r >^2 + (< \phi_r \mid \phi_r > -1)^2 \right) dsdr \tag{4.3}$$

Minima of $E(\phi)$ are solutions of $\nabla E(\phi) = 0$; this leads to the following differential system (where ∇^t stands for the transpose of the gradient operator):

$$\Delta \phi - \nabla^t(\mathcal{F} \nabla \phi) = 0 \, . \tag{4.4}$$

This equation expresses the evolution model. The surface must also verify boundary conditions. Because the surface interpolates the two original contours, one must take into account the two following boundary conditions concerning $r = 0$ and $r = 1$: $\phi(.,0) = C_1$ and $\phi(.,1) = C_2$.

Unfortunately, trying to solve this PDE results in intractable numerical problems. At this point, one may put the following observation: implicit interpolating surfaces never self-intersect (indeed, such surfaces always possess tubular neighborhoods [13]). Moreover, one can reasonably expects minimum curvature variations of implicit interpolating surfaces. Hence, instead of solving the last complicated PDE (4.4), one may try to find directly the motion of the control points and their radii of influence. The resulting implicit interpolating surface would then satisfy the preceding geometrical requirements. The aim of this paper is not to find explicitely an implicit interpolating surface but rather to perform the following experiment: given an image sequence

$\mathcal{I}_1, \mathcal{I}_2, \cdots, \mathcal{I}_n$, image $\mathcal{I}_1$ is first segmented using the method described in the last section. The resulting iso-contour is then given as an initialization for image $\mathcal{I}_2$. The minimization process is then applied over image $\mathcal{I}_2$ with such an initialization, producing an approximation contour which is then given as an initialization to image $\mathcal{I}_3$, and so on until last image $\mathcal{I}_n$. Then, by simply interpolating linearly between the geometrical attributes of the sequence of implicit control points and their radii of influence, we generate a modelling of the structure's motion which corresponds to temporal sections of an implicit interpolating surface between the segmented templates. This process is illustrated in the following section.

5. Experiments

The ideas presented in the previous sections are now applied on an infrared METEOSAT image sequence [11] of atmospheric temperature displayed in figure 5.1.This image sequence displays a vortex motion.

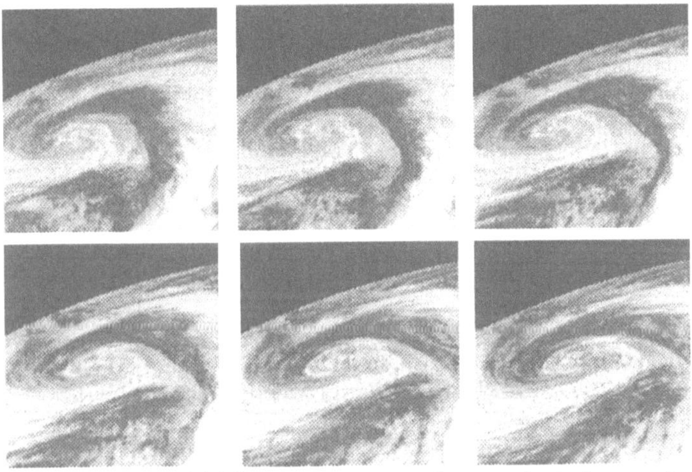

Fig. 5.1. *METEOSAT image sequence of atmospheric temperature.*

A standard snake model is firstly applied on these data to segment the structure of interest. The snake gives a parametrized curve which is then approximated by a C^2 spline curve as described in section 3. Then the method of the previous sections is applied to produce the results displayed in figure 5.2.

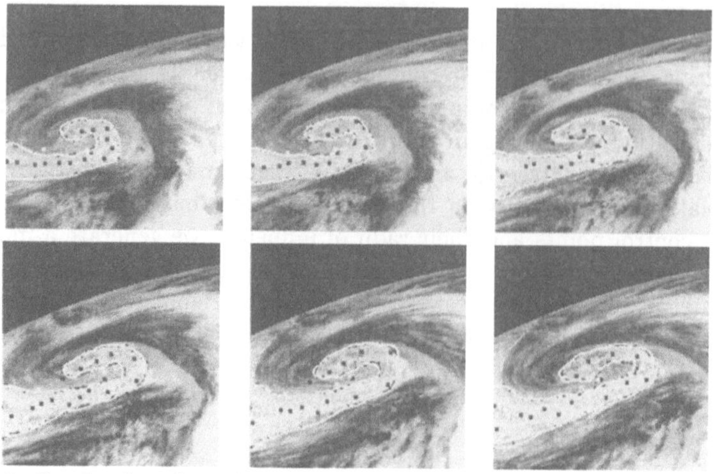

Fig. 5.2. *Result of tracking. Positive control points are depicted in black, negative in white, and the iso-contour is drawn in white.*

6. Conclusion and perspectives

The research presented in this paper is a first step towards a global study of complex structures undergoing deformation. Implicit functions seem to be a very valuable tool to track and model such deformation. For motion of deformable structures, we are working on translating the physical motion of a structure into simpler kinematic rules on particles and on the modelling of even larger deformation. Another challenging problem is to use implicit functions to solve the inverse problem for the motion of deformable structures in natural images.

7. Acknowledgments

We wish to thank A. Szantai and M. Desbois from Dynamical Meterology Laboratory (LMD) of Ecole Polytechnique (France) who provided the METEOSAT image sequence used in this work.

References

1. A. H Barr. Superquadrics and angle-preseving transforms. *IEEE Computer Graphics and Applications*, 1(1):11–23, 1981.
2. J. Blinn. A generalization of algebraic surface drawing. *ACM Transactions on Graphics*, pages 235–256, July 1982.
3. T. F. Cootes, C. J. Taylor, D. H. Cooper, and J. Graham. Active shape models–their training and application. *Computer Vision and Image Understanding*, 61(1):38–59, January 1995.

4. P. E. Gill, W. Murray, and M. H. Wright. *Practical Optimization*. Academic Press, New-York and London, 1981.
5. I.L. Herlin, C. Nguyen, and Ch. Graffigne. Stochastic segmentation of ultrasound images. In *ICPR Proceedings*, The Hague, Holland, August 30–September 3 1992.
6. S. Muraki. Volumetric shape description of range data using blobby model. *Computer Graphics*, 25(4):227–235, July 1991.
7. H. Nishimura, M. Hirai, T. Kawai, T. Kawata, I. Shirakawa, and K. Omura. Objects modeling by distribution function and a method of image generation (in japanese). *Trans. IEICE*, J68-D(4):718–725, 1985.
8. A. Pentland and K. Mase. Automatic visual recognition of spoken words. Technical Report 117, M.I.T. Media Lab Vision Science, August 1989.
9. C.H. Reinsch. Smoothing by spline functions I. *Numerische Mathematik*, 10:177–183, 1967.
10. S. Sclaroff and A. Pentland. Generalized implicit functions for computer graphics. *Computer Graphics*, 25(4):247–250, July 1991.
11. A. Szantai and M. Desbois. Construction of cloud trajectories to study the cloud life cycle. *Advanced Space Research*, 14(3):115–118, 1994.
12. J. A. Thorpe. *Elementary Topics In Differential Geometry*. Springer-Verlag, New-York Heidelberg Berlin, 1979.
13. L. Velho and J. Gomes. Approximate conversion of parametric to implicit surfaces. In *Implicit Surfaces'95*, pages 77–96, Grenoble, France, April 1995.
14. B. Wyvill and G. Wyvill. Field functions for implicit surfaces. *The Visual Computer*, 5:75–82, December 1989.
15. G. Wyvill, C. McPheeters, and B. Wyvill. Animating soft objects. *The Visual Computer*, 2:235–242, August 1986.
16. G. Wyvill, C. McPheeters, and Brian Wyvill. Data structures for soft objects. *The Visual Computer*, 2:227–234, August 1986.

An algorithm for generating motion by mean curvature

Steven J. Ruuth *

Department of Mathematics, University of British Columbia, Vancouver, Canada
E-Mail: ruuth@math.ubc.ca

Summary. A new algorithm based on the coupled level set method of Merriman, Bence and Osher [MBO] is proposed to track the curvature-dependent motion of surfaces. Similar to the usual finite difference implementation of the MBO-method, this algorithm uses the diffusion equation to generate curvature-dependent motion. Greatly improved efficiency is obtained, however, by using a higher order approximation to surfaces and by neglecting the contribution of rapidly decaying solution transients. Efficiency improvements, not possible with the original algorithm, are also made using extrapolation.

Results for several numerical experiments in two and three dimensions are also given. These demonstrate that the new algorithm is often more than 1000 times faster than the original implementation.

1. Introduction

In a variety of applications, one wants to track the motion of a front that moves with normal velocity equal to mean curvature or mean curvature plus a constant (affine velocity front motion). Such problems arise in the study of image enhancement (e.g., [15]), idealized crystal growth (e.g., [10]), the derivation of minimal surfaces (e.g., [4]), flame propagation (e.g., [17]) and grid generation[14]. Mathematicians are also interested in studying topological properties of mean curvature flow.

To study these phenomena, several numerical methods have been developed. These include front tracking methods (e.g., [3, 6]), Monte-Carlo methods such as Potts models (e.g., [7]), reaction-diffusion equations (e.g., [2]) and the Hamilton-Jacobi level set formulation of Osher and Sethian[11]. Of the various methods proposed, only a recent variational approach[16] and the MBO-method appear to be general enough to accurately treat multiple junctions in three dimensions. The main limitations for the usual implementation of the MBO-method are that it is exceedingly slow for 3-dimensional and affine velocity motions and that few theoretical results have been derived.

This article begins by describing the diffusion-generated motion by mean curvature algorithm given in [9]. Section 3 describes a new, spectral implementation for the MBO-method, which is typically much faster than the usual finite difference discretization. Section 4 describes several experimental results which demonstrate that extrapolation in conjunction with the new

* The work of the author was partially supported by an NSERC Canada Grant.

algorithm is a very effective technique for determining quantities of interest at arbitrary times. This section also includes results for a nonlocal curvature model. Section 5 concludes by summarizing the results of this article.

2. Diffusion-Generated Motion by Mean Curvature Algorithm

Suppose we wish to follow an interface moving with curvature dependent speed (e.g., Fig. 1). To carry out this motion over a domain, $\mathcal{D}$, the diffusion-generated motion by mean curvature algorithm of Merriman, Bence and Osher[9] can be used[1]:

MBO-Algorithm (Two Regions)
BEGIN
(1) Set U equal to the characteristic function for the initial region.
i.e., set $U(\boldsymbol{x},0) = \begin{cases} 1 & \text{if } \boldsymbol{x} \text{ belongs to the initial region} \\ 0 & \text{otherwise.} \end{cases}$

REPEAT for all steps, j, from 1 to the final step:
BEGIN
(2) Apply diffusion to U for some time, τ.
i.e., find $U(\boldsymbol{x}, j\tau)$ using $\begin{cases} U_t &= \Delta U, \\ \frac{\partial U}{\partial n} &= 0 \text{ on } \partial\mathcal{D}. \end{cases}$
(3) "Sharpen" the diffused region by setting
$U(\boldsymbol{x}, j\tau) = \begin{cases} 1 & \text{if } U(\boldsymbol{x}, j\tau) > \frac{1}{2} \\ 0 & \text{otherwise.} \end{cases}$
END
END

For any time t, the level set $\{\boldsymbol{x} : U(\boldsymbol{x},t) = \frac{1}{2}\}$ gives the location of the interface.

To illustrate this algorithm for the problem of Fig. 1, a grey-scaled representation of U after each of the steps (1) to (3) is given below:

After step (1) $U = 1$ for the black region of Fig. 2 and $U = 0$ elsewhere,
(2) U ranges between 0 and 1 as represented by the greyscale image of Fig. 3,
(3) $U = 1$ for the black region of Fig. 4 and $U = 0$ elsewhere.

By diffusing the characteristic function for each region, it is straightforward to extend the previous method to multiple regions[9]. Extensions to affine velocity motion are also possible[8] by tracking level sets other than $\frac{1}{2}$.

[1] See [5, 1] for convergence proofs for the case of two regions.

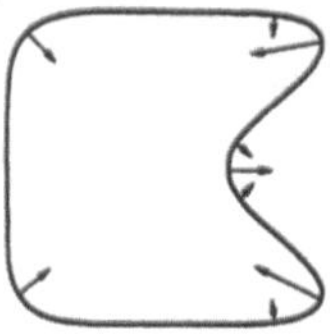

Fig. 1. *Initial Motion*

Fig. 2. *Initial Region*

Fig. 3. *After a Time τ*

Fig. 4. *Sharpened Region*

2.1 Selection of τ

It is important to select τ appropriately. We must select τ small enough to accurately resolve the motion of the interface[9]. Furthermore, when applying a finite difference approximation to the Laplacian term, diffusion must proceed long enough so that the level set $U = \frac{1}{2}$ moves at least one grid point, otherwise the interface remains stationary. This produces the restriction that

$$\text{(speed of motion of the interface)} \times \tau \gg \text{grid spacing}$$

Letting $\kappa(\boldsymbol{x})$ be the local curvature and h the grid spacing gives us

$$\kappa(\boldsymbol{x}) \times \tau \gg h \tag{2.1}$$

for the finite difference method.

3. A New, Spectral Method

Accurate computation of solutions using the usual finite difference implementation for the MBO-method can be expensive, even for simple 2-dimensional problems. Since we are mainly interested in 3-dimensional problems or problems involving more than two phases, a faster method is desired. This section describes a new, spectral method which is typically much faster than the usual implementation.

For notational simplicity, the method is described for the 2-dimensional case over the domain, $\mathcal{D} = [0,1] \times [0,1]$. Extensions to higher spatial dimensions or more phases is straightforward[13].

3.1 Discretization of the Heat Equation

As we have seen, carrying out the MBO-algorithm requires us to solve the heat equation

$$\begin{aligned} u_t &= \Delta u, \\ \frac{\partial u}{\partial n} &= 0 \text{ on } \partial\mathcal{D} \end{aligned} \tag{3.1}$$

repeatedly over time intervals of (possibly variable) length τ. Over any of these intervals, u may be approximated by the Fourier cosine tensor product,

$$U(x,y,t) = \sum_{i,j=1}^{n} c_{ij} \exp(-(i^2+j^2)\pi^2(t-t_{start})) \cos(\pi i x)\cos(\pi j y) \quad (3.2)$$

where t_{start} is the time when the most recent interval starts.

One might expect that a Fourier spectral approximation for u would be unwise because u is initially discontinuous at interfaces. We are only interested in the solution after a time τ, however. After a time τ, high frequency modes have dissipated[2]. Since the problem is linear, different modes do not interact and thus there is never a need to approximate high frequency modes (not even near t_{start}, when high frequency modes make an important contribution). For this reason, an accurate approximation to (3.1) at time τ can be obtained using far fewer basis functions than one might otherwise expect.

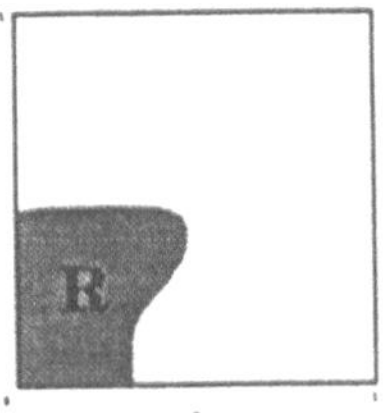

Fig. 5. *Initial Region*

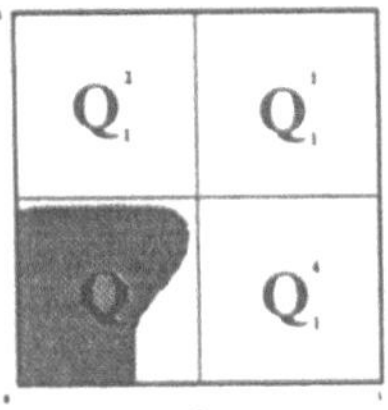

Fig. 6. *One subdivision*

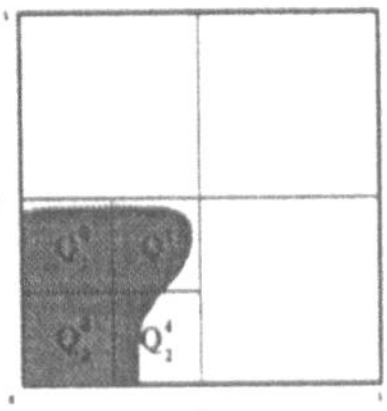

Fig. 7. *Two subdivisions*

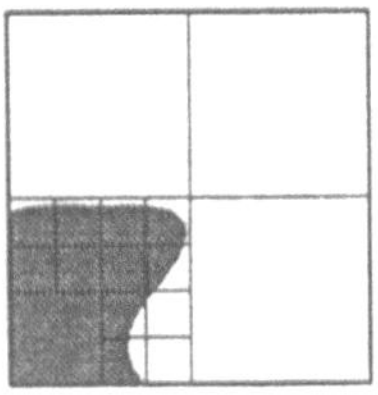

Fig. 8. *Four subdivisions*

3.2 Calculation of the Fourier Coefficients

The values of the Fourier coefficients, c_{ij}, of equation (3.2) must still be determined after each sharpening. These values are found using an adaptive quadrature method. Begin by defining

$$R_t = \{(x,y) : U(x,y,t) > \frac{1}{2}\}$$

as the approximation to the phase we are following. By multiplying equation (3.2) at time $t = t_{start}$ by $\cos(i\pi x)\cos(j\pi y)$, integrating over the domain and simplifying via the usual orthogonality conditions we find

$$\begin{aligned}
c_{00} &= \int_0^1\int_0^1 U(x,y,t_{start})\,dx\,dy, && \\
c_{i0} &= 2\int_0^1\int_0^1 U(x,y,t_{start})\cos(i\pi x)\,dx\,dy && \text{for } i \neq 0, \\
c_{0j} &= 2\int_0^1\int_0^1 U(x,y,t_{start})\cos(j\pi y)\,dx\,dy && \text{for } j \neq 0, \\
c_{ij} &= 4\int_0^1\int_0^1 U(x,y,t_{start})\cos(i\pi x)\cos(j\pi y)\,dx\,dy && \text{for } i,j \neq 0.
\end{aligned}$$

[2] Note that the required number of modes increases as τ decreases.

Immediately after sharpening,

$$U(x,y,t) = \begin{cases} 1 \text{ if } (x,y) \in R_t \\ 0 \text{ otherwise} \end{cases}$$

which gives us that

$$c_{ij} \;=\; \gamma_{ij} \iint\limits_{R_t} \cos(i\pi x)\cos(j\pi y)\; dA \tag{3.3}$$

where

$$\gamma_{ij} = \begin{cases} 1 & \text{if } i = j = 0 \\ 4 & \text{if } i \neq 0 \text{ and } j \neq 0 \\ 2 & \text{otherwise} \end{cases} \tag{3.4}$$

We thus need to integrate a simple function over a non-rectangular region, R_t. This may be accomplished by recursively subdividing the domain. We now give an illustration of this recursion for the region R given in Fig. 5.

Begin by determining the phase at each of the corner points. Because the phase at the origin is different than that at other corner points of the domain boundary we know that two phases are present. For this reason, we subdivide the domain into quadrants as shown in Fig. 6.

The phase color at the corner points of quadrants Q_1^1, Q_1^2 and Q_1^4 is white so we assume that $R \bigcap (Q_1^1 \bigcup Q_1^2 \bigcup Q_1^4) = \phi$ and that these subregions do not produce a contribution to the integral (3.3). The phases at the corners of Q_1^3 differ, however, so this region is subdivided further. See Fig. 7.

For each of Q_2^1, Q_2^2 and Q_2^4, the phases of the corner points differ so each of these subregions is subdivided. All corner points of the remaining subregion, Q_2^3, are grey. Thus we assume $Q_2^3 \bigcap R = Q_2^3$ and add a contribution

$$\gamma_{ij} \iint\limits_{Q_2^3} \cos(i\pi x)\cos(j\pi y)\; dA$$

to each of the Fourier coefficients, c_{ij}, for $0 \leq i,j \leq n$.

The domain is further subdivided recursively (see Fig. 8) until regions containing multiple phases can be safely approximated by a simple numerical technique. In this article, each c_{ij} is calculated directly (i.e. without transform methods) and the finest subdomains are approximated with piecewise linear approximations to the boundary [13]. We are currently developing fast implementations (see [13]) which apply an unequally spaced fast Fourier transform method to determine the c_{ij}.

3.3 Comparison to the Usual Finite Difference Implementation

There are several reasons why the spectral method outlined in this section is preferred over the usual finite difference implementation. These reasons are outlined below.

1. As has been discussed in §3.1, only low frequency modes need to be approximated provided τ is not taken very small. A large amount of computational work is saved by only treating these low frequency modes.
2. The new, spectral method does not require any time-stepping between t_{start} and $t_{start+\tau}$. This eliminates a possible source of error and produces large savings in computational work.
3. The new method is simpler to implement because there are no implicit systems to solve and local refinement is done in the context of a quadrature, rather than a discretization of a differential equation.
4. By using a spectral method, we can nearly eliminate the error arising from discretizing the heat equation. This is an attractive feature, because it makes extrapolation in τ practical (see next section), which in turn allows for larger τ-steps. When larger τ-steps are taken, even fewer basis functions are required to solve the heat equation to a given accuracy.
5. The original finite difference algorithm must satisfy (2.1) globally, or part of the front may erroneously remain stationary. By recursively refining near the interface, the new spectral method avoids this restriction.

These are indeed formidable advantages of the new, spectral method over the usual finite difference approach. Even for the straightforward implementation described above, large computational savings are typically observed.

To illustrate these savings, consider the motion by mean curvature of the kidney-shaped region displayed in Fig. 9. Using the new, spectral method and a finite difference implementation[3], we compare the area lost over a time $t = 0.0125$ with the exact answer, $0.0125 \times \pi = 0.0785398$ (see [10]). ¿From Table 1, we see that the new, spectral implementation is adequate for finding solutions to within a 1% error. The finite difference approach, however, becomes impractical when accurate solutions are sought (see Table 2). Numerical tests for the problems described in the next section also found that the new, spectral method required less than 1% of the computational time of the usual finite difference approach (results not shown).

τ	n	*Error*	*Time*[4]
0.003125	12	4%	0.4 s
0.00078125	23	1%	4 s

Table 1. New, Spectral Method

Δt	Δh	*Error*	*Time*[4]
1×10^{-4}	$\frac{1}{128}$	4%	85 s
2×10^{-5}	$\frac{1}{512}$	3%	10341 s

Table 2. Finite Difference Method

[3] The difference algorithm uses an adaptive τ-stepping method with backward Euler and a multigrid technique to solve the implicit equations.

[4] All timings were carried out on an HP735/100 workstation.

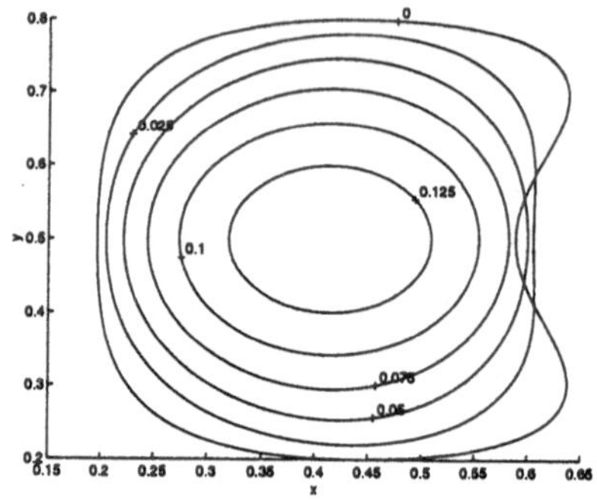

Fig. 9. *Smooth Interface at time t*

4. Numerical Experiments

In this section we report on various experiments using our algorithm.

4.1 Extrapolation for Smooth Interfaces

For smooth interfaces, it can be shown by an asymptotic expansion that the new, spectral implementation is first order in τ, and that the error is regular in τ [13]. This suggests that higher order results may be attainable using extrapolation in conjunction with the new, spectral method.

To test this possibility, we revisit the motion by mean curvature example displayed in Fig. 9. Using the new, spectral implementation (the direct method) and an extrapolation[5], the area lost over a time $t = 0.0125$ was compared to the exact answer for several τ. The results from a number of experiments are given in the table below.

	Direct Method		*Extrapolated Method*	
τ	Area Reduction	| Error |	Area Reduction	| Error |
0.00625	0.0840918	0.0055520	0.0770344	0.0015054
0.003125	0.0811712	0.0026314	0.0782506	0.0002892
0.0015625	0.0798342	0.0012944	0.0784972	0.0000426
0.00078125	0.0791813	0.0006415	0.0785284	0.0000106

These results support the conclusion that the MBO-algorithm is first order in τ and that extrapolation can be used with the new, spectral method to produce a second order result.

4.2 Extrapolation in the Presence of Triple Junctions

Extrapolation has also been used successfully for problems with multiple junctions. Consider, for example, the motion by mean curvature of the triple

[5] If we let S_τ be a result using the direct method with a step τ, then the extrapolation may be written $2S_\tau - S_{2\tau}$.

junction problem shown in Fig. 10. Using the new, spectral implementation (the direct method) and an extrapolated method[6], the time at which the smallest phase disappears was compared to the exact time, $t = 0.33$ (obtained using Brian Wetton's front tracking code, see [3]). The results are given in the table below.

	Direct Method		*Extrapolated Method*	
τ	T_f	| Error |	T_f	| Error |
0.04	0.1664	0.1636	0.2533	0.0767
0.02	0.2096	0.1204	0.3139	0.0161
0.01	0.2424	0.0876	0.3216	0.0084
0.005	0.2669	0.0631	0.3260	0.0040

These results suggest that the direct method is order $\sqrt{\tau}$ and that extrapolation can be used with the new, spectral method to produce higher order approximations when triple junctions are present. Indeed, even more accurate results have been produced using higher order extrapolations [13].

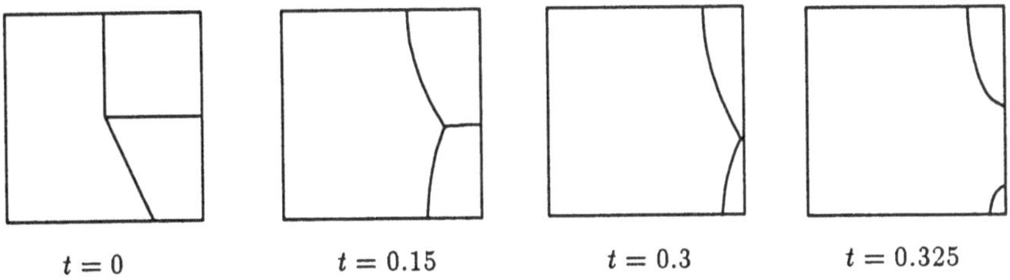

Fig. 10. *Triple Junction Results at Various Times, t*

4.3 Nonlocal Model

Implementations for affine velocity are not practical using the usual finite difference implementation. These may be carried out using the new, spectral method, however. Fig. 11, for example, shows the results of an experiment where the curve moves with normal velocity equal to the curvature minus the average curvature (see [13] for more details). Such a motion preserves phase areas and can be derived as a limit of a nonlinear model of binary alloys[12].

4.4 Three Dimensional Results

Three dimensional computations are also practical using the new, spectral implementation. Fig. 12, for example, shows a two-phase problem moving according to mean curvature motion.

[6] If we let S_τ be the result using the direct method with a step τ, then the extrapolation may be written $\frac{1}{\sqrt{2}-1}(\sqrt{2}S_\tau - S_{2\tau})$.

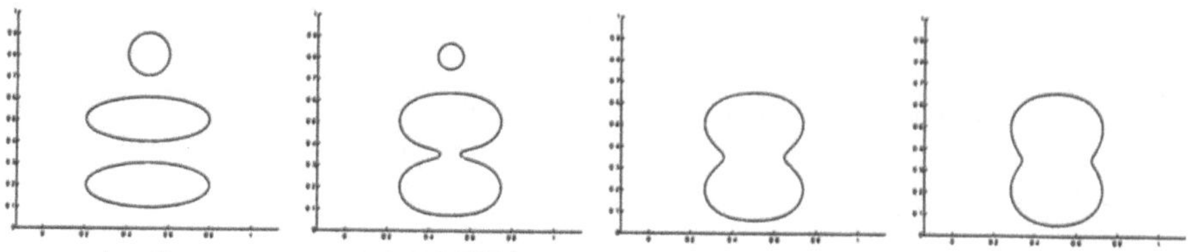

Fig. 11. *Nonlocal Model which Preserves Area*

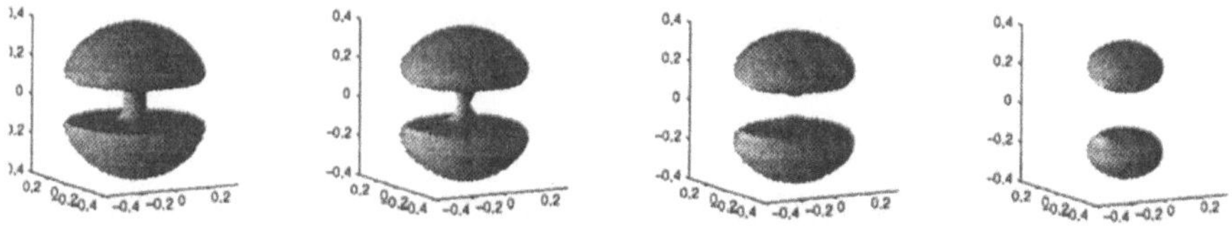

Fig. 12. *Three Dimensional Motion by Mean Curvature*

5. Conclusions

The new, spectral algorithm outlined in this article has several advantages over the usual finite difference approach for solving diffusion-generated motion by mean curvature. In particular, the method is easier to implement because no implicit systems need to be solved and the local refinement is handled by a simple quadrature method. It is also much more efficient because it gives a more accurate approximation to surfaces and it avoids approximating rapidly decaying transients. Furthermore, efficiency improvements, not possible with the original algorithm, can be made using extrapolation.

Numerical experiments show that extrapolation can be an effective method of improving the accuracy of the new, spectral method. Experiments also demonstrate that the method can be extended to approximate certain nonlocal curvature motions.

6. Acknowledgments

I would like to thank my supervisors Dr. Brian Wetton and Dr. Uri Ascher for their many helpful suggestions while working on this article.

References

1. Barles, G., Georgelin, C. (1995): *A simple proof of convergence for an approximation schemes for computing motions by mean curvature*, SIAM Journal of Numerical Analysis, **32**(2), 484-500.

2. Bronsard, L., Reitich, F. (1993): *On three-phase boundary motion and the singular limit of a vector-valued Ginzburg-Landau equation*, Arch. Rat. Mech., **124**, 355-379.
3. Bronsard, L., Wetton, B.T.R. (1995): *A numerical method for tracking curve networks moving with curvature motion*, Journal of Computational Physics, **120**(1), 66-87.
4. Chopp, D.L. (1993): *Computing minimal surfaces via level set curvature flow*, Journal of Computational Physics, **106**(1), 77-91.
5. Evans, L.C. (1993): *Convergence of an algorithm for mean curvature motion*, Indiana University Mathematics Journal, **42**, 553-557.
6. Fradkov, V.E., Glicksman, M.E., Palmera, M., Nordberg, J., Rajan, K. (1993): *Topological rearrangements during 2D normal grain growth*, Physica D, **66**, 50-60.
7. Glazier, J., Anderson, M., Grest, G. (1990): *Coarsening in the two-dimensional soap froth and the large-Q Potts model: a detailed comparison*, Philosophical Magazine B, **62**, 615-645.
8. Mascarenhas, P., *Diffusion generated by mean curvature*, Manuscript, University of California, Los Angeles.
9. Merriman, B., Bence, J., Osher, S. (1994): *Motion of multiple junctions: a level set approach*, Journal of Computational Physics, **112**(2), 334-363.
10. Mullins, W.W. (1956):, *Two-dimensional motion of idealized grain boundaries*, Journal of Applied Physics, **27**(8), 900-904.
11. Osher, S., Sethian, J.A. (1988): *Fronts propagating with curvature-dependent speed: algorithms based on Hamilton-Jacobi formulations*, Journal of Computational Physics, **79**, 12-49.
12. Rubinstein, J., Sternberg, P. (1992): *Nonlocal reaction-diffusion equations and nucleation*, IMA J. Appl. Math., **48**, 248-264.
13. Ruuth, S.J., Wetton, B.T.R. (1996): *Diffusion-generated motion by mean curvature.* In Preparation.
14. Sethian, J.A. (1994): *Curvature flow and entropy conditions applied to grid generation*, Journal of Computational Physics, **115**, 440-454.
15. Yuille, A. (1988): *The creation of structure in dynamic shape*, Proceedings of the Second International Conference on Computer Vision, Tampa, Florida, 685-689.
16. Zhau, H., Chan, T., Merriman, B., Osher, S. (1995): *A variational level set approach to multiphase motion*, CAM Report 95-36, University of California, Los Angeles.
17. Zeldovich, Y.B. (1981): *Structure and stability of steady laminar flame at moderately large Reynolds numbers*, Comb. Flame., **40**, 225-234.

A Numerical Model for Large Deformation on Meteorological Images

Jean-Paul BERROIR, Isabelle HERLIN, Isaac COHEN

INRIA, AIR project Rocquencourt
B.P. 105, 78153 Le Chesnay Cedex, France.
Email Jean-Paul.Berroir@inria.fr, Isabelle.Herlin@inria.fr,
Isaac.Cohen@inria.fr

Summary. The purpose of non-rigid motion in computer vision is the study of the geometrical deformation that may affect an object during its temporal evolution. One often assumes that this process is local and continuous; it is therefore the case for local geometrical features, which are considered as relevant for studying deformation. This assumption is no more valid when the structure is widely distorted between two successive temporal occurrences, and information about the underlying physical phenomenon is then required to analyze motion. In this paper, we propose a geometrical model of evolution that may be viewed as an approximation of the true physical deformation: we are reconstructing the surface generated by the evolution of the boundary of the structure. We present a stable numerical implementation of this model and apply it to the tracking of a vortex on a sequence of meteorological images.

1. Introduction

Many applications of motion study in computer vision require the analysis of non-rigid motion. This is for instance the case for medical applications [2], or satellite imagery, like clouds motion study. Most approaches of non-rigid motion share the same fundamental assumption, often called the "small deformation" hypothesis. This means that the motion is observed with a thin temporal resolution, so that successive occurrences of the structure are progressively deformed. The local geometrical features are then varying continuously. A well-known approach [1, 2, 5] is actually based on curvature extrema, that are the invariants of rigid motion. Temporal regularity is used by many other authors: for instance, Geiger's [7] approach for tracking considers high curvature points on the previous frame as the initialization of the current one. One of the characteristics of his matching approach is the search for a smooth displacement field. Serra and Berthod [11] are looking for a continuous matching function. To perform this task, they minimize a criterion based on the error measure between the different matches, that is, the difference between translation vectors: temporal regularity is assumed again.

The small deformation hypothesis can however not always be used. This is the case when the velocity of the observed phenomenon is too important with respect to the temporal resolution of the sensor. We are then dealing with large deformation, which often occurs with dynamic satellite imagery [4].

Since local features may have no relevance, matching or tracking structures in this case can not be done without adding exogenous knowledge about motion. This information should ideally be the physical evolution model which can yet be unavailable. We therefore propose to approximate it by a geometrical representation: we suppose we are given two temporal occurrences of the structure, namely, two 2D boundaries. Intermediate spatial positions generate during motion a continuous 3D surface, with the third dimension being time. The main objective is to compute the best possible surface, that is, that one interpolating the two contours and respecting the approximated evolution model. This approach presents several other advantages:

- The intermediate occurrences of the contour are cross sections of the surface, parallel to the original contours. This is the justification of this approach: the surface carries information about the evolution model and therefore enables to retrieve the continuous motion between two given temporal occurrences. This model does not rely on local geometrical features of the original contours.
- Points trajectories are included in the surface. It is possible to compute them, using the Hamilton-Jacobi formalism for instance.
- Topological changes may be expressed using geometrical considerations: for instance, the case of a structure splitting from one connected component into two different one can be solved by trying to find a surface with a branching.

The evolution model and the associated geometrical constraints are explained in section 2, with different numerical methods used to compute the surface. The problems, encountered by trying to define a stable numerical scheme, lead to a simplified and efficient model definition described in section 3. The adaptation of the model to the data is explained in section 4 with results obtained on METEOSAT images.

2. Definition of the surface model

The purpose of this section is to explicitly define the evolution model through the definition of the surface model, and to propose a direct implementation. As already mentioned, modelling is based on geometrical constraints. We propose to minimize the distortion of the surface, which gives it an elastic behavior.

Two 2D contours C_1 and C_2 are given and placed in two parallel planes $z = t_1$ and $z = t_2$, t_i being the date of acquisition of the contour C_i. Since the model only deals with these two given dates, the temporal axis has an arbitrary scale: assuming that $t_1 = 0$ and $t_2 = 1$ does not influence further computation. In order to focus this study on deformation, we first apply the rigid motion that performs the best matching between C_1 and C_2. This enables to manage the global translation motion that is often observed on meteorological and oceanographic image sequences. We then look for an interpolating surface ϕ, between C_1 and C_2, with an elastic behavior. The surface ϕ is described by two parameters $(s, r) \in \Omega = [0,1]^2$. The parameter r represents in fact the time. A constant value of r generates a cross section of the surface parallel to the original contours and described by s. The intermediate temporal occurrences of the contour are then described by the different values of r between 0 and 1.

Elasticity may be modelled by control of the first fundamental form. This form, $\mathcal{F}$ (see equation 2.1, where $< .,. >$ is the dot product in $\mathbb{R}^3$), expresses the distortion produced by ϕ on an elementary square of the parameters' space.

$$\mathcal{F} = \begin{pmatrix} < \frac{\partial \phi}{\partial s}, \frac{\partial \phi}{\partial s} > & < \frac{\partial \phi}{\partial s}, \frac{\partial \phi}{\partial r} > \\ < \frac{\partial \phi}{\partial s}, \frac{\partial \phi}{\partial r} > & < \frac{\partial \phi}{\partial r}, \frac{\partial \phi}{\partial r} > \end{pmatrix} \tag{2.1}$$

If $\mathcal{F}$ is equal to identity, the local basis $(\frac{\partial \phi}{\partial s}, \frac{\partial \phi}{\partial r})$ of the tangent plane of ϕ is orthonormal. In this case the distortion is minimal. Moreover, the closer to identity is $\mathcal{F}$, the smaller is the distortion. We then measure the local distortion of the surface by the expression (2.2), where Id stands for identity matrix:

$$||\mathcal{F} - Id||^2 \tag{2.2}$$

The elastic behavior is obtained by minimizing the integral of this criterion over the surface. This corresponds to the following energy functional:

$$E(\phi) = \int\int_{\Omega} \left((< \phi_s, \phi_s > -1)^2 + 2 < \phi_s, \phi_r >^2 + (< \phi_r, \phi_r > -1)^2\right) dsdr \tag{2.3}$$

Minima of $E(\phi)$ are solutions of $\nabla E(\phi) = 0$, and this leads to the following differential system (where ∇^t stands for the transpose of the gradient operator):

$$\triangle \phi - \nabla^t(\mathcal{F} \nabla \phi) = 0 \tag{2.4}$$

This equation expresses the evolution model. The surface must also verify boundary conditions. Because the surface, that we are looking for, interpolates the original contours, we obtain the two following boundary conditions concerning $r = 0$ and $r = 1$: $\phi(.,0) = C_1$ and $\phi(.,1) = C_2$. The conditions concerning $s = 0$ and $s = 1$ depend on the topology of the contours. We previously mentioned that topological changes are not discussed in this paper and the study is restricted to the case where the two contours have only one connected component. If the contours are closed, the surface has then the topology of a cylinder, and the boundary conditions become those of equation (2.5). If they are opened, with no more information, we do not apply constraints on $s = 0$ and $s = 1$.

$$\begin{cases} \phi(.,0) = C_1 \\ \phi(.,1) = C_2 \\ \phi(0,.) = \phi(1,.) \end{cases} \tag{2.5}$$

The energy is of course not quadratic. To achieve numerical resolution, we use a relaxation model (eq. 2.6):

$$\frac{\partial \phi}{\partial \tau} + \Delta\phi - \nabla^t(\mathcal{F}\nabla\phi) = 0 \tag{2.6}$$

where τ is the relaxation time of the system. After temporal discretization and linearization of equation 2.6, ϕ at time $\tau + d\tau$ may be computed knowing ϕ at time τ. We therefore need an initial estimate, ϕ at time $\tau = 0$, that can be a bilinear interpolation of the two original contours. Moreover, ϕ must be constrained, at each step of the relaxation process, to respect the boundary conditions.

This process has the advantage of being easily implemented. It has yet several drawbacks:

- As the energy functional is not convex, each solution is in fact associated to the initial estimate. It is therefore crucial to find a sufficiently good estimate of the surface.
- Getting the first fundamental form close to identity tends to produce parallel cross sections of length 1. But the lengths of the original contours (L_1 for C_1 and L_2 for C_2) are given and that of the cross section must vary along the r-axis between these two values. This can be easily solved by minimizing the norm of $\mathcal{F} - K(r)$, instead of $\mathcal{F} - Id$, where $K(r)$ is:

  $$K(r) = \begin{pmatrix} rL_2 + (1-r)L_1 & 0 \\ 0 & 1 \end{pmatrix} \tag{2.7}$$

  Remaining calculation is unchanged. This scheme tends to produce a linear variation of the length of the cross sections between L_1 and L_2.
- The most important problem concerns regularity. We saw that we must impose the surface to respect the boundary conditions during the relaxation process. Irregularities may occur in the neighborhood of the two original contours. Furthermore, the energy functional does not prevent local discontinuities. Therefore, the regularity of the surface is not assured. This point will be further discussed in the next section.

3. Restricted model

The main difficulty of our approach is to find a surface that optimizes a criterion that does not constraint regularity. A simple method to achieve this task is to restrict the space of available surfaces. In other words, we define a parametric model of the surface: we consider the bilinear interpolations of the two contours as the set of available surfaces.

We now consider two opened contours, which is the case for vortex tracking. These contours are described by their curvilign abscissa s. A bilinear interpolation of C_1 and C_2 is generated by the matching function f, that links their curvilign abscissa. If the lengths of the two contours are respectively L_1 and L_2, each function f mapping $[0, L_1]$ into $[0, L_2]$ generates the following bilinear interpolation (see eq. 3.1):

$$S(s,r) = \begin{cases} x(s,r) & = & (1-r)x_1(s) + rx_2(f(s)) \\ y(s,r) & = & (1-r)y_1(s) + ry_2(f(s)) \\ z(s,r) & = & r \end{cases} \tag{3.1}$$

where $s \in [0, L_1]$, $r \in [0, 1]$ and $(x_i(s), y_i(s))$ is the point of contour C_i with curvilign abscissa s. The set of available surfaces is in fact too restricted to allow computation of realistic points trajectories. But it enables a simple formulation of the matching between two contours: the matched points have respectively the curvilign abscissa s and $f(s)$. The computation of the energy functional (equation 2.3) also gets easier: first, we are now looking for the matching function f, which is a one-dimensional problem; then, integrating the energy functional along the r-axis can be approximated by the area of the quadrangle formed by the matched points, X_1 and $X_2(f)$, and the tangents of the contours at these points, $\overrightarrow{\Gamma_1}$ and $\overrightarrow{\Gamma_2(f)}$. The associated energy is then defined by:

$$e(f) = \int_s (\overrightarrow{X_1 X_2(f)} \wedge \overrightarrow{\Gamma_1})^2 + (\overrightarrow{X_1 X_2(f)} \wedge f'\overrightarrow{\Gamma_2(f)} + f'\overrightarrow{\Gamma_2(f)} \wedge \overrightarrow{\Gamma_1})^2 \tag{3.2}$$

The gradient of this functional, with respect to the matching function, can be computed, and this leads to the following differential equation, where N_i is the normal vector of the contour C_i at X_i:

$$\begin{aligned} &< \overrightarrow{X_1 X_2(f)} \wedge \overrightarrow{\Gamma_1}, \overrightarrow{\Gamma_1} \wedge \overrightarrow{\Gamma_2(f)} > + f''((\overrightarrow{X_1 X_2(f)} - \overrightarrow{\Gamma_1}) \wedge \overrightarrow{\Gamma_2(f)})^2 + \\ &f'^2 < (\overrightarrow{X_1 X_2(f)} - \overrightarrow{\Gamma_1}) \wedge \overrightarrow{\Gamma_2(f)}, (\overrightarrow{X_1 X_2(f)} - \overrightarrow{\Gamma_1}) \wedge \overrightarrow{N_2(f)} > \quad = 0 \end{aligned} \tag{3.3}$$

But this criterion does not express the fact that the contours have to be entirely matched, and that the solution f must be non-decreasing. Moreover, regularity problems may still occur. We therefore propose to use a parametric model for the matching function f. This represents an adaptation of the model to the applicative structures. In the context of vortex tracking, we suppose that one can perform a static segmentation of a fully developed vortex into n meaningful regions. These regions may grow at different speeds, but we still want them to be matched one to each other during the temporal evolution. The matching function should be approximately linear on those regions. We then use piecewise linear functions, smoothed by a C^2 filter at the $n-1$ junction points. The coordinates $(s_i, f(s_i))$ of these points are the curvilign abscissa of the separation between the different regions on the two contours. The energy functional depends now on the $(n-1)$ junction points of the matching function. Optimization is performed using a hierarchical scheme: the intervals $[0, L_1]$ and $[0, L_2]$ are first discretized into few admissible values (20 for instance) and the minimum is obtained using an exhaustive search on these values. Result is then progressivly refined, by increasing the number of discretization points.

4. Results

Classical approaches of non-rigid motion are not adapted to remote sensed data for two main reasons:

- the observed structures are often not related to a physical object, but they express physical measurements of fluids: meteorologic images measure the water vapor density in the atmosphere, oceanographic images may for example concern the temperature of the sea. A vortex is a specific configuration of the fluid in motion. Thus, it is uneasy to define its shape and its boundary. As a consequence, one can not expect to compute local features accurately. It is in particular the case for curvature (and a fortiori curvature extrema) measure, which definition involves differential characteristics of order 2.
- The temporal resolution of the sensor is related to the trajectory of the satellite, to the time it spends to perform an acquisition, and to atmospheric conditions. If the evolution of the phenomenom is fast, there may be not enough intermediate occurrences to observe a progressive evolution.

We present results concerning vortex tracking on a sequence of METEOSAT infra-red images. On figure 4.1 are displayed three occurrences of a vortex over northern atlantic, together with a static segmentation performed using *snakes* modelling.

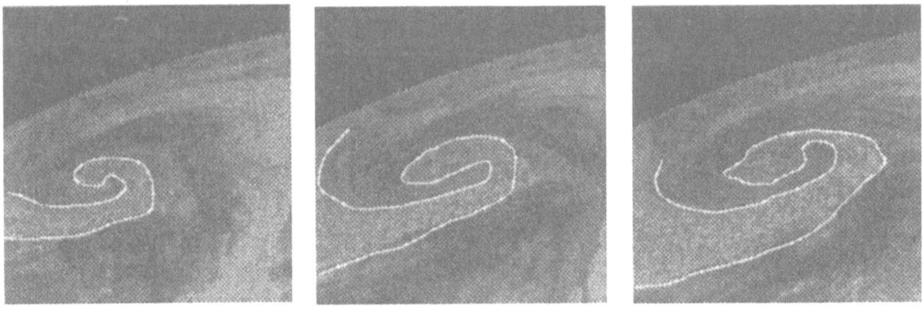

Fig. 4.1. Three consecutive occurrences of a vortex on METEOSAT infra-red data.

The vortex is in fact made of several bands of homogeneous temperature winding around its center. We have chosen to model the structure by the contour of its brightest band, that we split into three regions: the high-curvature area around the center and its extension on both sides. These regions are defined by two separation points and we then have to find four values, corresponding to the curvilign abscissa of these points on each contour, in order to perform the matching.

Results are shown on figure 4.2, where we display the matching between early occurrences of the vortex with the last one.

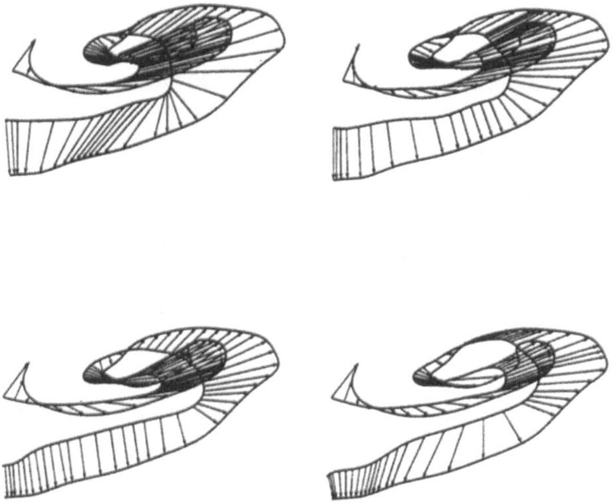

Fig. 4.2. Matching of the early occurrences of the vortex with the last one. Small deformation hypothesis does not hold.

5. Conclusion

Modelling deformation is a crucial problem when studying motion. Large deformation often occur with remote sensed data, and more generally with application concerning environment. Our approach to this problem is to model the evolution by geometrical constraints when physical ones are not available. That model allows tracking and matching of natural structures without relying on local features. It is designed accordingly to the geometry of the structure but can easily be adapted to other kind of structures.

Several improvements are possible. First, we have restricted the set of surfaces to bilinear interpolations. Consequently, point to point trajectories are included in a straight line. Finding realistic trajectories supposes to use a larger set of surface. Introducing an evolution model, that is closer to the physical one, could help restricting that set in a more appropriate way. At last, this approach, based on a parameterization of the contours, supposes an a priori known topology, and handling topological changes requires further research.

Acknowledgments. We thank M. Desbois and A. Szantai from the LMD (Laboratoire de Météorologie Dynamique) for providing the atmospheric image sequence.

References

1. A.A. Amini, R.L. Owen, and J.S. Duncan. Non-rigid motion for tracking left-ventrical wall. In *IPMI*, 1992.
2. N. Ayache, I. Cohen, and I.L. Herlin. Medical image tracking. In Andrew Blake and Alan Yuille, editors, *Active Vision*, chapter 17. MIT Press, Cambridge, Mass, 1992. December.
3. I.A. Bachelder and S. Ullman. Contour matching using local affine transformations. In *Image Understanding Workshop*, Jan 1992.
4. J.P. Berroir, I.L. Herlin, and I. Cohen. Tracking highly deformable structures: a surface model applied to vortex evolution within satellite oceanographic images. In *EOS-SPIE Satellite Remote Sensing II*, 1995.
5. I. Cohen, N. Ayache, and P. Sulger. Tracking points on deformable objects using curvature extrema. In *European Conference on Computer Vision*, Santa Margherita Ligure, Italy, May 1992.
6. J.S. Duncan, R.L. Owen, L.H. Staib, and P. Anandan. Measurement of non-rigid motion using contour shape descriptors. In *Computer Vision and Pattern Recognition*, Hawaii, June 1991.
7. D. Geiger, A. Gupta, L.A. Costa, and J. Vlontzos. Dynamic programming for detecting, tracking, and matching deformable contours. *IEEE Transactions on Pattern Analysis and Machine Intelligence*, 17(3):294–302, March 1995.
8. I.L. Herlin, I. Cohen, and S. Bouzidi. Detection and tracking of vortices on oceanographic images. In *SCIA'95*, 1995.
9. B. Horowitz and A. Pentland. Recovery of non-rigid motion and structure. In *Computer Vision and Pattern Recognition*, Hawaii, June 1991.
10. D. Metaxas and D. Terzopoulos. Shape and non-rigid motion estimation through physics-based synthesis. *IEEE Transactions on Pattern Analysis and Machine Intelligence*, 15(6):580–591, 1993.
11. B. Serra and M. Berthod. Subpixel contour matching using continuous dynamic programming. In *Computer Vision and Pattern Recognition*, 1994.
12. A. Szantai and M. Desbois. Construction of cloud trajectories to study the cloud life cycle. *Advanced Space Research*, 14(3):115–118, 1994.
13. A. Yuille and P. Hallinan. Deformable templates. In Andrew Blake and Alan Yuille, editors, *Active Vision*. MIT Press, Cambridge, Mass, 1992.

Image Enhancement and Restoration, Scale-Space

Understanding the structure of diffusive scale-spaces

Nicolas ROUGON and Françoise PRÊTEUX

Institut National des Télécommunications, Département Signal & Image
9, Rue Charles Fourier - 91011 Evry, France
Nicolas.Rougon@int-evry.fr, Francoise.Preteux@int-evry.fr

Abstract. This paper investigates structural properties of diffusive scale-spaces and develops a Riemannian description based on electromagnetic (EM) field theory. The generalized diffusion equation defining photometric transitions is interpreted as a Lorentz gauge condition expressing the trace Lorentz-invariance of an EM quadripotential with covariant scalar and contravariant vector components, respectively related to photometric and geometric image properties. This gauge condition determines EM quadrifield and quadricharge which satisfy Maxwell equations. Deriving their general expressions as functions of scale-space geometric or energetic features yields Lorentz-invariants which synthetize intrinsic multiscale image properties.
Keywords: multiscale analysis, geodesic flows, deformable manifolds, variational methods, gauge theory.

1. Introduction

Anisotropic diffusion is an efficient nonlinear filtering technique for deriving deterministic multiscale image descriptions. Extensive studies based on differential geometry, group theory and PDE theory have led to characterize its essential properties, understand its limitations and derive extented models [7, 11]. In particular, deep connections have been established with regularized, invariant and morphological formulations of low-level vision, and with multiscale shape theory [7]. Relationships with such stochastic theories as Bayesian estimation, Gibbs fields and mean field approximation have also been clarified [3, 5, 10].

Let images be modeled as elements within the space $\mathcal{F}^2(\mathcal{E}_2, \mathbb{R})$ of a.e. $\mathcal{C}^2$-continuous real mapping with compact support Ω in the Euclidean plane $\mathcal{E}_2$. A diffusive scale-space (DSS) is defined as the solution set of a Neumann problem associated with a 1st-order non-linear parabolic PDE of the form :

$$\gamma \frac{\partial I}{\partial t} = \nabla \cdot \vec{J} \qquad \text{on } \Omega\times]0,T]. \tag{1}$$

with initial condition $I(\cdot,0) = I_0$. Equation (1), called *generalized (Euclidean) diffusion equation*, models the diffusion of the scalar field I within an inhomogeneous medium. $\vec{J}(I)$ is a regular vector field on $\mathcal{E}_2$, called *diffusion flux*, and γ denotes a non-negative function on $\Omega\times]0,T]$ controlling the diffusion kinetics.

When $\vec{J} = c\nabla I$ for a constant conduction $c > 0$ and $\gamma \equiv \gamma_0 > 0$, equation (1) is a linear isotropic heat equation generating the Gaussian scale-space [8, 16]. Gaussian filtering induces oversmoothing and delocation artefacts along edges at large scales. Both effects are avoided by inhibiting diffusion along edge direction while preserving its isotropy within stationary luminance regions. This is achieved via a non-linear process defined by $\vec{J} = g(\frac{\|\nabla I\|}{K})\,\nabla I$ where g is a gradient-dependent conduction function and $K > 0$ a stationarity threshold [13]. Admissible conduction functions satisfy [3, 5] : (i) g non-increasing, non-negative, $\mathcal{C}^1$-continuous on $\mathbb{R}$; (ii) $g(0) = 1$; (iii) $\lim_{u \to +\infty} g(u) = 0$. Overcoming such drawbacks as local inversion of heat flow and asymptotic degeneracy has led to many extensions [7], requiring a consistent theory to assess their properties.

This paper attempts to elaborate a unified mathematical perspective based on field theory. Section 2. recalls classical geometric and energetic characterizations of DSS. In Section 3., a Riemannian structural description is developed : equation (1) is interpreted as a Lorentz gauge condition expressing the trace Lorentz-invariance of an EM quadripotential with covariant scalar and contravariant vector components, respectively related to photometric and geometric image properties. This gauge determines EM quadrifield and quadricharge which satisfy Maxwell equations. Deriving their expressions in terms of DSS features yields Lorentz-invariants which synthetize intrinsic multiscale image properties.

2. Diffusive scale-space representations

Elaborating representations of DSS requires a local morphological description of images. Classical approaches rely on differential geometry and define salient geometric features as differential invariants of the luminance field w.r to a Lie group of transforms [7]. Relevant groups are symmetry groups for the PDE (1). Here, we focus on level set-based image local descriptions which are invariant w.r. to the Euclidean group on $\mathcal{E}_2$, denoted by $ISO(\mathcal{E}_2)$ [7].

The level sets of an image define a partition of $\mathcal{E}_2$ which is invariant w.r. to invertible luminance transforms. Their boundaries are non-intersecting a.e. $\mathcal{C}^2$-continuous curves, called *level lines* (LL). The LL orthogonal trajectories, called *stream lines* (SL), determine a dual family consisting of integral curves of $\|\nabla I\|$. Orientation along LL is specified by the orthonormal local frame $\left(\vec{t} = \frac{\nabla^{\perp} I}{\|\nabla I\|}, \vec{n} = \frac{\nabla I}{\|\nabla I\|}\right)$ where $\nabla^{\perp} = \left(\frac{\partial}{\partial y}, -\frac{\partial}{\partial x}\right)^T$. The $ISO(\mathcal{E}_2)$-invariant local coordinate transform $\mathcal{T} : (x, y) \to (v, w) = (\vec{t}\cdot\vec{X}\,,\, \vec{n}\cdot\vec{X})$ induces Euclidean arclengths v (w) along LL (SL). Euclidean local image structure is then described by $ISO(\mathcal{E}_2)$ differential invariants w.r. to the (v, w) system (see Table 1).

	Metric properties	Curvature properties
Differential invariants	$\nabla_{v,w} = [\vec{t},\vec{n}]^T \cdot \nabla$ $\nabla^{\perp}_{v,w} = [\vec{t},\vec{n}]^T \cdot \nabla^{\perp}$	$H_{v,w} = [\vec{t},\vec{n}]^T \cdot H \cdot [\vec{t},\vec{n}]$ with : $H = \nabla \cdot \nabla^T$
Photometric invariants	$I_v = 0$ $I_w = \|\nabla I\|$	$I_{vv} = \vec{t}^T \cdot H \cdot \vec{t} = -I_w k_I$ $I_{vw} = \vec{t}^T \cdot H \cdot \vec{n} = -I_w k_S$ $I_{ww} = \vec{n}^T \cdot H \cdot \vec{n} = -I_w \delta_S$
Invariance relationships	$\|\nabla_{v,w}\| = \|\nabla\|$	$\mathrm{tr}(H_{v,w}) = \mathrm{tr}(H)$ $\det(H_{v,w}) = \det(H)$
Connection formulas	$\left[\vec{t},\vec{n}\right]_v = k_I \left[\vec{n}, -\vec{t}\right]$ $\left[\vec{t},\vec{n}\right]_w = k_S \left[\vec{n}, -\vec{t}\right]$	$\nabla \cdot \left[\vec{t},\vec{n}\right] = [k_S, -k_I]$ $\nabla \wedge \left[\vec{t},\vec{n}\right] = [k_I, k_S]\,\vec{u}_z$

Table 1 : Level set-based $ISO(\mathcal{E}_2)$-invariant local geometry. Here, k_I (k_S) denotes LL (SL) curvature, and δ_S is the relative variation of $\|\nabla I\|$ along SL.

2.1. Geometric representation of diffusive scale-space flows

Equation (1) generates a motion flow $\vec{\mathcal{V}} = \left[\vec{X}_t, \frac{\partial I}{\partial t}\right]$. An $ISO(\mathcal{E}_2)$-invariant geometric description of DSS is derived by analyzing $\vec{\mathcal{V}}$ along the LL-SL network:

$$\vec{X}_t(v,w,t) = \tau(v,t)\,\vec{t} + \eta(v,t)\,\vec{n} \quad . \tag{2}$$

Here, η and τ are a.e. $\mathcal{C}^1$-continuous functions on $[0,1]\times]0,T]$ that respectively specify shape and metric properties along LL. This connection between luminance flows and curve evolution theory can in fact be developed within the general framework of scale-space models described by a 1st-order PDE : $\frac{\partial I}{\partial t} = \mathcal{L}I$. The LL equation $\frac{\partial I}{\partial t} + \vec{X}_t \cdot \nabla I = 0$ yields :

$$I_w\,\eta(v,t) = -\mathcal{L}I \quad . \tag{3}$$

A closed-form expression for τ is obtained by considering the evolution of spatial differentials of photometry. Indeed, differenciating equation (3) leads to :

$$\frac{\partial I_w}{\partial t} = -I_w\,(\eta_w - \eta\,\delta_S) \quad (4) \qquad\qquad \frac{\partial \vec{n}}{\partial t} = -(\eta_v - \eta\,k_S)\,\vec{t} \quad (5)$$

Using the general property : $\frac{\partial \vec{n}}{\partial t} = -(\eta_v - \tau k_I)\,\vec{t}$ [14], we obtain :

$$I_w\,k_I\,\tau(v,t) = -\mathcal{L}I\;k_S \quad . \tag{6}$$

From these relationships and general results on deformation of 1-codimensional manifolds w.r. to 1st-order kinematics [14], complete analytic and geometric characterizations of classical scale-spaces can be achieved (see Table 2).

2.2. Weak formulation of diffusive scale-space flows

Equation (1) can be seen as a descent method for quasi-linear elliptic variational problems arising in $ISO(\mathcal{E}_2)$-invariant discontinuity-preserving regularization [3, 12]. For example, Perona-Malik's models derive from functionals of the forms :

$$E_1(I) = \int_\Omega \varphi(\|\nabla I\|)\, d\Omega\,, \quad E_2(I,b) = \int_\Omega \left[\, b\, \|\nabla I\|^2 + \lambda\, \psi(b) \,\right] d\Omega\,.$$

Here, E_1 specifies a *generalized membrane* defined by an adaptive potential

Orientation	$\frac{d}{dt}\left[\vec{t}, \vec{n}\right] = (\eta_v - \eta k_S)\left[\vec{n}, -\vec{t}\right]$	Perimeter	$\frac{d\mathcal{L}}{dt} = -\int_0^1 \eta k_I\, dv$
Curvature	$\frac{dk_I}{dt} = \eta_{vv} + \tau \frac{\partial k_I}{\partial v} + \eta k_I^2$	Area	$\frac{d\mathcal{A}}{dt} = -\int_0^1 \eta\, dv$

Table 2: Kinematics of the level-stream lines network.

$\varphi \in \mathcal{F}^2(\mathbb{R}^+, \mathbb{R}^+)$, and E_2 determines a *weak membrane*[1] with stretching properties controlled by an auxiliary line variable $b \in \mathcal{F}^1(\mathcal{E}_2, \mathbb{R})$ and penalized by a line density $\psi \in \mathcal{F}^1(\mathcal{E}_2, \mathbb{R})$ conditionally to a critical parameter $\lambda > 0$. The two models are asymptotically equivalent iff. : (i') $u \longrightarrow \varphi(\sqrt{u})$ *strictly concave*, and $u \longrightarrow \frac{\varphi'(u)}{u}$ bounded; (i'') ψ strictly convex, non-increasing and bounded [3, 5]. Then : $\varphi = \inf_b \left[\, bu^2 + \lambda\psi(b) \,\right]$, and steady-states verify [3, 5, 10, 12] :

$$\left\{ \begin{array}{lcl} \nabla \cdot (b \nabla I) & = & 0 \\ b(\vec{X}) & \stackrel{\text{def}}{=} & g(\|\nabla I\|) = \left(\psi'|_\Omega\right)^{-1}\left(-\frac{\|\nabla I\|^2}{\lambda}\right) = \frac{\varphi'(\|\nabla I\|)}{2\|\nabla I\|} \end{array} \right. . \tag{7}$$

Consistency with conduction functions requires: (i'') $\lim_{u \to 0} \frac{\varphi'(u)}{u} = \lim_{u \to 0} \varphi''(u) = 1$; (ii'') $\lim_{|u| \to \infty} \frac{\varphi'(u)}{u} = \lim_{|u| \to \infty} \varphi''(u)$. In addition, diffusion inhibition along $\vec{n}$ in edge vicinity occurs iff. (iii'') $\lim_{|u| \to \infty} \frac{\varphi''(u)}{\frac{\varphi'(u)}{u}} = 0$, which holds whenever φ is *convex*[3].

3. Structural scale-space description

The curvilinear nature of DSS geodesics (*e.g.* zero-crossings of k_I) suggests that DSS structure is Riemannian, i.e. multiscale analysis relies on hyperbolic metrics. Moreover, for uniform dynamics, equation (1) is formally similar to the Lorentz gauge condition introduced in EM [9]: $\nabla \cdot \vec{A} + \frac{1}{c^2}\frac{\partial V}{\partial t} = 0$ for EM quadripotentials $\mathcal{V} = \left(\frac{V}{c}, \vec{A}\right)$ with $\frac{V}{c} = I$ and $\vec{A} = -\left(\alpha\,\vec{n} + \beta\,\vec{t}\right)$. Hence, the idea of casting anisotropic diffusion into the framework of EM field theory. Equation (1) then determines EM quadrifield $\mathcal{E} = \left(\frac{\vec{E}}{c}, \vec{B}\right)$ and quadricharge $\mathcal{J} = (\rho c, \vec{\jmath})$ which satisfy Maxwell equations. Deriving their expressions in terms of scale-space geometric or energetic features is therefore to yield *Lorentz-invariants* reflecting deep structural properties. Here, the constant c sets up an upper bound on spatial variations during scale transitions. Introducing this bound formalizes the intuition that scale transitions cannot be arbitrary large in order to preserve topological and geometrical consistency of the luminance field.

Using Euclidean connection identities, equation (1) can be rewritten as :

$$\frac{1}{c}\frac{\partial I}{\partial t} = -\alpha\, k_I + \alpha_w + \beta\, k_S + \beta_v \stackrel{\text{def}}{=} \mathcal{L} I \tag{8}$$

In anisotropic DSS theories (ADSST), $\beta \equiv 0$ and $\alpha = \varphi'(I_w)$ derives from a potential φ satisfying (i'', ii''). Additionally, in convex ADSST, φ verifies (iii''). In both cases, we have :

$$\eta = -\frac{c}{I_w}\left(\frac{\varphi'(I_w)}{I_w}\, I_{vv} + \varphi''(I_w)\, I_{ww}\right) \stackrel{\text{def}}{=} -\frac{c}{I_w}\,\mathcal{L}_\varphi I \quad . \tag{9}$$

3.1. Magnetic field : luminance stationarity model

From definition $\vec{B} = \nabla \wedge \vec{A}$, the magnetic field $\vec{B}$ is given by :

$$\vec{B} = -[\alpha\, k_S + \alpha_v + \beta\, k_I - \beta_w]\, \vec{u}_z \quad . \tag{10}$$

The vector field $\vec{B}$ accounts for the implicit discontinuity model encoded in the conduction function. For ADSST, equation (10) can be simplified as :

$$\vec{B} = \left[\frac{\varphi'(I_w)}{I_w} - \varphi''(I_w)\right] I_{vw}\, \vec{u}_z \quad . \tag{11}$$

Hence : $\lim_{I_w \to 0} \vec{B} = \vec{0}$. Moreover, $\vec{B} \underset{I_w \to +\infty}{\sim} \frac{\varphi'(I_w)}{I_w}\, I_{vw}\, \vec{u}_z = -\alpha k_S\, \vec{u}_z$ for convex ADSST. Specifically, fields of the form $\vec{B} = \varepsilon\, \alpha\, k_S\, \vec{u}_z$, $\varepsilon \in \mathbb{R}$ derive from vector potentials $\vec{A}$ such that $\alpha = \alpha_0\, (I_w)^{\varepsilon+1}$, $\alpha_0 \in \mathbb{R}$. Conversely, since $\vec{B}$ depends linearly on $\vec{A}$, potentials defined as $\alpha = \sum_{i=0}^{N} \alpha_i\, (I_w)^i$ determine vector fields of the form $\vec{B} = k_S \sum_{i=0}^{N} \alpha_i\, (i-1)\, (I_w)^i\, \vec{u}_z$. The case $\alpha = \alpha_1\, I_w$, i.e. $\vec{A} = -\alpha_1 \nabla I$, corresponds to *purely electric fields* (i.e. $\vec{B} = \vec{0}$) and yields Gaussian scale-space. The case $\varepsilon \in [-1, 0[$ leads to generalized Gaussian models [2].

3.2. Electric field : 1st-order scale-space properties

From definition $\vec{E} = -\nabla V - \frac{\partial \vec{A}}{\partial t}$, the electric field $\vec{E}$ is given by :

$$\vec{E} = -c\nabla I + \frac{\partial}{\partial t}(\alpha\, \vec{n} + \beta\, \vec{t}) \stackrel{\text{def}}{=} \vec{E}_0 + \vec{E}_d \quad . \tag{12}$$

The field $\vec{E}$ consists of a monoscale photometric component $\vec{E}_0$, independent of scale transition characteristics, and of a multiscale component $\vec{E}_d$ which

Model	$\varphi(u)$	$\lim\limits_{I_w \to +\infty} \vec{B} \cdot \vec{u}_z$	$\lim\limits_{I_w \to +\infty} c^{-1} \vec{E}_d$
Gaussian [8, 16]	$\frac{1}{2} u^2$	0	$\nabla_{v,w} \Delta I$
Perona-M. [13]	$\frac{1}{2}\left(1 - e^{-u^2}\right)$	$-2 I_w^3 e^{-I_w^2} k_S$	$-4 I_w^2 e^{-I_w^2} \frac{\partial}{\partial w} \left(I_w^3 e^{-I_w^2} \delta_S\right) \vec{n}$
Perona-M. [13]	$\frac{1}{2} \ln\left(1 + u^2\right)$	$-2 I_w^{-1} k_S$	$\frac{1}{I_w^2} \left[-\vec{t} \frac{\partial}{\partial v} + \vec{n} \frac{\partial}{\partial w}\right] \left(\frac{k_I - \delta_S}{I_w}\right)$
Geman-R. [5]	$1 - (1 + u)^{-1}$	$-k_S$	$\frac{1}{I_w^3} \left[-\vec{t} \frac{\partial}{\partial v} + 2\vec{n} \frac{\partial}{\partial w}\right] \left(\frac{k_I - 2\delta_S}{I_w^2}\right)$
Geman-M. [4]	$\frac{u^2}{2} \left(1 + u^2\right)^{-1}$	$-4 I_w^3 k_S$	$\frac{1}{I_w^4} \left[-\vec{t} \frac{\partial}{\partial v} + 3\vec{n} \frac{\partial}{\partial w}\right] \left(\frac{k_I - 3\delta_S}{I_w^3}\right)$
Green [6]	$\ln\cosh(u)$		
Charbonnier [3]	$\left(1 + u^2\right)^{\frac{1}{2}} - 1$	$-k_S$	$-\frac{1}{I_w} \frac{\partial k_I}{\partial v} \vec{t}$
Osher-R. [15]	u		

Table 3: Structural description of some classical DSS.

can be related to LL kinematics. Assuming $\alpha = \alpha(I_w)$ and $\beta = \beta(I_w)$, we obtain :

$$\frac{\vec{E}_d}{c} = \left[-\vec{t} \left[\beta' \frac{\partial}{\partial w} + \frac{\alpha}{I_w} \frac{\partial}{\partial v} \right] + \vec{n} \left[-\alpha' \frac{\partial}{\partial w} + \frac{\beta}{I_w} \frac{\partial}{\partial v} \right] \right] (\alpha\, k_I - \alpha_w) \tag{13}$$

The vector field $\vec{E}$ accounts for edge information carried by 1st-order variations of scale-space features : $\vec{E}_0$ measures the stationarity of the luminance field, and $\vec{E}_d$ describes the shape of geometric structures. The latter component can be understood in two ways. First, the geometric relevance of gradient information is measured as a function of DSS characteristics. Stated otherwise, shape analysis is modeled as an intrinsic multiscale mechanism. Second, the problem of devising 1st-order edge estimators is solved in a model-dependent perspective. Indeed, provided that the diffusion dynamics is sufficiently slow, the field $\vec{E}$ can be considered as a.e. model-independent when $t \to 0$. When $t \to +\infty$, however, vector fields $\vec{E}$ corresponding to different DSS models are *not* equivalent. Equation (12) then defines *the* covariant, Lorentz-invariant vector field consistent with the model at hand (see Table 3). For ADSST, we have indeed :

$$\frac{\vec{E}_d}{c} = \frac{\varphi'(I_w)}{I_w} \frac{\partial (\mathcal{L}_\varphi I)}{\partial v} \vec{t} + \varphi''(I_w) \frac{\partial (\mathcal{L}_\varphi I)}{\partial w} \vec{n} \quad . \tag{14}$$

so that : $\lim\limits_{I_w \to 0} \frac{\vec{E}_d}{c} = \nabla_{v,w} \Delta I$. Moreover : $\frac{\vec{E}_d}{c} \underset{I_w \to +\infty}{\sim} \frac{\varphi'(I_w)}{I_w} \frac{\partial}{\partial v} \left(\frac{\varphi'(I_w)}{I_w} I_{vv} \right) \vec{t}$ for convex ADSST that prove to be asymptotically equivalent to the total variation [15] for which $\vec{B} = -k_S\, \vec{u}_z$ and $\frac{\vec{E}_d}{c} = -\frac{1}{I_w} \frac{\partial k_I}{\partial v} \vec{t}$ holds *everywhere*.

3.3. Energy density : global 1st-order stationarity

The energetic structure of the EM quadrified is represented in a local form by the EM *energy density* : $w = \frac{1}{2}\left[\varepsilon_0 \|E^2\| + \frac{\|B\|^2}{\mu_0}\right]$. Its expression is readily deduced from equations (10,12,13). The scalar field w provides a global multiscale representation of 1st-order edge information. It generalizes classical potentials used in discontinuity-preserving regularization.

3.4. Source densities : higher-order scale-space properties

From Maxwell-Gauss equation $\nabla \cdot \vec{E} = \frac{\rho}{\varepsilon_0}$, the free charge density ρ is given by:

$$\varepsilon_0\,\rho \;=\; -c\Delta I \,+\, \frac{\partial}{\partial t}\left[-\alpha\,k_I \,+\, \alpha_w \,+\, \beta\,k_S \,+\, \beta_v\right] \;\stackrel{\text{def}}{=}\; \varepsilon_0\,(\rho_0 \,+\, \rho_d) \qquad (15)$$

The scalar field ρ accounts for edge information carried by 2nd-order variations of scale-space features. It comprises a monoscale photometric component ρ_0, independent of scale transition properties and crossing zero at edge points, and a multiscale geometric component. Again, geometric edge information is made dependent on the scale-space model at hand, and the previous arguments apply.

Using Maxwell-Ampère equation $\nabla \wedge \vec{B} = \mu_0\,\vec{\jmath} + \frac{1}{c^2}\frac{\partial \vec{E}}{\partial t}$, the current density $\vec{\jmath}$ is shown to be the sum of two scale-space components $\vec{\jmath}_0$ and $\vec{\jmath}_d$ such that :

$$\mu_0\,\vec{\jmath}_0 \;=\; -\nabla^{\perp}\,(\alpha\,k_S \,+\, \alpha_v \,+\, \beta\,k_I \,-\, \beta_w\,) \qquad (16)$$

$$\mu_0\,\vec{\jmath}_d \;=\; \nabla\,(\,-\alpha\,k_I + \alpha_w + \beta\,k_S + \beta_v\,) \,-\, \frac{1}{c^2}\frac{\partial^2}{\partial t^2}\,(\alpha\,\vec{n} \,+\, \beta\,\vec{t}\,) \qquad (17)$$

The vector field $\vec{\jmath}$ provides a multiscale description of higher-order geometric properties and can be interpreted as a junction map. Note that junction modeling depends on an *a priori* stationarity model for luminance. Moreover, junctions are defined as singularities of k_S rather than k_I only as usually stated [7]. This is consistent with the classical source/well description of the EM field. The general expressions of ρ and $\vec{\jmath}$ are complex and will not be given in this paper.

4. Conclusion

We have developed a field-based structural description of DSS. Our main result is the identification in terms of DSS features of a set of Lorentz-invariants expressing intrinsic multiscale image properties. The resulting formulation is in complete accordance with classical theories of low-level vision. We therefore conjecture that it underlies the deep (hyperbolic) structure of DSS. Extensions to non-degenerate and non-Euclidean scale-space models are currently investigated.

References

[1] A. Blake, A. Zisserman, *Visual reconstruction*, MIT Press, Cambridge, MA, 1987.

[2] C. Bouman, K. Sauer, "A generalized gaussian image model for edge-preserving MAP estimation", *IEEE Image Processing*, 2(3), July 1993, pp. 296–310.

[3] P. Charbonnier, "Reconstruction d'image : Régularisation avec prise en compte des discontinuités", *Ph.D. Dissertation - Univ. de Nice, France*, Sept. 1994.

[4] S. Geman, D.E. McClure, "Bayesian image analysis : An application to single photon emission tomography", *Proc. Statistical Computational Section, American Statistical Association, Washington, DC*, 1985, pp. 12–18.

[5] D. Geman, G. Reynolds, "Constrained restoration and the recovery of discontinuities", *IEEE PAMI*, 14(3), March 1992, pp. 367–383.

[6] P. Green, "Bayesian reconstructions from emission tomography data using a modified EM algorithm", *IEEE Medical Imaging*, 9(1), March 1990, pp. 84-93.

[7] B.M. Ter Haar Romeny Ed., *Geometry-driven diffusion in computer vision*, Kluwer Academic Publishers, Dordrecht, The Netherlands, 1994.

[8] J.J. Koenderink, "The structure of images", *Biol. Cyber.*, 50, 1984, pp. 363–370.

[9] L. Landau, E. Lifchitz, *Physique théorique – Tome II : Théorie des champs*, MIR, Moscou, Russia, 1989.

[10] S.Z. Li, "On discontinuity-adaptive smoothness priors in computer vision", *IEEE PAMI*, 17(6), pp. 576–586.

[11] T. Lindeberg, *Scale-space theory in computer vision*, Kluwer Academic Publishers, Dordrecht, The Netherlands,1995.

[12] N. Nordström, "Biased anisotropic diffusion-A unified regularization and diffusion approach to edge detection", *Image & Vis. Comp.*, 8(4), Nov. 1990, pp. 318–327.

[13] P. Perona, J. Malik, "Scale-space and edge detection using anisotropic diffusion", *IEEE PAMI*, 12(7), July 1992, pp. 629–639.

[14] N. Rougon, "On mathematical foundations of local deformations analysis", *Proc. SPIE Math. Meth. Med. Imag. II, San Diego, CA*, Vol. 2035, July 1993, pp. 1–12.

[15] L.I. Rudin, S. Osher, E. Fatemi, "Nonlinear total variation-based noise removal algorithms", *Physica*, D60, 1992, pp. 259–268.

[16] A. Witkin, "Scale-space filtering", *Proc. IJCAI'83, Karlsruhe, Germany*, 1983, pp. 1019-1021.

Nonlinear Diffusion Scale-Spaces: From the Continuous to the Discrete Setting

Joachim Weickert

RWC, Imaging Center Utrecht,
Utrecht University Hospital, Heidelberglaan 100,
Room E01.334, NL-3584 CX Utrecht.
Tel.: +31 30 250 8377, Fax: +31 30 251 3399.
E-mail: Joachim.Weickert@cv.ruu.nl

Summary. A survey on continuous, semidiscrete and discrete well-posedness and scale-space results for a class of nonlinear diffusion filters is presented. This class does not require any monotony assumption (comparison principle) and, thus, allows image restoration as well. The theoretical results include existence, uniqueness, continuous dependence on the initial image, maximum-minimum principles, average grey level invariance, smoothing Lyapunov functionals, and convergence to a constant steady state.

Keywords. scale-space, nonlinear diffusion, discrete smoothing transformations.

1. Introduction

In the last years nonlinear diffusion filtering has been established as a successful tool for image smoothing and restoration. Strict scale-space results have been found recently for the continuous case [16]. The goal of the present paper is to outline how they can be extended to the semidiscrete and discrete setting. This is of significant practical importance, since a scale-space representation cannot perform better than its discrete realization.

The paper is organized as follows: Section 2 reviews well-posedness and scale-space results in the continuous framework and interprets the meaning of the obtained results. Section 3 introduces the requirements that we need in order to establish similar properties in a semidiscrete setting. Finally, Section 4 gives related results for the fully discrete framework which utilizes a finite number of scales. We conclude with a summary in Section 5.

2. Continuous Case

We are concerned with a class of nonlinear diffusion processes using a diffusion tensor D. This symmetric positive definite matrix can be adapted to the local image structure by means of the so-called structure tensor (scatter matrix, second-moment matrix) $J_\rho(\nabla u_\sigma) := K_\rho * (\nabla u_\sigma \, \nabla u_\sigma^T)$, where $\nabla u_\sigma := \nabla K_\sigma * u$, K_σ denotes a Gaussian with standard deviation σ, and $*$ is the convolution product [17]. Matrices of this type play an important role in the local structure analysis of textures, corners and T-junctions, shape cues and spatio-temporal images [14, 3, 12, 8, 11, 2]. The regularization parameter $\sigma > 0$

makes the structure tensor insensitive to noise of order σ, while the integration scale ρ gives the window size over which the orientation information is averaged. The structure tensor generalizes the edge detector ∇u_σ and provides us with additional information, for instance about corners and coherent structures. Its use for steering nonlinear diffusion filters in order to enhance coherent flow-like structures has been demonstrated in [17].

Let us consider an open rectangle $\Omega \subset I\!R^2$ as image domain and let a (monochromatic) image be represented by a function $f \in L^\infty(\Omega)$. The initial boundary value problem we are concerned with is as follows (n denotes the normal vector and $\langle .,. \rangle$ is the Euclidean scalar product):

Assume that $f \in L^\infty(\Omega)$, $\rho \geq 0$, and $\sigma, T > 0$.
Let $a := ess\,inf_\Omega\, f$, $b := ess\,sup_\Omega\, f$, and consider the problem

$$\begin{aligned} \partial_t u = \operatorname{div}(D(J_\rho(\nabla u_\sigma))\, \nabla u) \quad & \text{on} \quad \Omega \times (0,T], \\ u(x,0) = f(x) \quad & \text{on} \quad \Omega, \\ \langle D(J_\rho(\nabla u_\sigma))\nabla u,\, n\rangle = 0 \quad & \text{on} \quad \partial\Omega \times (0,T], \end{aligned}$$

where the diffusion tensor $D = (d_{ij})$ satisfies

(C1) Smoothness:
$D \in C^\infty(I\!R^{2\times 2}, I\!R^{2\times 2})$.

(C2) Symmetry:
$d_{12}(J) = d_{21}(J)$ for all symmetric matrices $J \in I\!R^{2\times 2}$.

(C3) Uniform positive definiteness:
For all $w \in L^\infty(\Omega, I\!R^2)$ with $|w(x)| \leq K$ on $\bar{\Omega}$, there exists a positive lower bound $\nu(K)$ for the eigenvalues of $D(J_\rho(w))$.

(P_c)

We observe that (P_c) may comprise linear diffusion filters [9] as well as certain nonlinear models [5, 19, 15, 16, 17]. This class reveals the subsequent properties ([18], see also [5] for a proof of the major part of (a)):

Theorem 2.1. ***(Properties of the continuous filter class)***
For the continuous filter class (P_c) the following statements are valid:

(a) (Well-posedness and regularity results)
For any $T > 0$, the problem (P_c) has a unique solution $u(x,t)$ in the distributional sense. This solution satisfies $u \in C([0,T]; L^2(\Omega)) \cap L^2(0,T; H^1(\Omega))$ and $\partial_t u \in L^2(0,T; (H^1(\Omega))')$. Moreover, $u \in C^\infty(\bar{\Omega} \times (0,T])$, and it depends continuously on f with respect to the $L^2(\Omega)$ norm.

(b) (Extremum principle)
Let $a := ess\,inf_{x\in\Omega}\, f(x)$ and $b := ess\,sup_{x\in\Omega}\, f(x)$. Then, $a \leq u(x,t) \leq b$ on $\Omega \times [0,\infty)$.

(c) (Average grey level invariance)
The average grey level $\mu := \frac{1}{|\Omega|}\int_\Omega f(x)\,dx$ is not affected by nonlinear diffusion filtering: $\frac{1}{|\Omega|}\int_\Omega u(x,t)\,dx = \mu$ for all $t > 0$.

(d) (Lyapunov functionals)
$V(t) := \Phi(u(t)) := \int_\Omega r(u(x,t))\,dx$ is a Lyapunov function for all $r \in C^2[a,b]$ with $r'' \geq 0$ on $[a,b]$: $V(t)$ is decreasing and bounded from below by $\Phi(Mf)$, where $(Mf)(y) := \mu$ for all $y \in \Omega$.
(e) (Convergence to a constant steady state)
$\lim_{t\to\infty} \|u(t) - Mf\|_{L^p(\Omega)} = 0$ for $1 \leq p < \infty$.

The well-posedness results in (a) have significant practical impact, as they guarantee the stability with respect to perturbations of the initial images. This is of importance when considering stereo images, image sequences or slices from medical CT or MRT sequences, since we know that similar images remain similar after filtering.

Many smoothing scale-space properties are closely related to extremum principles: Hummel [7] for instance shows that under certain conditions the maximum principle for parabolic operators is equivalent to the property that the corresponding scale-space never creates additional level-crossings for $t > 0$.

Average grey level invariance is a property which distinguishes diffusion filters from morphological scale-spaces. In addition to this invariance it is evident that (P_d) satisfies classical scale-space invariances like grey level shift invariance, reverse contrast invariance, translation invariance and isometry invariance (see [1] for precise definitions). Usual architectural properties of scale-spaces (e.g. the semi-group property) are satisfied as well.

The Lyapunov functionals introduced in (d) show that the considered evolution equation is a simplifying, information-reducing transform with respect to many aspects: Indeed, special choices for r immediately imply that all L^p norms with $2 \leq p \leq \infty$ are decreasing (e.g. the energy $\|u(t)\|^2_{L^2(\Omega)}$), all even central moments are decreasing (e.g. the variance), and the entropy $S[u(t)] := -\int_\Omega u(x,t)\,\ln(u(x,t))\,dx$ is increasing with respect to t [16].

The result (e) tells us that, for $t \to \infty$, this simplifying scale-space representation tends to the most global image representation that is possible: a constant image with the same average grey level as f.

Interestingly, (P_c) does not need any monotony assumption (comparison principle) [1]. This is in contrast to linear diffusion and morphological scale-spaces and allows nonlinear diffusion scale-spaces to reveal contrast-enhancing properties leading to segmentation-like results (see [16, 18] for examples). In this sense we have a scale-space framework which does not contradict certain image restoration demands.

3. Semidiscrete Case

Let us now establish conditions under which comparable well-posedness and scale-space results can be proved for the semidiscrete framework. This case is of special interest since it involves the spatial discretization which is characteristic for digital images but it keeps the scale-space idea of using a continuous scale parameter. It leads to nonlinear systems of ordinary differential equations.

A discrete image can be regarded as a vector $f \in I\!R^N$, $N \geq 2$, whose components f_j, $j=1,...,N$ represent the grey values at the pixels. We denote the index set $\{1, ..., N\}$ by J. The semidiscrete problem class (P_s) we are concerned with is defined as follows:

Let $f \in I\!R^N$. Find a function $u \in C^1([0,\infty), I\!R^N)$ which satisfies the initial value problem

$$\frac{du}{dt} = A(u)\,u,$$
$$u(0) = f,$$

where $A = (a_{ij})$ has the following properties: (P_s)

(S1) Lipschitz-continuity of $A \in C(I\!R^N, I\!R^{N\times N})$ for every bounded subset of $I\!R^N$,

(S2) symmetry: $a_{ij}(u) = a_{ji}(u) \quad \forall\, i,j \in J,\ \forall\, u \in I\!R^N$,

(S3) vanishing row sums: $\sum_{j\in J} a_{ij}(u) = 0 \quad \forall\, i \in J,\ \forall\, u \in I\!R^N$,

(S4) nonnegative off-diagonals: $a_{ij}(u) \geq 0 \quad \forall\, i \neq j,\ \forall\, u \in I\!R^N$,

(S5) irreducibility for all $u \in I\!R^N$.

Under these prerequisites we obtain the subsequent well-posedness and scale-space results [18]:

Theorem 3.1. ***(Properties of the semidiscrete filter class)***
For the semidiscrete filter class (P_s) the following statements are valid:

(a) (Well-posedness)
For every $T > 0$ the problem (P_s) has a unique solution $u(t) \in C^1([0,T], I\!R^N)$. This solution depends continuously on the initial value and the right-hand side of the ODE system.

(b) (Extremum principle)
Let $a := \min_{j\in J} f_j$ and $b := \max_{j\in J} f_j$. Then, $a \leq u_i(t) \leq b$ for all $i \in J$ and $t \in [0,T]$.

(c) (Average grey level invariance)
The average grey level $\mu := \frac{1}{N}\sum_{j\in J} f_j$ is not affected by the semidiscrete diffusion filter: $\frac{1}{N}\sum_{j\in J} u_j(t) = \mu$ for all $t > 0$.

(d) (Lyapunov functionals)
$V(t) := \Phi(u(t)) := \sum_{i\in J} r(u_i(t))$ is a Lyapunov function for all $r \in C^1[a,b]$ with increasing r' on $[a,b]$: $V(t)$ is decreasing and bounded from below by $\Phi(c)$, where $c := (\mu, ..., \mu)^\top \in I\!R^N$.

(e) (Convergence to a constant steady state)
$\lim_{t\to\infty} u(t) = c$.

These results allow the same interpretation as their continuous counterparts. Not all of the requirements (S1)–(S5) are necessary for each of the theoretical results above. (S1) is needed for well-posedness, the proof of a maximum–minimum principle involves (S3) and (S4), while average grey value invariance uses (S2) and (S3). The existence of Lyapunov functionals can be established by means of (S2)–(S4), and convergence to a constant steady state requires (S5) in addition to (S2)–(S4).

It can be shown [18] that there exist finite difference approximations of (P_c) fulfilling the requirements (S1)–(S5) of (P_s). However, (P_c) is not the only family which leads to semidiscrete filters satisfying (S1)–(S5): Interestingly, a semidiscrete version of the Perona–Malik filter [13] (which is claimed to be ill-posed in the continuous setting [5]) on a fixed grid also satisfies (S1)–(S5) and, thus, reveals all the beforementioned well-posedness and scale-space properties. This is due to the fact that the extremum principle limits the modulus of discrete gradient approximations. Hence, the spatial discretization implicitly causes a regularization.

4. Discrete Case

In practice, scale-spaces are always approximated with a finite number of scales. This corresponds to the fully discrete case which shall be treated now. In this setting we impose the subsequent requirements:

(P_d):

Let $f \in I\!R^N$. Calculate a sequence $(u^{(k)})_{k \in I\!N_0}$ of processed versions of f by means of

$$u^{(0)} = f,$$
$$u^{(k+1)} = Q(u^{(k)})\, u^{(k)}, \qquad \forall\, k \in I\!N_0,$$

where $Q = (q_{ij})$ has the following properties:

(D1) continuity in its argument: $Q \in C(I\!R^N, I\!R^{N \times N})$,
(D2) symmetry: $q_{ij}(v) = q_{ji}(v) \quad \forall\, i,j \in J,\ \forall\, v \in I\!R^N$,
(D3) unit row sum: $\sum_{j \in J} q_{ij}(v) = 1 \quad \forall\, i \in J,\ \forall\, v \in I\!R^N$,
(D4) nonnegativity: $q_{ij}(v) \geq 0 \quad \forall\, i,j \in J,\ \forall\, v \in I\!R^N$,
(D5) irreducibility for all $v \in I\!R^N$,
(D6) positive diagonal: $q_{ii}(v) > 0 \quad \forall\, i \in J,\ \forall\, v \in I\!R^N$.

This gives similar results as in the continuous and semidiscrete case [18]:

Theorem 4.1. ***(Properties of the discrete filter class)***
For the discrete filter class (P_d) the following statements are valid:
(a) (Continuous dependence on initial image)
For every $k > 0$ the unique solution $u^{(k)}$ of (P_d) depends continuously on the initial image f.

(b) (Extremum principle)
Let $a := \min_{j \in J} f_j$ and $b := \max_{j \in J} f_j$. Then, $a \le u_i^{(k)} \le b$ for all $i \in J$ and $k \in I\!N_0$.

(c) (Average grey level invariance)
The average grey level $\mu := \frac{1}{N} \sum_{j \in J} f_j$ is not affected by the discrete diffusion filter: $\frac{1}{N} \sum_{j \in J} u_j^{(k)} = \mu$ for all $k \in I\!N_0$.

(d) (Lyapunov functionals)
$V^{(k)} := \Phi(u^{(k)}) := \sum_{i \in J} r(u_i^{(k)})$ is a Lyapunov sequence for all convex $r \in C[a,b]$: $V^{(k)}$ is decreasing and bounded from below by $\Phi(c)$, where $c := (\mu, ..., \mu)^{\top} \in I\!R^N$.

(e) (Convergence to a constant steady state)
$\lim_{k \to \infty} u^{(k)} = c.$

It can be shown [18] that the semi-implicit scheme

$$\frac{u^{(k+1)} - u^{(k)}}{\tau} = A(u^{(k)})\, u^{(k+1)}$$

with A satisfying (S1)–(S5) fulfils the prerequisites (D1)–(D6) for discrete diffusion models for every positive time step size τ. Hence, (P_d) arises in a natural way from (P_s).[1] On the other hand, the assumptions (S1)–(S5) are sufficient conditions for the semi-implicit scheme to fulfil (D1)–(D6), but they are not necessary: Nonnegativity of $Q(u^{(k)})$ may also be achieved using spatial discretizations where $A(u^{(k)})$ violates nonnegativity (see [4] for examples).

5. Summary and Conclusions

We have investigated a complete scale-space framework for nonlinear diffusion filtering in the continuous, semidiscrete and discrete setting. This is of special importance, since besides a few exceptions [10, 6] the practically relevant problem of how to design (semi-)discrete scale-spaces has hardly been addressed in the literature.

We have established conditions under which one can prove well-posedness and scale-space results for nonlinear diffusion filtering. In all three settings we have used similar assumptions: smoothness ((C1), (S1), (D1)), symmetry ((C2), (S2), (D2)), nonnegativity ((C4), (S4), (D4)), requirements ensuring a nonvanishing diffusion at all locations ((C3), (S5), (D5)–(D6)) and assumptions expressing the conservation of the average grey level (divergence form and boundary condition in the continuous case, (S3), (D3)). However, it should be observed that the smoothness assumptions can be weakened during the transition from the coontinuous to the discrete framework. This is

[1] Explicit and α-semi-implicit schemes satisfy (D1)–(D6) as well, provided that some time step size restrictions are imposed (see [18] for more details).

also true for the convex function r which generates the smoothing Lyapunov functionals.

References

1. Alvarez, L., Guichard, F., Lions, P.-L., Morel, J.-M.: Axioms and fundamental equations in image processing. Arch. Rat. Mech. Anal. **123** (1993) 199–257
2. Ballester Nicolau, C.: An affine invariant model for image segmentation: Mathematical analysis and applications. Ph.D. thesis, University of Illes Baleares, Palma de Mallorca, Spain, 1995
3. Bigün, J., Granlund, G.H., Wiklund, J.: Multidimensional orientation estimation with applications to texture analysis and optical flow. IEEE Trans. Pattern Anal. Mach. Intell. **13** (1991) 775–790
4. Bramble, J.H., Hubbard, B.E.: New monotone type approximations for elliptic problems. Math. Comp. **18** (1964) 349–367
5. Catté, F., Lions, P.-L., Morel, J.-M., Coll, T.: Image selective smoothing and edge detection by nonlinear diffusion. SIAM J. Numer. Anal. **29** (1992) 182–193
6. Cohignac, T., Eve, F., Guichard, F., Lopez, C. Morel, J.-M.: Numerical analysis of the fundamental equation of image processing. Preprint No. 9254, CEREMADE, Université Paris IX – Dauphine, Place du Maréchal de Lattre de Tassigny, 75775 Paris Cedex 16, France, 1992
7. Hummel, R.A.: Representations based on zero-crossings in scale space. Proc. IEEE Comp. Soc. Conf. Computer Vision and Pattern Recognition (CVPR '86, Miami Beach, June 22–26, 1986), IEEE Computer Society Press, Washington, 204–209, 1986
8. Jähne, B.: Spatio-temporal image processing. Lecture Notes in Comp. Science, Vol. 751, Springer, Berlin, 1993
9. Koenderink, J.J.: The structure of images. Biol. Cybern. **50** (1984) 363–370
10. Lindeberg, T.: Scale-space for discrete signals. IEEE Trans. Pattern Anal. Mach. Intell. **12** (1990) 234–254
11. Lindeberg, T.: Scale-space theory in computer vision. Kluwer, Boston, 1994
12. Nitzberg, M., Shiota, T.: Nonlinear image filtering with edge and corner enhancement. IEEE Trans. Pattern Anal. Mach. Intell. **14** (1992) 826–833
13. Perona, P., Malik, J.: Scale space and edge detection using anisotropic diffusion. IEEE Trans. Pattern Anal. Mach. Intell. **12** (1990) 629–639
14. Rao, A.R., Schunck, B.G.: Computing oriented texture fields. CVGIP: Graphical Models and Image Processing, **53** (1991) 157–185
15. Weickert, J.: Anisotropic diffusion filters for image processing based quality control. In Fasano, A., Primicerio, M. (Eds.): Proc. Seventh European Conf. on Mathematics in Industry. Teubner, Stuttgart, 355–362, 1994
16. Weickert, J.: Scale-space properties of nonlinear diffusion filtering with a diffusion tensor. Report No. 110. Laboratory of Technomathematics, University of Kaiserslautern, P.O. Box 3049, 67653 Kaiserslautern, Germany, 1994 (submitted)
17. Weickert, J.: Multiscale texture enhancement. V. Hlaváč, R. Šára (Eds.), Computer analysis of images and patterns, Lecture Notes in Comp. Science, Vol. 970, Springer, Berlin, 230–237, 1995
18. Weickert, J.: Anisotropic diffusion in image processing. Ph.D. thesis, Dept. of Mathematics, University of Kaiserslautern, Germany, January 1996

19. Whitaker, R.T., Pizer, S.M.: A multi-scale approach to nonuniform diffusion. CVGIP: Image Understanding **57** (1993) 99–110

Nonlinear diffusions and hyperbolic smoothing for edge enhancement

Satyanad Kichenassamy

School of Mathematics, University of Minnesota, 127 Vincent Hall, 206 Church St., S. E., Minneapolis, MN 55455-0487, U. S. A.

Abstract

We give a mathematical foundation for the interpretation of numerical results for the Perona-Malik equation and introduce a reversible smoothing procedure, based on a second-order hyperbolic equation. These results are briefly compared with other edge-detection techniques, including mean curvature flow.

Key-Words:

Segmentation, computer vision, ill-posed problems, multi-scale analysis.

1 Introduction

The purpose of this paper is (1) to report on a new theory of generalized solutions of the Perona-Malik equation, (2) to present a reversible, local, and causal smoothing procedure for images, based on a second-order hyperbolic equation.

It is well-known that the smoothing of digital images by Gaussian smoothing, which is equivalent to the solution of the initial-value problem for the heat equation, has led to difficulties which brought about proposals involving nonlinear diffusions. The first and best known is the Perona-Malik equation

$$u_t = \operatorname{div} \rho(|\nabla u|^2)\nabla u,$$

where ρ is nearly one for small gradients, but is small for large gradients. This equation is parabolic if $\rho(s) + 2s\rho'(s)$ is positive for all s, but becomes ill-posed in the usual sense otherwise. The surprising discovery of Perona and Malik is that a straightforward discretization of this equation is numerically stable in the uniform norm even in cases when the equation is ill-posed. In fact, they specifically considered the cases

$$\rho(s^2) = 1/(1 + s^2/K^2),$$

and

$$\rho(s^2) = \exp(-s^2/K^2),$$

where K is a constant. (These numerical results are reminiscent of those of Höllig and Nohel (1983) for equations which are nonlinear for small gradients only.) The rationale was that if we identify, as usual, features with large-gradient regions, this procedure would make regions of small gradient more uniform, while sharpening the "edges." The unexpected result is that the ill-posedness of the equation does not result in violent oscillations.

The Perona-Malik equation is more than a curiosity, because it can be thought of as a limiting, or special, model case of most current segmentation procedures. Thus, they already observed that procedures based on the minimization of a functional by statistical-mechanical methods could be viewed as discretized versions of the stationary Perona-Malik equation. Mumford noticed that the conformal motion by mean curvature, which was introduced (see Kichenassamy *et al.*, (1995)) as a natural extension of the method recently proposed by Malladi *et al.*, and Caselles *et al.*, leads to the Perona-Malik equation with an additional factor of $|\nabla u|$, if we assume the 'stopping function' to be defined from the gradient of u instead of that of the original image. Finally, a mean field approximation applied to a discrete version of the Mumford-Shah functional leads again in the first approximation to an equation similar to the Perona-Malik equation. See e.g. Morel and Solimini (1995) for a recent survey of the variational approach to segmentation.

We discuss below (1) earlier attempts at resolving the paradox raised by the numerical success of this equation; (2) our results which include (a) a nonexistence theorem for weak solutions; (b) a new notion of generalized solutions such that the equation does admit of solutions without the need for regularization; (c) a stability result for the difference scheme considered by Perona and Malik, and which we also extensively tested.

2 Earlier results

Earlier results on this problem deal with regularizations of the equation. Note that although it can be formulated as a gradient flow, the corresponding functional has an identically vanishing convex envelope, which prevents us from making use of most of the extensive literature on non-convex variational problems.

A first result is that if one regularizes the nonlinear term by a Gaussian, the equation becomes well-posed, but still remains numerically efficient (see Catté *et al.*, and also Alvarez *et al.*).

A second result is that a time-delay regularization combined with some smoothing of the gradient term also gives satisfactory results. This regularization has the advantage that if no gradient smoothing is allowed, the equation has the same stationary solutions as the PM equation itself (see the discussion in Mumford, Nitzberg and Shiota).

Also, it seems to be a consensus that if there is a theory of generalized solutions for this equation, it must allow step functions as stationary solutions; we note that this implies that the gradient of such a solution is a measure and not a function—this is the starting point of our definition of generalized solutions. In addition, we should not expect any uniqueness for the associated stationary problem, since there are infinitely many step functions with the same boundary values.

We restrict ourselves in the following to the 1D case, not only because it is the one most often considered in earlier papers, as containing already the salient difficulties of the problem, but also because in that case, the equation is either forward or backward parabolic only.

3 Results

The proofs of the following results are barely outlined and a more detailed account will be given elsewhere (see Kichenassamy (1995b)). We write the equation as

$$u_t = R(u_x)_x.$$

3.1 Nonexistence of weak solutions

Let us define a weak solution as a function in say, $L^\infty(0,T;W^{1,2}_{\text{loc}})$ which solves the equation in the sense of distributions. The main restriction is here that the gradient be a function, since otherwise we will need to deal with a nonlinear function of a measure, as in 3.2 below.

Theorem 1 *There is no weak solution of the Perona-Malik equation which is defined on a rectangle $(a,b)\times(0,T)$, where $a<b$ and $T>0$, and which has large gradient throughout this rectangle.*

In view of this result, we see that, in order to go further, we must either consider solutions which oscillate in every subinterval of the interval of definition, or else deal only with piecewise smooth solutions with small slopes, separated by sharp discontinuities. The former is clearly unacceptable for real images—it would lead to the creation of infinitely many edges. We therefore turn to the second.

3.2 Generalized solutions

Definition. An integrable function u of (x, t) will be said to be a generalized solution of the PM equation if its space derivative is a measure for every t, and if the equality

$$u_t = (R(r))_x$$

holds in the sense of distributions, where r is the regular part of u_x in the sense of the Radon-Nikodym theorem. We say that such a solution is *admissible* if its total variation does not increase with t.

We can construct large classes of generalized solutions using the following patching theorem:

Theorem 2 *Let u_l and u_t be two C^1 solutions defined on either side of a smooth curve $\Gamma : x = x(t)$. Then they define a generalized solution across Γ iff*

$$dx/dt = -[R(u_x)]/[u],$$

where brackets denote jumps across Γ.

Typical applications of this result include solutions derived from piecewise linear data with intervening jumps. The above rule provides an algorithm for computing the solution for all time, provided we specify how to evolve the solution between two jumps, as well as the rule for interaction of jumps. It turns out that the admissibility condition discourages the formation of jumps in continuous piecewise smooth solutions *only* in the well-posed region, and encourages them otherwise, which is satisfactory.

The simplest interaction rule requires that colliding jumps merge. It suffices to define the evolution of piecewise smooth data. In particular, the number of jumps does not increase.

The sensitivity of the solution to initial data is also preserved in this formalism: for instance, a piecewise constant function provides a constant generalized solution; this is as it should be, since such solutions appear as long-time limits in numerical calculations. However, if we tilt slightly one of the steps by increasing or decreasing its slope, one can show easily that the resulting solution tends to a single wide ramp (with small slope), if the slope is made positive. But if it is made negative, however small, two jumps approach each other and merge into one, after which we have a new stationary solution. This shows that the large-time limit is extremely sensitive to the data. However, the process by which solutions evolve is extremely stable over fixed times.

3.3 Stability of the difference scheme

We include below a few simple results on the difference scheme originally proposed by Perona-Malik, which consists in forward differencing in time, and using $\delta_+ R(\delta_- u)$ for the second derivative term. The scheme takes therefore the form

$$u_i^{n+1} - u_i^n = \lambda h[R(v_{i+1}^n) - R(v_i^n)],$$

where $\lambda = k/h^2$, h and k are the space and time steps, and

$$v_i^n = (u_i^n - u_{i-1}^n)/h.$$

We neglect the treatment of boundaries here, for simplicity. The approximation is therefore defined at times nk, $n = 0, 1, \ldots$

The observed stability for increasing data can be accounted for thanks to

Theorem 3 *If λ is small enough, then, if the approximation is nondecreasing at time nk, it is still so at the next step.*

This implies that large gradients can occur only in the form of 'staircasing,' which is exactly what is observed. It also implies that the maximum and the total variation of the computed solution remain finite, and therefore accounts for the stability of the scheme.

We have further tested this algorithm numerically, and found in particular that the presence of a sharp discontinuity is equivalent to the introduction of an artificial Neumann condition on each side of the jump. This is consistent with one of the possible classes of generalized solutions.

3.4 Hyperbolic smoothing.

A common preprocessing step in many segmentation procedures is to perform some smoothing. This introduces another ill-posed problem which comes from the fact that if smoothing is performed by means of a parabolic equation, it is not possible to undo its effects. This can be a problem if smoothing is viewed in a scale-space context. We propose a reversible smoothing procedure based on a hyperbolic equation, namely the Euler-Poisson-Darboux equation, which in 2D takes the form

$$u_{tt} + \frac{u_t}{t} - u_{xx} - u_{yy} = 0.$$

The usefulness of this equation comes from the following results.

Theorem 4 *This equation has precisely one bounded solution if $u(x, 0)$ is given and smooth. The solution operator maps H^s to $H^{s+1/2}$ for every s.*

Theorem 5 *The solution $u(x, t)$ can be computed by pixel averaging of the initial image over the circle of radius t.*

This latter result is familiar in the construction of the solution of the wave equation. It is also motivated by the fact that the Euler-Posson-Darboux equation is a special case of a Fuchsian equation, and such equations are now known to admit of existence results in H^s even if rather general nonlinear perturbations are allowed (see Kichenassamy (1995a)). These results were actually motivated by the problem of description of solutions of nonlinear wave equations near their singularities. We omit here the few available results on exact reconstruction from such pixel averages.

This procedure is 'causal' in the sense of scale-space theory, and is stably reversible for $t > 0$, with a mild instability if we wish to go as far back as $t = 0$. Note that if the image is smooth enough, $u \approx u(x,0) + t^2 \Delta u(x,0)/4$, which shows that this is in fact close to the solution of the heat equation at time $t^2/4$ for very small t.

The next issue is to examine whether the gain of half a derivative might be too little. We have tested the procedure on real images and shown that it is good enough for edge enhancement, by constructing the associated high-pass filter.

Acknowledgement. Computations reported in this paper were performed at the Geometry Center.

References

L. Alvarez, P.-L. Lions and J.-M. Morel (1992): Image selective smoothing and edge detection by nonlinear diffusion, II, *SIAM J. Num. Anal.*, **29**(3) 845–866.

F. Catté, P.-L. Lions, J.-M. Morel and T. Coll (1992): Image selective smoothing and edge detection by nonlinear diffusion, SIAM J. Num. Anal., **29**(1) 182–193.

V. Caselles, F. Catté, T. Coll, and F. Dibos, (1992): A geometric model for active contours in image processing, Technical Report #9210, CEREMADE, Université Paris Dauphine.

V. Caselles, R. Kimmel and G. Sapiro (1995): Geodesic active contours, Proc. ICCV.

K. Höllig and J. A. Nohel (1983): A diffusion equation with a nonmonotone constitutive function, in *Systems of Nonlinear Partial Differential Equations*, J. Ball ed., D. Reidel, Boston, MA, 409–422.

S. Kichenassamy, A. Kumar, P. Olver, A. Tannenbaum, and A. Yezzi, (1995): Conformal curvature flows: from phase transitions to active vision, to appear in Archive for Rational Mechanics and Analysis.

S. Kichenassamy, (1995a): *Nonlinear Wave Equations*, Texts and Monographs in Pure and Applied Mathematics, vol. 194, Marcel Dekker, New York.

S. Kichenassamy, (1995b): The Perona-Malik paradox, preprint.

R. Malladi, J. Sethian, and B. Vemuri (1995): Shape modeling with front propagation: a level set approach, IEEE Trans. Pattern Anal. Machine Int. **17**, 158-175.

D. Mumford, M. Nitzberg and T. Shiota (1993): *Filtering, Segmentation and Depth*, Lect. Notes in Comp. Sci., **662**, Springer.

J.-M. Morel and S. Solimini, (1995): *Variational Methods in Image Segmentation*, Birkhäuser, Boston.

P. Perona and J. Malik (1990) Scale-space and edge detection using anisotropic diffusion, IEEE Trans. Pattern Anal. Machine Int. **12**, 629-639.

Convergence of iterated affine and morphological filters by nonlinear semi-group theory

Francine Catté

Ceremade - (URA CNRS 749)
Université Paris 9-Dauphine
Place du Maréchal de Lattre de Tassigny
75775 Paris Cedex 16

1. Introduction

A numerical image can be modelled as a real function $u_0(x)$ defined in IR^N (in practice, $N = 2$ ou 3). A main concept of vision theory and image analysis is multiscale analysis (or "scale space"). Multiscale analysis associates with $u(0) = u_0$ a sequence of pictures $u(t,x) = (T_t u_0)(x)$ which depend upon an abstract parameter $t > 0$, the scale. The image $u(t,x)$ is called analysis of the image u_0 at scale t.

Alvarez and all [1] proved that under reasonable assumptions on a multiscale analysis T_t, all sequence of pictures $u(t,x) = (T_t u_0)(x)$ are solutions of a partial differential equation

$$\frac{\partial u}{\partial t} = |Du| \, F \, (div \, \frac{Du}{|Du|}, \, t) \, , \quad u(0) = u_0 \tag{1}$$

A particular case is when F is a constant function : if $F = +1$ or $F = -1$, then the equation becomes $\frac{\partial u}{\partial t} = |Du|$ (*resp.* $\frac{\partial u}{\partial t} = -|Du|$) which corresponds to the so-called morphological erosion when the sign is $-$ and to a morphological dilation when the sign is $+$ (see [8]). Dilation and erosion are the basic operators of the Morphologie Mathématique, founded by Matheron (1975). The dilation at scale t is defined by

$$D_t \; u_0(x) = \sup_{y \in B(x,t)} \; u_0(y)$$

where $B(x,t)$ is a set centered at x. For the erosion, on simply replaces "sup" by "inf".
Among the particular instances of (1), we find the "mean curvature equation"

$$(MCM) \qquad \frac{\partial u}{\partial t} = |Du| \, div \left(\frac{Du}{|Du|} \right) = |Du| \, curvu \tag{2}$$

This equation comes from a reformulation by Osher and Sethian [14] of a differential geometry model studied by Grayson [13], Gage and Hamilton [12].
If the multiscale analysis also satisfies the so called contrast invariance and affine invariance, the only equation is, upon a rescaling, see [1] :

$$(AMSS) \qquad \frac{\partial u}{\partial t} = |Du| \left(tdiv \frac{Du}{|Du|}\right)^{1/3} \tag{3}$$

(AMSS means affine morphological scale space).
Sapiro and Tannenbaum [9] have proposed an equivalent model for curve evolution. Guichard, Lasry and Morel [10] recently proposed a mathematical morphology interpretation of this equation : Let IB be a family of convex sets, symmetric with respect to 0, all having area equal to 1 and invariant by the special affine group that is $AB \in IB$ whenever $det(A) = 1$ and $B \in IB$. We set $IB_s = \{s^{1/2}B, \quad B \in IB\}$. The elements of IB_s have area s. We consider a scaling parameter $s = h^{3/2}$. We also set

$$S\ I_h\ u(x) = \sup_{B \in IB_{h^{3/2}}} \inf_{y \in x+B} u(y) \tag{4}$$

$$I\ S_h\ u(x) = \inf_{B \in IB_{h^{3/2}}} \sup_{y \in x+B} u(y) \tag{5}$$

$S\ I_h$ is understood as an "affine erosion" of u and $I\ S_h$ as an "affine dilation" and we also consider the alternate scheme $= SI_h(IS_h(u))$. Denote by T_h one of the schemes IS_h, SI_h or $SI_h\ IS_h$, we set by induction

$$\begin{cases} u_h(x,(n+1)h) = T_h\ u_h\ (x+nh) \\ \qquad u_h(x,0) = u_0 \end{cases} \tag{6}$$

It is proved in [10] that if u_0 is a Lipschitz function of IR^2, $u_h(x, nh)$ tends uniformly of every compact set to $u(x,t)$, where $u(x,t)$ is the unique viscosity solution of

$$\frac{\partial u}{\partial t} = C_{IB}\ |Du|\ g(curv\ u), \quad u(0) = u_0 \tag{7}$$

where C_{IB} is a constant only depending on IB,

$$\begin{aligned} g(\sigma) &= (\sigma^-)^{1/3} \quad if \quad T_h = SI_h \\ &= (\sigma^+)^{1/3} \quad if \quad T_h = IS_h \\ &= (\sigma)^{1/3} \quad\ \ if \quad T_h = SI_h\ IS_h. \end{aligned}$$

A variant of the schemes IS_h, SI_h or $SI_h\ IS_h$ which is consistent with MCM is studied in Catté and all [3], [4], [11].

In this note, using nonlinear semi-group theory, we propose an alternative proof for the convergence of the algorithm T_h :

$$T^m_{\frac{t}{m}}\ u_0 \to u(t), \qquad m \to +\infty$$

uniformly for times t belonging to compact subsets of $[0, \infty[$, where u is the unique viscosity solution of (7). Our proof is inspired from Evans [7] and Barles-Georgelin [11], but is in fact simpler because of the good properties of equation [7]. In Section 2, we survey the good properties of the schemes (4) and (5) which justify their use. In Section 3, we define an m-accretive operator and a nonlinear semigroup associated with equation (3). Section 4 is devoted to our main convergence theorem.

2. Properties of the discrete scheme

Let us list some simple and useful properties of the operators T_h . Their proof is straightforward.

- $W^{1,\infty}$-stability :

If u_0 is Lipschitz on IR^2 , then, for all h and n, the functions $u_h(\cdot, nh)$ are Lipschitz with the same Lipschitz constant.

- Monotonicity :

If $u(x) \geq v(x)$ for all $x \in IR^2$, then $(T_h u)(x) \geq T_h v)(x)$.
Evident with the definition of $IS_h, S_h I$.

- L^∞ stability :

If u_0 is bounded on IR^2, then for all h and all n, the functions $u_h(\cdot, nh)$ are bounded and

$$u_h(\cdot, nh)||_\infty \;\leq\; ||u_0||_\infty$$

Evident by monotonicity and the fact that $T_h C \;=\; C$ when C is a constant function.

- Morphological invariance :

$$g \; o \; T_h \;=\; T_h \; o \; g$$

for any change of contrast g, that is, any non decreasing and continuous function g.

- Affine invariance :

$$T_t(Au) \;=\; AT_t u$$

where A is an affine map with determinant equal to 1.
With the above properties of the scheme T_h, Guichard, Lasry and Morel [10] proved that this morphological scheme is consistent with the AMSS equation.

Theorem 2.1 There exists a constant $C_{IB} \; > \; 0$ such that for every u which is C^3 in a neighbourhood of x_0,

$$\lim_{h \to 0} \frac{(T_h u)(x_0) - u(x_0)}{h} \;=\; C_{IB} |Du(x_0)| \; g \; (curv \; u(x_0))$$

where

$$\begin{aligned} g(\sigma) &= (\sigma^-)^{1/3} \quad if \quad T_h = SI_h \\ &= (\sigma^+)^{1/3} \quad if \quad T_h = IS_h \\ &= (\sigma)^{1/3} \quad if \quad T_h = SI_h\, IS_h. \end{aligned}$$

3. Nonlinear semi-groups and the AMSS equation

Let us fix some closed square Q in IR^2 and hereafter take X to be the space of real-valued, continuous, Q-periodic functions on IR^2, taken with supremum norm. Fix $f \in X$, $t > 0$, and consider the nonlinear equation

$$\begin{cases} u - \lambda\, |Du|\, (curv\ u)^{1/3} = f \\ u\ Q\text{−}periodic. \end{cases} \tag{8}$$

Definitions

A bounded, continuous, Q-periodic function $u : IR^2 \to IR$ is a viscosity subsolution of (9) provided for each $\phi \in C^\infty(IR^2)$, if $u - \phi$ has a maximum at a point $x_0 \in IR^2$ then

$$u(x_0) - \lambda\, |D\phi(x_0)|\, (curv\ \phi(x_0))^{1/3} \leq f(x_0). \tag{9}$$

Similarly, u is a viscosity supersolution provided for each $\phi \in C^\infty(IR^2)$, if $u - \phi$ has u minimum at a point $x_0 \in IR^2$ then

$$u(x_0) - \lambda |D\phi\,(x_0)|\, (curv\ \phi(x_0))^{1/3} \geq f(x_0). \tag{10}$$

We say that u is a viscosity solution if u is both a viscosity subsolution and supersolution. Let us define a nonlinear operator A associated with Equation (9). Given $f \in X$, we say that $u \in D(A)$ and write

$$u + \lambda A u \ni f \tag{11}$$

provided u is a viscosity solution of (9).

Theorem 3.1
The operator A defined by (11) is m - accretive on X, that is (see [2, 6]):

1. If $u \in D(A)$, $\hat{u} \in D(A)$, $f \in u + \lambda A\, u$, $\hat{f} \in \hat{u} + \lambda\, A\, \hat{u}$, then $||u - \hat{u}|| \leq ||f - \hat{f}||$.

2. $\mathcal{R}(I + \lambda A) = X$, $\forall \lambda > 0$, where $\mathcal{R}(I + \lambda A)$ is the range of A.

Proof.

1. Select $u \in D(A)$, $\hat{u} \in D(A)$, $f \in u + \lambda A u$, $\hat{f} \in \hat{u} + \lambda A \hat{u}$.

We must first show

$$||u - \hat{u}|| \le ||f - \hat{f}||.$$

We define $\phi(x, y) = u(x) - \hat{u}(y) - \frac{1}{4\epsilon}||x - y||^4$. Let (x_0, y_0) a maximum point of ϕ.
According to the main approximation theorem 3.2 of [5], for any $\mu > 0$, there exist symmetric 2×2 matrices X and Y such that

$$\begin{pmatrix} X & 0 \\ 0 & Y \end{pmatrix} \le \begin{pmatrix} B + \mu B^2 & -B - \mu B^2 \\ -B - \mu B^2 & B + \mu B^2 \end{pmatrix}$$

and

$$\begin{aligned} u(x_0) &- F(X, \epsilon^{-1}||x_0 - y_0||^2(x_0 - y_0)) \le f(x_0) \\ \hat{u}(y_0) &- F(Y, \epsilon^{-1}||x_0 - y_0||^2(x_0 - y_0)) \ge \hat{f}(y_0) \end{aligned}$$

where $B = \overset{-1}{\epsilon}\, ||x_0 - y_0||^2 I_2 + 2\epsilon^{-1}(x_0 - y_0) \otimes (x_0 - y_0)$
and $F(D^2u, Du) = |Du|\ (curv\ u)^{1/3}$.

We define

$$\Gamma = \begin{pmatrix} I_2 & I_2 \\ I_2 & I_2 \end{pmatrix}.$$

Since $\Gamma \ge 0$, we have

$$\begin{pmatrix} X & 0 \\ 0 & -Y \end{pmatrix} \Gamma \le \begin{pmatrix} B + \mu B^2 & -B - \mu B^2 \\ -B - \mu B^2 & B + \mu B^2 \end{pmatrix} \Gamma = \begin{pmatrix} 0 & 0 \\ 0 & 0 \end{pmatrix}$$

$$\Longrightarrow \begin{pmatrix} X - Y & 0 \\ 0 & X - Y \end{pmatrix} \le 0 \Longrightarrow X \le Y.$$

Since F is decreasing with respect to its first argument, we obtain that

$$\begin{aligned} u(x_0) - f(x_0) &\le F(X, \epsilon^{-1}\, ||x_0 - y_0||^2\, (x_0 - y_0)) \\ &\le F(Y, \epsilon^{-1}\, ||x_0 - y_0||^2 (x_0 - y_0)) \le \hat{u}(y_0) - \hat{f}(y_0). \end{aligned}$$

So we have

$$\begin{aligned} u(x) - \hat{u}(x) &\le \sup_{(x,y)} (u(x) - \hat{u}(y) - \frac{1}{4\epsilon}||x - y||^4) \\ &= u(x_0) - \hat{u}(y_0) - \frac{1}{4\epsilon}||x_0 - y_0||^4 \\ &\le f(x_0) - \hat{f}(y_0) - \frac{1}{4\epsilon}||x_0 - y_0||^4 \\ &\le \sup_{(x,y)} (f(x) - \hat{f}(y) - \frac{1}{4\epsilon}||x - y||^4). \end{aligned}$$

Since this last relation is true for any $\epsilon > 0$ and since f and $\hat{f}$ are continuous, we obtain

$$u(x) - \hat{u}(x) \le ||f - \hat{f}||$$

2. Next, we verify $\mathcal{R}(I + \lambda A) = X$ for each $\lambda > 0$.
Since the function $F(D^2u, Du) = |Du|\,(curv\ u)^{1/3}$ is continous with respect to Du and D^2u, we can use Perron's Method [5] :

a) If u is a subsolution of (9) and v a supersolution of (9), then $u \le v$.

b) Suppose also that there is a subsolution $\underline{u}$ and a supersolution $\bar{u}$ of (9), then

$$u(x) = \sup\{w(x) : \underline{u} \le w \le \bar{u} \textit{ and } w \textit{ is a subsolution of } (9)\}$$

is a solution of (9). We have a) in the proof of the first section of Theorem 3.1. Next, $\underline{u} = ||f||_\infty$ is a subsolution and $\bar{u} = -||f||_\infty$ is a supersolution, and so $\mathcal{R}(I + \lambda A) = X$ for each $\lambda > 0$. □

Theorem 3.2
The domain $D(A)$ of the operator defined in (12) is dense in X.

Proof.
Fix $\phi \in C^\infty \cap X$. Take $\lambda > 0$ and let $u^\lambda \in D(A)$ be the solution of

$$u^\lambda - \lambda\,|Du^\lambda|\,(curv\ u^\lambda)^{1/3} = \phi.$$

Assume that $u^\lambda - \phi$ attains a maximum at x_0. Then, noting $O(\lambda)$ a function which tends to zero as $\lambda \to 0$,

$$u^\lambda(x_0) - \lambda D\phi(x_0)|\,(curv\ \phi(x_0))^{1/3} \le \phi(x_0) \Longrightarrow u^\lambda(x_0) - \phi(x_0) \le O(\lambda)$$

If $u^\lambda - \phi$, instead, has a minimum at x_0, we similarly obtain

$$u^\lambda(x_0) - \phi(x_0) \ge -O(\lambda).$$

Thus $|u^\lambda - \phi| \le O(\lambda) \quad (\lambda > 0)$. Since $u^\lambda \in D(A)$ and $C^\infty(IR^2) \cap X$ is dense in X, we conclude that $D(A)$ is dense in X. □

Since A is m-accretive, by Crandall-Liggett lemma, [6] :

$$\lim_{\substack{m \to \infty \\ \lambda m \to t}} J_\lambda^m\, u_0 = u(t) \qquad \forall u_0 \in \overline{D(A)}$$

exists in X, uniformly in time on compact subsets of $[0, \infty)$, where $J_\lambda = (I + \lambda A)^{-1}$, $\lambda > 0$ is the nonlinear resolvent. We usually write

$$u(t) = S(t)u_0 \quad (t \ge 0,\ u_0 \in \overline{D(A)})$$

Then $\{S(t)\}_{t \ge 0}$ is the nonlinear semigroup generated by A . Next, we verify that semi-group techniques generate the unique viscosity solution of the AMSS equation :

Theorem 3.3

Let $u_0 \in X$ and $S(t)$ be the semi-group generated by A.
Then $u(x,t) = S(t)u_0(x)$ $(x \in IR^2,\ t \geq 0)$ is the unique viscosity solution of the AMSS equation

$$\begin{cases} \frac{\partial u}{\partial t} = |Du|\ (curv\ u)^{1/3} \\ u(x,0) = u_0 \\ u\ Q - periodic \end{cases} \tag{12}$$

Proof.

1. Guichard [10] shows the uniqueness for AMSS equation in Appendix C of [10].

2. The existence of viscosity solutions is proved by the same argument, but simpler, since the second member of (12) is a continuous fonction of Du and D^2u as in the proof of Theorem 2.5 in Evans [7].

4. Main Convergence Theorem

Theorem 4.1

Let $u_0 \in X$ and define the family of non linear operators $(T_t)_t \geq 0$ as in Section 1. Then

$$\lim_{m \to +\infty} T^m_{\frac{t}{m}}\ u_0 = u(t)$$

uniformly for t in compact subsets of $[0, \infty)$ where u is the unique viscosity solution of the equation

$$\begin{aligned} \frac{\partial u}{\partial t} &= C_{IB}\ |Du|\ (curv^+ u)^{1/3} \quad if\ \ T_h = IS_h \\ \frac{\partial u}{\partial t} &= C_{IB}\ |Du|\ (curv^- u)^{1/3} \quad if\ \ T_h = SI_h \\ \frac{\partial u}{\partial t} &= C_{IB}\ |Du|\ (curv\ u)^{1/3} \quad if\ \ T_h = (IS_h\ SI_h) \end{aligned} \tag{13}$$

Proof.

Let A_{IB} be the nonlinear operator associated with Equation (13). Since A_{IB} is m-accretive, $S_{IB}(t)$ is a nonlinear semi-group. It suffices to prove that

i) $||T_t u - T_t \hat{u}|| \leq ||u - \hat{u}|| \quad \forall u \in X,\ \forall \hat{u} \in X,\ \forall t \geq 0,$

ii) $\lim_{t \to 0}(I + \lambda(\frac{I - T_t}{t}))^{-1} u = (I + \lambda A_{IB})^{-1} u, \quad \forall u \in \overline{D(A_{IB})},\ \forall \lambda > 0.$

Then, with Chernoff's formula for non linear semi-groups (see [7], [2]), we have

$$\lim_{m\to\infty} T^m_{\frac{1}{m}} u_0 = S_{IB}(t)u_0, \quad \forall u_0 \in X.$$

In order to prove i) and ii), we take for example, $T_t = SI_t\ IS_t$. It is easy to prove i) and it suffices to prove ii) with $\lambda = 1$. We set

$$u^t = \left(I + \frac{I - T_t}{t}\right)^{-1} u_0, \quad u_0 \in X,$$

so that

$$u^t + \frac{u^t - T_t u^t}{t} = u_0.$$

We also set

$$A_t u = \frac{u - T_t u}{t}.$$

Remarking that T_t is a contraction , which maps X into X, we deduce that A_t is m-accretive on X. By accretivity, we have

$$||u^t|| \leq ||u_0|| \quad \forall t > 0 \tag{14}$$

and

$$||u - \hat{u}|| \leq ||u - \hat{u} + (A_t u - A_t \hat{u})|| \qquad \forall u \in X,\ \forall \hat{u} \in X.$$

Setting $u = u^t(\cdot)$ and $\hat{u} = u^t(\cdot + y)$, we have

$$||u^t(\cdot) - u^t(\cdot + y)|| \leq ||u_0(\cdot) - u_0(\cdot + y)||, \qquad \forall t > 0. \tag{15}$$

From (13) and (14), we deduce that $(u^t)_{0<t<1}$ is bounded and equicontinuous in Q. Then, there exists a sequence $t_k \to 0$ and a function $u \in X$ such that

$$u^{t_k} \to u \quad in\ X$$

We must show that if $u = (I + A_{C_{IB}})^{-1} u_0$; then u is solution of $u - C_{IB}\ |Du|\ (curv\ u)^{1/3} = u_0$ in the viscosity sense.
Let us prove that u is a viscosity subsolution. The proof that u is a supersolution is similar. So, let us fix a smooth test function $\phi \in C^\infty(IR^2)$ and suppose that $u - \phi$ has a strict positive maximum at a point x_0. We must verify that

$$u(x_0) - C_{IB}\ |D\phi(x_0)|\ (curv\ \phi(x_0))^{1/3} \leq u_0(x_0) \tag{16}$$

Since $u^{t_k} \to u$ uniformly and $u - \phi$ has a strict maximum at x_0, there exist points $x_k \to x_0$ such that

$$u^{t_k} - \phi \textit{ has a positive maximum at } x_k.$$

We next observe using i) that

$$T_{t_k}\ u^{t_k}(x_k) - T_{t_k}\ \phi(x_k) \leq u^{t_k}(x_k) - \phi(x_k),$$

hence

$$A_{t_k}\phi(x_k) \leq A_{t_k}u^{t_k}(x_k),$$

and so

$$u^{t_k}(x_k) + \frac{\phi(x_k) - T_{t_k}\phi(x_k)}{t_k} \leq u_0(x_k).$$

Using the consistency of the scheme T_t, we must now let $x_k \to x_0$, $t_k \to 0$ in this inegality, then (16) is true.

Acknowledgements : We thank Jean Michel Morel , Francoise Dibos and Guy Barles for all valuable conversations we had.

References

[1] L. Alvarez, F. Guichard, P.L. Lions, J.M. Morel, *Axiomatisation et nouveaux opérateurs de la morphologie mathématique*, CRAS. Acad. Sci. Paris., t. 315, série I, (1992), 265-268.

[2] H. Brezis and A. Pazy, *Convergence and approximate of semi groups of non linear operators in Banach spaces,"* J. Func. Analysis, 9 , (1972), 63-74.

[3] F. Catté and F. Dibos, *A morphological approach of mean curvature motion*, Report 9310, Ceremade, Université Paris-Dauphine.

[4] F. Catté, F. Dibos and G. Koepfler, *A morphological scheme for mean curvature motion and applications to anisotropic diffusion and motion of level sets*, Report 9338, Ceremade, Université Paris-Dauphine.

[5] M.G. Crandall, H. Ishii and P.L. Lions, *User's guide to viscosity solutions of second order partial differential equations*, Bull. Amer. Math. Soc., 27, (1992), 1-67.

[6] M.G. Grandall and T. Liggett, *Generation of semi groups of non linear transformations on general Banach spaces*, Amer. J. Math., 93, (1971), 265-298.

[7] L.C. Evans, *Convergence of an algorithm of the mean curvature*, Indiana Univ. Math. J., 42, (1993), 533-557.

[8] P. Maragos, *Tutcrial on advances in morphological image processing and analysis* Optical engineering, July 1987, Vol. 26 N° 7.

[9] G. Sapiro and A. Taunenbaum, *On affine plane curve evolution*, Department of electrical engineering. Technion Israel Institute of Technology. Haifa Israel. EE Pub 821. February 1992, submitted.

[10] F. Guichard, *Axiomatisation des analyses multiéchelles d'images et de films*, Thèse Université Paris-Dauphine, 1994.

[11] G. Barles et Ch. Georgelin, *A simple proof for the convergence of an approximation scheme for computing mean curvature motion.*
To appear in SIAM J. on Numerical Analysis (1995).

[12] M. Gage and R.S. Hamilton, *The heat equation shrinking convex plane*, J. Differential Geometry, 23, 69-96, (1986).

[13] M. Grayson, *The heat equation shrinks embedded plane curves to round points*, J. Differential Geometry 26, 285-314, (1987).

[14] S. Osher and J. Sethian, *Fronts propagating with curvature dependant speed : algorithms based on the Hamilton-Jacobi formulation.*, J. Comp. Physics, 79, 12-49, 1988.

Anisotropic Diffusion of Multivalued Images

Guillermo Sapiro[1] and Dario L. Ringach[2]

[1] Hewlett-Packard Labs, 1501 Page Mill Road, Palo Alto, CA 94304
[2] Center for Neural Science, New York University, New York, NY 10003.

Summary. A general framework for anisotropic diffusion of multivalued images is presented. We propose an evolution equation where at each point in time the directions and magnitudes of the maximal and minimal rate of change in the image are first evaluated. These are given by eigenvectors and eigenvalues of the first fundamental form in the given image metric. Then, the image diffuses in the direction of minimal change, while the diffusion "strength" is controlled by a function that measures the degree of dissimilarity between the eigenvalues. The proposed framework can be applied to the filtering of color, texture, and space-frequency representations of images.

1. Introduction

Anisotropic diffusion for image denoising and enhancement of images has been studied by a number of authors in recent years [1, 13, 17, 18, 19]. Briefly, the idea is to smooth the image in a direction *parallel* to object boundaries, and prevent, as much as possible, diffusion *across* edges.[1] Traditionally, "edges" are localized in those regions where the gradient of the image luminance is high. However, a measure based solely on luminance contrast does not always produce satisfactory results. Looking at other image attributes, such as color, texture, or motion may provide additional information that can help in localizing image discontinuities. The measurement of several image attributes at a single location leads to a multivalued representation of the image. In this paper, we present a framework that extends the ideas presented in [1, 17] for real images to the multivalued case.

2. Edges in multivalued images

We present a definition of edges in vector-valued images based on classical Riemannian geometry [10]. Early approaches to detecting discontinuities in multivalued images attempted to combine the response of single-valued edge detectors applied separately to each of the image components (see for example [12]). The way the responses for each component are combined is in general heuristic, and has no theoretical basis. A principled way to look at gradients in multivalued images, and the one we adopt in this paper, has been described in [4].

[1] Anisotropic diffusion is also connected with the theory of curve evolution, studied in a number of recent works [5, 6, 8, 9, 14, 18, 22].

The idea is the following. Let $\Phi(u_1, u_2) : \mathbb{R}^2 \to \mathbb{R}^m$ be a multi-valued image with components $\Phi_i(u_1, u_2) : \mathbb{R}^2 \to \mathbb{R}$, $i = 1, 2, ..., m$. For color images, for example, we have $m = 3$ components. The value of the image at a given point (u_1^0, u_2^0) is a vector in $\mathbb{R}^m$, and the difference of image values at two points $P = (u_1^0, u_2^0)$ and $Q = (u_1^1, u_2^1)$ is given by $\Delta\Phi = \Phi(P) - \Phi(Q)$. When the (Euclidean) distance $d(P, Q)$ between P and Q tends to zero, the difference becomes the arc element

$$d\Phi = \sum_{i=1}^{2} \frac{\partial \Phi}{\partial u_i} du_i, \tag{2.1}$$

and its squared norm is given by

$$d\Phi^2 = \sum_{i=1}^{2} \sum_{j=1}^{2} \frac{\partial \Phi}{\partial u_i} \frac{\partial \Phi}{\partial u_j} du_i du_j. \tag{2.2}$$

This quadratic form is called the *first fundamental form* [10]. Let us denote $g_{ij} := \frac{\partial \Phi}{\partial u_i} \cdot \frac{\partial \Phi}{\partial u_j}$, then

$$d\Phi^2 = \sum_{i=1}^{2} \sum_{j=1}^{2} g_{ij} du_i du_j = \begin{bmatrix} du_1 \\ du_2 \end{bmatrix}^T \begin{bmatrix} g_{11} \, g_{12} \\ g_{21} \, g_{22} \end{bmatrix} \begin{bmatrix} du_1 \\ du_2 \end{bmatrix}. \tag{2.3}$$

The first fundamental form allows the measurement of changes in the image. For a unit vector $\hat{v} = (v_1, v_2) = (\cos\theta, \sin\theta)$, $d\Phi^2(\hat{v})$ is a measure of the rate of change of the image in the $\hat{v}$ direction. The extrema of the quadratic form (2.3) are obtained in the directions of the eigenvectors of the matrix $[g_{ij}]$, and the values attained there are the corresponding eigenvalues. Simple algebra shows that the eigenvalues are

$$\lambda_{\pm} = \frac{g_{11} + g_{22} \pm \sqrt{(g_{11} - g_{22})^2 + 4g_{12}^2}}{2}, \tag{2.4}$$

and the eigenvectors are $(\cos\theta_{\pm}, \sin\theta_{\pm})$, where the angles $\theta_{\pm}$ are given (modulo π) by

$$\theta_+ = \frac{1}{2} \arctan \frac{2g_{12}}{g_{11} - g_{22}}, \quad \theta_- = \theta_+ + \pi/2. \tag{2.5}$$

Thus, the eigenvectors provide the direction of maximal and minimal changes at a given point in the image, and the eigenvalues are the corresponding rates of change. We call θ_+ the *direction of maximal change* and λ_+ the *maximal rate of change*. Similarly, θ_- and λ_- are the *direction of minimal change* and the *minimal rate of change* respectively. Note that for $m = 1$, that is, for grey-level images, $\lambda_+ \equiv \| \nabla\Phi \|^2$, $\lambda_- \equiv 0$, and $(\cos\theta_+, \sin\theta_+) = \nabla\Phi / \|\nabla\Phi\|$.

In contrast to grey-level images ($m = 1$), the minimal rate of change λ_- may be different than zero. In the single-valued case, the gradient is always perpendicular to the level-sets, and $\lambda_- \equiv 0$. As a consequence, the "strength" of an edge in the multi-valued case is not simply given by the rate of maximal change, λ_+, but by how λ_+ compares to λ_-. For example, if $\lambda_+ = \lambda_-$, we know that the image changes at an equal rate in all directions. Image discontinuities can be detected by defining a function $f = f(\lambda_+, \lambda_-)$ that measures the dissimilarity between λ_+ and λ_-. A possible choice is $f = f(\lambda_+ - \lambda_-)$, which has the nice property of reducing to $f = f(\|\nabla\Phi\|^2)$ for the one dimensional case, $m = 1$.

Cumani [3] extended some of the above ideas. He analyzed the directional derivative of λ_+ in the direction of its corresponding eigenvector, and looked for edges by localizing the zero crossings of this function. Note that this approach entirely neglects the behavior of λ_-. As we already mentioned, it is the relationship between λ_- and λ_+ that is important.

A noise analysis for the above vector-valued edge detector has been presented in [11]. It was shown that, for correlated data, this scheme is more robust to noise than the simple combination of the gradient components.

3. Anisotropic diffusion of multivalued images

Based on the results in previous section, we proceed to derive the anisotropic flow of multivalued images in an analogous way to the single-valued case presented in [1].

Diffusion of the image occurs normal to the direction of maximal change θ_+ (i.e., in the direction perpendicular to the multivalued "edge") which, in our case, is given by θ_-. Thus, we obtain

$$\frac{\partial \Phi}{\partial t} = \frac{\partial^2 \Phi}{\partial \theta_-^2}, \tag{3.1}$$

which is a system of coupled PDE's. The coupling results from the fact θ_- depends on *all* the components of the image Φ.

In order to have control over the local diffusion coefficient we add a factor g similar to the one proposed in [1] for single-valued images. As we want to reduce the amount of diffusion near the edges, we require that g will be close to zero when $\lambda_+ \gg \lambda_-$. The general evolution equation reads,

$$\frac{\partial \Phi}{\partial t} = g(\lambda_+, \lambda_-)\frac{\partial^2 \Phi}{\partial \theta_-^2}, \tag{3.2}$$

A suitable choice for g is any decreasing function of the difference $(\lambda_+ - \lambda_-)$.

Since a function of the form $f = f(\lambda_+ - \lambda_-)$ is the analogue multivalued extension of $f = f(\|\nabla\Phi\|^2)$ for single-valued images ($m = 1$), image processing algorithms for single-valued images based on $\|\nabla\Phi\|^2$ can be extended to multivalued images by replacing the square magnitude of the gradient with

$\lambda_+ - \lambda_-$. For example, an extension to the *total variation* algorithm developed by Rudin *et al.* [19] (see also [7]) for image denoising can be suggested. Rudin *et al.* propose to minimize (under certain constraints) the functional $\int_\Omega \| \nabla \Phi \| \, dxdy$. The analogue functional for vector-valued images would be $\int_\Omega \sqrt{\lambda_+ - \lambda_-} \, dxdy$. The variational approach to vector-valued image denoising is similar to the anisotropic diffusion described above in the sense that it leads to a system of coupled PDE's when the gradient-descent approach is used. We are currently investigating in this direction.

3.1 Related works

Little has been done regarding anisotropic diffusion of vector-valued images. Recently, Chambolle [2] presented preliminary results on anisotropic diffusion of color images. Chambolle's evolution equation reads

$$\Phi_t = \Phi_{\xi\xi}, \tag{3.3}$$

where ξ is the direction normal to η given by

$$\eta := \arg \max_{\eta \in \mathbb{R}^2, |\eta|=1} \| (< \nabla\Phi_1, \eta >, < \nabla\Phi_2, \eta >, < \nabla\Phi_3, \eta >) \| \tag{3.4}$$

Equation (3.3) is different from our model since there is no control over the diffusion coefficient. Diffusion is independent of the relationship between λ_+ and λ_-, which does not prevent edges from moving during the evolution (see [21] for specific examples). In our scheme, control over the diffusion coefficient solves this problem. It is easy to show that the direction dictated by (3.4) is exactly the same as the one obtained by our approach. This can be seen by writing $\eta = (\cos\theta, \sin\theta)$ and showing that maximizing (3.4) is equivalent to maximizing (2.3). Actually, no specific way of computing η is reported in [2]. While (3.4) can be generalized to other metrics, we consider that the approach based on Riemannian geometry is more intuitive and allows the use of well known results.

Chambolle also proposed ξ to be the direction perpendicular to the vector

$$(\Phi_1 - \Phi_2)\nabla\Phi_3 + (\Phi_2 - \Phi_3)\nabla\Phi_1 + (\Phi_3 - \Phi_1)\nabla\Phi_2. \tag{3.5}$$

This choice has severe problems. If any two planes are constant, as expected for example in a black and white image, the direction given by (3.5) will always be zero. This is also true if two planes are locally constant, and the image resides in the third component. We expect a consistent multi-valued model to automatically reduce to the existing ones for single-valued images when $m = 1$. This requirement holds for our approach and equation (3.4). It fails for equation (3.5).

Whitaker and Gerig [23] presented a general framework for the diffusion of vector-valued images. In contrast with our scheme, which is motivated by extending $\Phi_t = \Phi_{\xi\xi}$, where ξ is parallel to the edge, their approach is based on a straightforward extension of Perona and Malik's work to a system of coupled diffusion equations. The coupling is given by a conductance matrix. Their work does not discuss the extension of gradient computations for multi-valued images, and the importance of anisotropic diffusion perpendicular to the direction of maximal rate of change is not explicitly addressed.

We conclude this section pointing out that in [16], the authors present a number of systems of coupled PDE's for different computer vision applications like stereo and motion.

4. Experimental results

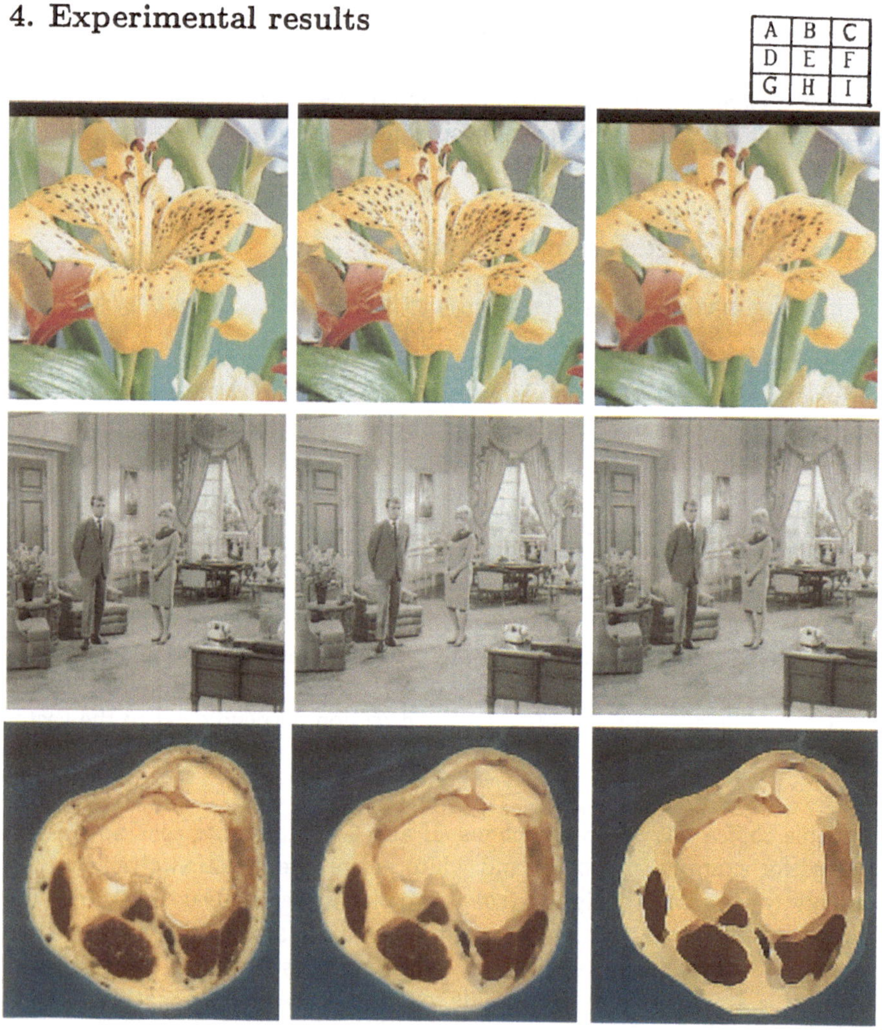

Fig. 4.1. (A–C) Anisotropic diffusion of a noisy color image. (D–F) Image processing in a wavelet-like decomposition. (G–I) Color self-snakes for image simplification.

An example of vector-valued anisotropic diffusion for color data is given in Figure 4.1(**A**–**C**). We represented the image in the CIE 1976 $L^*a^*b^*$-space [24], which is an approximately uniform color space. In (**A**) we see the original image, the noisy one is shown in (**B**), and the filtered image is depicted in (**C**). Additional examples can be found in [21].

In Figure 4.1(**D**–**F**) we present an example where the image was first locally represented in 4 frequency bands (low-low, low-high, high-low, high-high). This yields a vector-valued representation of the image. We processed the multivalued image in this space, and subsequently reconstructed the associated real-valued image. The original image is shown in (**D**), the noisy one in (**E**), and processed one in (**F**). Similar decompositions can be applied to process textured images [20].

Figure 4.1(**G**–**I**) shows an example of a flow introduced in [20]. The evolution is governed by the equation $\frac{\partial \Phi_i}{\partial t} = |\nabla \Phi_i| \mathrm{div}\left(g(\lambda_+, \lambda_-)\frac{\nabla \Phi_i}{|\nabla \Phi_i|}\right)$. This flow moves each one of the level-sets according to the *geodesic color snakes* introduced in [20]. Besides an anisotropic diffusion term, it contains a shock one. The original image is shown in (**G**), and the result of two different steps in the algorithm are shown in (**H**) and (**I**). Note the piecewise constant form of the image on the right, which is the steady state of the flow. For details on this flow and its relations with previous works, see [20].

5. Concluding remarks

A new approach for anisotropic diffusion of multivalued images was proposed. The approach was exemplified by extending known flows on real images to the multi-valued case. Results for the filtering of color and multiscale data were presented.

Given a number of image "attributes" one can measure *psychophysically* the degree to which the human visual system is sensitive to small changes in each of the components. Psychophysical studies, therefore, can help in defining an image metric space that is relevant to human vision. Finding ways to assess the performance of the algorithm for different choices of the metric is of importance. It is not clear how the performance of our algorithm depends on the choice of different metrics. We are currently investigating in this direction.

Since most of the research on PDE's for image processing has been done for single-valued data, not many theoretical results for systems of coupled equations are presently available. The rigorous analysis of the equations proposed in this paper opens an interesting theoretical area.

Acknowledgement. The image in Figure 4.1(**D**) was obtained from the image archive of the National Library of Medicine (The Visible Human Project).

References

1. L. Alvarez, P. L. Lions, and J. M. Morel, "Image selective smoothing and edge detection by nonlinear diffusion," *SIAM J. Numer. Anal.* **29**, pp. 845-866, 1992. anisotropic
2. A. Chambolle, "Partial differential equations and image processing," to appear in *Proc. IEEE International Conference on Image Processing*, Texas, Austin, November 1994.
3. A. Cumani, "Edge detection in multispectral images," *CVGIP-GMIP* **53**, pp. 40-51, 1991.
4. S. Di Zenzo, "A note on the gradient of a multi-image," *CVGIP* **33**, pp. 116-125, 1986.
5. O. Faugeras, "On the evolution of simple curves of the real projective plane," *Comptes rendus de l'Acad. des Sciences de Paris* **317**, pp. 565-570, September 1993.
6. M. Gage and R. S. Hamilton, "The heat equation shrinking convex plane curves," *J. Differential Geometry* **23**, pp. 69-96, 1986.
7. D. Geman and G. Reynolds, "Constrained restoration and the recovery of discontinuities," *IEEE-PAMI* **14**, pp. 367-383, 1992.
8. M. Grayson, "The heat equation shrinks embedded plane curves to round points," *J. Differential Geometry* **26**, pp. 285-314, 1987.
9. B. B. Kimia, A. Tannenbaum, and S. W. Zucker, "Shapes, shocks, and deformations, I," *International Journal of Computer Vision* **15**, pp. 189-224, 1995.
10. E. Kreyszig, *Differential Geometry*, University of Toronto Press, Toronto, 1959.
11. H-C. Lee and D. R. Cok, "Detecting boundaries in a vector field," *IEEE Trans. Signal Proc.* **39**, pp. 1181-1194, 1991.
12. R. Nevatia, "A color edge detector and its use in scene segmentation," *IEEE Trans. Syst. Man, Cybern.*, **7**, pp. 820-826, 1977.
13. M. Nitzberg and T. Shiota, "Nonlinear image filtering with edge and corner enhancement," *IEEE-PAMI* **14**, pp. 826-833, 1992.
14. S. J. Osher and J. A. Sethian, "Fronts propagation with curvature dependent speed: Algorithms based on Hamilton-Jacobi formulations," *Journal of Computational Physics* **79**, pp. 12-49, 1988.
15. S. Osher and L. I. Rudin, "Feature-oriented image enhancement using shock filters," *SIAM J. Numer. Anal.* **27**, pp. 919-940, 1990.
16. E. J. Pauwels, P. Fiddelaers, and L. J. Van Gool, "Coupled geometry-driven diffusion equations for low-level vision," in [18].
17. P. Perona and J. Malik, "Scale-space and edge detection using anisotropic diffusion," *IEEE-PAMI* **12**, pp. 629-639, 1990.
18. B. Romeny (Ed.), *Geometry Driven diffusion in Computer Vision*, Kluwer, 1994.
19. L. I. Rudin, S. Osher, and E. Fatemi, "Nonlinear total variation based noise removal algorithms," *Physica D* **60**, pp. 259-268, 1992.
20. G. Sapiro, "Color snakes," *HPL-TR*, September 1995, submitted.
21. G. Sapiro and D. Ringach, "Anisotropic diffusion in color space," *HP-TR*, September 1994.
22. G. Sapiro and A. Tannenbaum, "Affine invariant scale-space," *International Journal of Computer Vision* **11:1**, pp. 25-44, 1993.
23. R. T. Whitaker and G. Gerig, "Vector-valued diffusion," in [18].
24. G. Wyszecki and W. S. Stiles, *Color Science: Concepts and Methods, Qualitative Data and Formulae*, (2nd edition), John Wiley & Sons, 1982.

Wavelets

Creating And Comparing Wavelets

Gilbert Strang

Department of Mathematics, Massachusetts Institute of Technology
Cambridge, Massachusetts 02139, USA
E-Mail: gs@math.mit.edu

Summary. This paper emphasizes two points about the design and application of filters and filter banks and wavelets:

- The algebra behind wavelet design is now quite simple.
- The comparison of two competing wavelets is still experimental an d empirical.

As example we consider two particular 9/7 constructions (nine coefficients in analysis and seven in synthesis). Both are symmetric, so neither is orthogonal. Both have important advantages. We are unable to say which construction is better. The paper begins with the conditions on the filter coefficients for perfect reconstruction. That part is a brief summary of the exposition in Strang and Nguyen [1].

1. Introduction

This paper discusses two-channel filter banks, leading to wavelets. The structure involves four filters (which are just convolutions, see below). The analysis bank has a lowpass filter H_0 and a highpass fi lter H_1. The outputs y_0 and y_1 from those filters are "downsampled" by keeping only the even-numbered components:

$$\begin{aligned} y_0 &= H_0 x \quad \text{and} \quad v_0(n) = (\downarrow 2) y_0(n) = y_0(2n) \\ y_1 &= H_1 x \quad \text{and} \quad v_1(n) = (\downarrow 2) y_1(n) = y_1(2n). \end{aligned}$$

The full-length input vector x yields two half-length vectors v_0 and v_1. *That analysis step is inverted by the synthesis step.* The v's are "upsampled" to put zeros into their odd-numbered components. The results are filtered by F_0 and F_1, and their sum is the output $\hat{x}$:

$$\hat{x} = F_0(\uparrow 2)v_0 + F_1(\uparrow 2)v_1.$$

The intention is that $\hat{x} = x$.

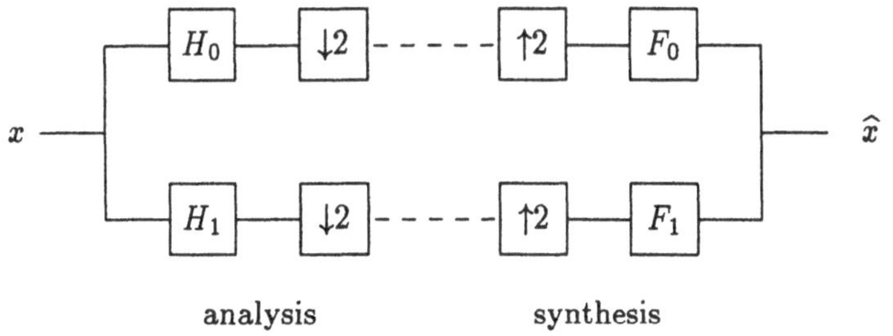

In that case the filter bank gives *perfect reconstruction*. The analysis wavelets and synthesis wavelets will be *biorthogonal*. So are the scaling functions (biorthogonality means $\int \tilde{\phi}(t-k)\phi(t-\ell)dt = \delta_{k\ell}$). All these functions come from solving two-scale equations, which involve t and $2t$. This second scale $2t$ corresponds to the $2n$ in downsampling.

We emphasize particularly how the properties of these functions in continuous time follow from the properties of the filters in discrete time. The heart of the construction is the choice of H_0 and F_0. These are given by the impulse responses h_0 and f_0, which are the vectors of filter coefficients: $h_0 = (h_0(0), \ldots, h_0(8))$ and $f_0 = (f_0(0), \ldots, f_0(6))$. Those choices determine H_1 and F_1 by a pattern of "alternating signs", $h_1(n) = (-1)^n f_0(n)$ and $f_1(n) = (-1)^{n+1} h_0(n)$.

We first determine the conditions on H_0 and F_0 (thus on h_0 and f_0) for perfect reconstruction. Then we explain the special importance of "zeros at $z = -1$" in the transfer functions, which are simply polynomials built from the filter coefficients:

$$H_0(z) = \sum_0^8 h_0(n) z^{-n} \qquad \text{and} \qquad F_0(z) = \sum_0^6 f_0(n) z^{-n}.$$

The algebra is all straightforward. We have a change of basis, produced by a *wavelet transform*. The components of v_0 and v_1 express the input vector x in the new basis. This transform can be applied again to the lowpass output v_0 that is normally most important. Scaling functions and wavelets appear in the limit of an infinite iteration. Four or five levels give a typical tree, in practice.

If successful, many components of the v's will be small. The signal is *compressed* by setting small coefficients to zero (not invertible!). The reconstructed output from the synthesis bank will no longer agree exactly with x. But if x and $\hat{x}$ are close, the input signal is now represented by a small number of components. In the new basis, the signal can be efficiently transmitted and stored. One important and unresolved difficulty is the meaning of the word "close".

2. The Conditions for Perfect Reconstruction

A filter is a convolution: $y(n) = \sum h(k)x(n-k)$. This linear transformation is represented by a Toeplitz matrix (meaning constant diagonals). The coefficient $h(k)$ appears along the kth subdiagonal. The input vector x is very long in practice and infinitely long in theory— thus the filter matrix is doubly infinite:

$$Hx = \begin{bmatrix} h(3) & h(2) & h(1) & h(0) & & \\ & h(3) & h(2) & h(1) & h(0) & \\ & & h(3) & h(2) & h(1) & h(0) \end{bmatrix} \begin{bmatrix} x(-1) \\ x(0) \\ x(1) \end{bmatrix}$$

$$= \begin{bmatrix} y(-1) \\ y(0) \\ y(1) \end{bmatrix}.$$

Downsampling removes $y(-1)$ and $y(1)$. In the product $(\downarrow 2)H$, all the odd-numbered rows of H are removed:

$$(\downarrow 2)H = \begin{bmatrix} h(3) & h(2) & h(1) & h(0) & & \\ & & h(3) & h(2) & h(1) & h(0) \end{bmatrix}.$$

Notice the *double shift* between rows. $(\downarrow 2)H$ is the fundamental operator in wavelet analysis (1×2 block Toeplitz matrix). When the two analysis filters $(\downarrow 2)H_0$ and $(\downarrow 2)H_1$ are combined, by interleaving rows of the two matrices, we get the block Toeplitz matrix (with 2×2 blocks) that represents the analysis bank:

$$H_b = \begin{bmatrix} h_0(3) & h_0(2) & h_0(1) & h_0(0) & & \\ h_1(3) & h_1(2) & h_1(1) & h_1(0) & & \\ & & h_0(3) & h_0(2) & h_0(1) & h_0(0) \\ & & h_1(3) & h_1(2) & h_1(1) & h_1(0) \end{bmatrix}$$

The inverse of H_b is the synthesis matrix F_b. *The key featu re of these matrices is that both are banded.* In the language of signal processing, all filters are FIR (finite impulse response). Banded Toeplitz matrices with banded inverses are possible only because these are *block* matrices. The inverse of a polynomial $1 + z^{-1}$ is not a polynomial. But a matrix polynomial can have a polynomial inverse: for example

$$H_p(z) = \frac{1}{2} \begin{bmatrix} 1 + z^{-1} & 1 - z^{-1} \\ 1 - z^{-1} & 1 + z^{-1} \end{bmatrix}$$

has determinant z^{-1} and inverse

$$\frac{1}{2} \begin{bmatrix} z + 1 & -z + 1 \\ -z + 1 & z + 1 \end{bmatrix}.$$

This example illustrates two further points. First, the determinant is a monomial. Since we divide by the determinant, this monomial is the key to a polynomial inverse. Second, the inverse is anticausal (powers of z) when the original filters are causal (powers of z^{-1}). To keep all filters causal and all matrices lower triangular, the convention is to separate out the monomial determinant:

$$\begin{aligned} F_p(z) &= z^{-1}H_p^{-1}(z) && \text{is causal,} \\ F_p(z)H_p(z) &= z^{-1}I && \text{is a one-step delay.} \end{aligned}$$

This means that the output $\hat{x}(n)$ from the filter bank agrees with $x(n-1)$, not with $x(n)$. We still call this perfect reconstruction! In general the filters produce ℓ delays and the output is $\hat{x}(n) = x(n-\ell)$. The product of the lower triangular block Toeplitz matrices F_b and H_b is the shift matrix that has identity blocks on the ℓth subdiagonal. Our real task is to find the conditions on the coefficients $f_0(n)$ and $h_0(n)$ for this to happen.

The algebra of convolution is summed up in the convolution rule. This transforms the matrix equation $y = Hx$ into a simpler equation for the associated polynomials. These polynomials are just multiplied: $Y(z) = H(z)X(z)$ is

$$\left(\sum y(n)z^{-n}\right) = \left(\sum h(k)z^{-k}\right)\left(\sum x(\ell)z^{-\ell}\right).$$

The terms z^{-k} and $z^{-(n-k)}$ in the factors give z^{-n} in the product. Thus the coefficient in $y(n)z^{-n}$ is the convolution $\sum h(k)x(n-k)$ that comes from matrix multiplication.

The algebra of $(\downarrow 2)$ and $(\uparrow 2)$ is almost as ne at. For vectors,

$$(\uparrow 2)(\downarrow 2)\begin{bmatrix} y(-2) \\ y(-1) \\ y(0) \\ y(1) \\ y(2) \end{bmatrix} = (\uparrow 2)\begin{bmatrix} y(-2) \\ y(0) \\ y(2) \end{bmatrix} = \begin{bmatrix} y(-2) \\ 0 \\ y(0) \\ 0 \\ y(2) \end{bmatrix}.$$

For the z-transform $Y(z) = \sum y(n)z^{-n}$, only even powers remain. The result of downsampling and upsampling is the even part

$$\frac{1}{2}\Big(Y(z) + Y(-z)\Big) = \frac{1}{2}\Big(H(z)X(z) + H(-z)X(-z)\Big).$$

That term with $-z$ reflects aliasing. Two inputs can give the same output. The constant vector $y(n) \equiv 1$ and the alternating vector $y(n) = (-1)^n$ have the same even components, and therefore they look the same after downsampling.

The conditions for perfect reconstruction $\hat{x}(n) = x(n-\ell)$ come by following the signal through the filter bank. We do it in the z-domain, starting with $X(z)$. The lowpass channel yields $Y_0(z) = H_0(z)X(z)$. Then it takes the even part. Then it multiplies by $F_0(z)$. The highpass channel has 1 in place of 0, and we add:

$$\hat{X}(z) = \begin{matrix} \frac{1}{2}F_0(z)\Big(H_0(z)X(z) + H_0(-z)X(-z)\Big) \\ + \\ \frac{1}{2}F_1(z)\Big(H_1(z)X(z) + H_1(-z)X(-z)\Big) \end{matrix} = z^{-\ell}X(z).$$

The coefficient of $X(z)$ is $z^{-\ell}$ (no distortion, only a delay). The coefficient of $X(-z)$ must be zero (no aliasing in the final output). These are the PR conditions:

$$F_0(z)H_0(z) + F_1(z)H_1(z) = 2z^{-\ell} \quad (2.1)$$

$$F_0(z)H_0(-z) + F_1(z)H_1(-z) = 0. \quad (2.2)$$

It is the anti-aliasing equation (2.2) that leads to the "alternating sign" constructions, $h_1(n) = (-1)^n f_0(n)$ and $f_1(n) = (-1)^{n+1}h_0(n)$. In terms of polynomials, this is $H_1(z) = F_0(-z)$ and $F_1(z) = -H_0(-z)$. Then (2.2) is automatically satisfied, and (2.1) reduces to an equation for the *product filter* $P_0(z) = F_0(z)H_0(z)$:

$$P_0(z) - P_0(-z) = 2z^{-l}. \quad (2.3)$$

This is the key equation for perfect reconstruction.

Note that the left side of (2.3) is an odd function, so ℓ must be odd. The equation says that *the only odd term in* $P_0(z)$ *is* $z^{-\ell}$ *with coefficient 1*. We can separate the design of a PR filter bank into three steps:

1. Choose a polynomial $P_0(z)$ that satisfies (2.3).
2. Factor $P_0(z) = F_0(z)H_0(z)$.
3. Choose $H_1(z) = F_0(-z)$ and $F_1(z) = -H_0(-z)$.

This simplicity is what was meant in our first point in the abstract. It is deceptive, because it does not indicate what makes one design better than another. Part of the answer (only part!) is in the number of zeros at $z = -1$.

3. Zeros at −1: Approximation and Vanishing Moments

For a lowpass filter, the polynomial

$$H(z) = \sum h(n)z^{-n} = \sum h(n)e^{-in\omega}$$

is near zero at the highest frequency $\omega = \pi$. In the z-plane, with $z = e^{i\omega}$, this is the point $z = e^{i\pi} = -1$. Then the multiplication $Y(z) = H(z)X(z)$ ensures that high frequencies in x (at or near $\omega = \pi$) are nearly annihilated in y. The lowest frequency $\omega = 0$ (or $z = +1$) passes through the filter provided $H(1) = 1$. Hence the name "lowpass".

This condition $H(-1) = 0$ is fundamental in wavelet theory. It must hold exactly, not just approximately, to have any chance of continuous scaling functions and wavelets. The scaling function $\phi(t)$ solves the *dilation equation* or *refinement equation*:

$$\phi(t) = 2\sum h(k)\phi(2t-k). \tag{3.1}$$

The Fourier transform of this equation is

$$\hat{\phi}(\omega) = \sum h(n)e^{-i\omega n/2}\hat{\phi}(\omega/2) = H(e^{i\omega/2})\hat{\phi}(\omega/2). \tag{3.2}$$

Periodicity gives $H(e^{ik\pi}) = 0$ for all odd integers k. Then equation (3.2) yields $\hat{\phi}(2\pi n) = 0$ for every $n \neq 0$. This is the first of the so-called "Strang–Fix conditions", implying that the translates of $\phi(t)$ add to a constant, which we may normalize to 1:

$$\sum_{-\infty}^{\infty} \phi(t-n) \equiv 1. \tag{3.3}$$

This conclusion could also be reached directly from the dilation equations

$$\phi(t-n) = 2\sum h(k)\phi(2t-2n-k)$$

by summing on n. The sum $S(t)$ is 1-periodic, and we restrict to $0 \leq t < 1$. Separating even from odd k leads to

$$S(t) = \left(2\sum_{\text{even } k} h(k)\right) S(2t) + \left(2\sum_{\text{odd } k} h(k)\right) S(2t-1). \tag{3.4}$$

The coefficients are $H(1) \pm H(-1)$. Thus both coefficients are 1. The sum satisfies the "Haar equation" $S(t) = S(2t) + S(2t-1)$, whose solution on the period interval $[0,1)$ is $S(t) \equiv 1$.

We often use the simplified notation $H(\omega)$ to replace $H(e^{i\omega})$. Note that the transform of $\phi(2t)$ is $\frac{1}{2}\hat{\phi}(\omega/2)$. Then (3.1) applies recursively to $\omega/2, \omega/4, \ldots$ and leads to an infinite product formula for $\hat{\phi}(\omega)$:

$$\hat{\phi}(\omega) = H(\frac{\omega}{2})H(\frac{\omega}{4})\hat{\phi}(\frac{\omega}{4}) = \cdots = \prod_{1}^{\infty} H(\frac{\omega}{2^j}). \tag{3.5}$$

This product converges for each ω. But is it the transform of a smooth $\phi(t)$?

Summary

The special zero $H(-1) = 0$ leads to $S(t) \equiv 1$. Constant polynomials can be produced from translates of $\phi(t)$. By standard results in approximation theory, these translates give at least first-order approximation to any smooth function $f(t)$:

$$\|f(t) - \sum a_k\phi(t-k)\| \leq C\|f'(t)\|,$$

for suitable a_k. The particular choice $a_k = f(k)$ is the "quasi-interpolate". When the mesh size changes from 1 to h, by rescaling t, the familiar factor h appears on the right side and the approximation error is $O(h)$.

All this followed from a simple zero $H(-1) = 0$. *Suppose that the zero at $z = -1$ has higher multiplicity p.* Then the corresponding st eps lead to the Strang–Fix condition of order p: the derivatives are $\hat{\phi}^{(j)}(2\pi n) = 0$ for $n \neq 0$ and $j < p$. This is equivalent to pth-order approximation by the translates of $\phi(t)$.

Theorem 3.1. *A pth-order zero of $H(z)$ at $z = -1$ implies these properties, provided $\phi(t)$ exists in L^2:*

1. *The translates $\phi(t-n)$ can reproduce all polynomials of degree less than p.*
2. *The translates give pth-order approximation of a smooth $f(t)$:*

$$\|f(t) - \sum a_k \phi(t-k)\| \leq C_p \|f^{(p)}(t)\| \tag{3.6}$$

for suitable a_k. Again the rescaling of t produces the factor h^p in (3.6).

3. *The wavelets that are orthogonal to the scaling functions have p vanishing moments.*

Orthogonality to the scaling functions $\phi(t-k)$ means orthogonality to their combinations $1, t, \ldots, t^{p-1}$. This gives the vanishing moments:

$$\int_{-\infty}^{\infty} t^m \tilde{w}_{jk}(t)dt = \int_{-\infty}^{\infty} t^m 2^{j/2} \tilde{w}(2^j t - k)dt = 0 \qquad for\ m < p.$$

Note! The convention is that $\phi(t)$ without the tilde is the *synthesis* function. Thus the coefficients in (4) and (5) should be written $f_0(k)$ instead of $h(k)$. The *analysis* functions $\tilde{\phi}(t)$ and $\tilde{w}(t)$ are constructed from

$$\begin{aligned} \tilde{\phi}(t) &= 2\sum h_0(k)\tilde{\phi}(2t-k) \\ \tilde{w}(t) &= 2\sum h_1(k)\tilde{\phi}(2t-k). \end{aligned}$$

Biorthogonality between tilde and non-tilde follows from the perfect reconstruction conditions, at every step of the iteration that solves the dilation equation. This iteration is the "cascade algorithm":

$$\tilde{\phi}^{(i+1)}(t) = 2\sum h_0(k)\tilde{\phi}^{(i)}(t). \tag{3.7}$$

The initial $\tilde{\phi}^{(0)}(t)$ is the box function on [0,1]. *But the iteration may not converge.* There is a condition on the eigenvalues of the matrix $T = (\downarrow 2)2HH^T$: *Condition E*: All eigenvalues satisfy

$$|\lambda(T)| < 1$$

except for a simple eigenvalue $\lambda = 1$.

This gives L^2 convergence of the cascade algorithm [1]. The scaling function basis and the wavelet basis are stable [2]. Furthermor e the smoothness of $\tilde{\phi}(t)$ and $\tilde{w}(t)$ are determined [3] by the spectral radius $\rho = |\lambda_{max}(T)|$, when we exclude the special eigenvalues $\lambda = 1, \frac{1}{2}, \frac{1}{4}, \ldots, (\frac{1}{2})^{2p-1}$ that are automatic from the pth order zero of $H(z)$ at $z = -1$. The number of derivatives of $\tilde{\phi}(t)$ and $\tilde{w}(t)$ in L^2 is given by

$$s_{max} = -\log \rho / \log 4. \tag{3.8}$$

Each additional factor $(1 + z^{-1})/2$ in $H(z)$ increases p by 1 and s_{max} by 1. The new $\tilde{\phi}(t)$ is just the convolution of the old $\tilde{\phi}(t)$ with the box function. The perfect examples are *B-splines*, which come from the "pure" filter

$$H(z) = \left(\frac{1 + z^{-1}}{2}\right)^p.$$

The B–spline of degree $p - 1$ has $s_{max} = p - 1/2$. Of course the B–splines are not biorthogonal to themselves (except for the box function when $p = 1$). The product $F_0(z)H_0(z)$ is allowed only one odd power $z^{-\ell}$. We now create filters with this perfect reconstruction property, and compare the functions that come out of the cascade algorithm.

4. Three Choices of 9/7 Symmetric Biorthogonal Filters

The most popular filters are constructed by factoring a Daubechies polynomial of degree $4p - 2$:

$$D_{4p-2}(z) = (1 + z^{-1})^{2p} Q_{2p-2}(z).$$

$Q(z)$ is needed so that $D(z)$ will have only one odd power (with coefficient 1). The unique polynomial Q_{2p-2} of lowest degree $2p - 2$ is the Daubechies choice. Examples are

$$\begin{aligned} D_6(z) &= (1 + z^{-1})^4 \, (-1 + 4z^{-1} - z^{-2})/32 \\ D_{14}(z) &= (1 + z^{-1})^8 \, (-5 + 40z^{-1} - 131z^{-2} + 208z^{-3} \\ &\qquad -131z^{-4} + 40z^{-5} - 5z^{-6})/2^{12}. \end{aligned}$$

The polynomial Q_{2p-2} comes directly from the binomial expansion of $(1 - y)^{-p}$, truncated after p terms [1]. This polynomial of degree $(p - 1)$ in $y = (2 - z - z^{-1})/4$ becomes a polynomial of degree $2p - 2$ in z^{-1} (after shifting by z^{1-p}).

Here are three filter banks with interesting properties. Many others are interesting too!

1. The FBI 9/7 filters were constructed by Daubechies and chosen by the FBI in digitizing fingerprints. $H_0(z)$ and $F_0(z)$ are factors of $D_{14}(z)$, each with $p = 4$ zeros at $z = -1$. (Thus $(1 + z^{-1})^8$ is split down the middle.) The other factors of degree 4 in H_0 and 2 in F_0 are chosen to preserve the symmetry of $Q_6(z)$. Real reciprocal roots z and $1/z$ go into F_0, and complex reciprocal roots $z, \bar{z}, 1/z, 1/\bar{z}$ go into H_0. *The coefficients are not rational*, and we give (inadequately) two decimals of the filter coefficients:

$$\begin{aligned} h_0 &= [\ \ 0.03\ \ -0.02\ \ -0.08\ \ 0.27\ \ 0.60\ \ 0.27\ \ -0.08\ \ -0.02\ \ 0.03\,] \\ f_0 &= [\ -0.05\ \ -0.03\ \ 0.30\ \ 0.56\ \ 0.30\ \ -0.03\ \ -0.05\,]. \end{aligned}$$

These filters are frequently used in image compression. Our normalizati on is

$$\sum h_0(k) = \sum f_0(k) = 1.$$

Then an extra $\sqrt{2}$ is needed in all four filters H_0, H_1, F_0, F_1, by Eq.(1).

2. The spline 9/7 filters come from a different factorization of $D_{14}(z)$, in which $F_0(z) = (1 + z^{-1})^6/64$. Then $\phi(t)$ is the extremely smooth B–spline of degree 5 (with 4 continuous derivatives). But this leaves only two zeros at -1 for $H_0(z)$, which must swallow $Q_6(z)$ whole:

$$\begin{aligned} h_0 &= [\ -5\ \ 30\ \ -56\ \ -14\ \ 154\ \ -14\ \ -56\ \ 30\ \ -5]/64 \\ f_0 &= [\ \ 1\ \ 6\ \ 15\ \ 20\ \ 15\ \ 6\ \ 1]/64. \end{aligned}$$

This is a foolish choice. We will see that there is no L_2 solution to the dilation equation involving h_0. Condition E fails.

3. The "binary" 9/7 filter bank selects $F_0(z) = D_6(z)$. This choice (by the physicist Tomas Arias of M.I.T.) surprised the author. In itself it has one odd power z^{-3}—which leads to the useful interpolating property $\phi(n) = \delta(n-3)$. But it is the product $F_0 H_0$ that must have one odd power, and what is H_0? The biorthonormal filter is needed in compression, if not in physics.

The lowest degree is 8 for a symmetric H_0 with a zero at $z = -1$. A direct calculation gives

$$\begin{aligned} h_0 &= [\ \ 1\ \ 0\ \ -8\ \ 16\ \ 46\ \ 16\ \ -8\ \ 0\ \ 1]/64 \\ f_0 &= [\ -1\ \ 0\ \ 9\ \ 16\ \ 9\ \ 0\ \ -1]/32. \end{aligned}$$

Notice that all coefficients are integers divided by powers of 2. This means perfect arithmetic, and fast execution on a chip.

We communicated this construction to Wim Sweldens who responded that he had already found the same binary filters. His method of "lifting" is extremely useful[5]. Starting from one admissible pair, in this case $H_0 \equiv 1$ and $F_0 = D_6$, the choice $H_{new}(z) = H_0(z) + F_0(-z)S(z^2)$ also gives perfect reconstruction. The unrestricted S allows our filter (Wim's filter) to have two zeros at $z = -1$. With four zeros the lengths become 13/7, and our compression experiments were less satisfactory—we don't know why.

5. Smoothness and Behavior of the Filters

We come now to a *comparison* of the three examples: FBI, spline and binary. Various measures are easy to compute. They give partial information:

Zeros at $z = -1$: 4/4, 2/6, 2/4 (thus binary loses)
Smoothness s_{max} of $\tilde{\phi}(t)$ and $\phi(t)$: 1.4/2.1, −2.2/5.5, 0.59/2.44.

The largest non–special eigenvalues of the matrix T for the binary h_0 and f_0 were 0.4394 and 0.0339. The smoothness $s_{max} = 0.59$ and 2.44 came direct ly from Eq. (3.8).

The largest non–special eigenvalue for the foolish choice h_0, biorthonormal to the quintic B–spline filter, was 21.314. The function $\tilde{\phi}(t)$ is a wild distribution (−2.2 derivatives in L_2). This example is *not* to show that splines are a poor choice—they are often very good. But we must keep enough zeros at $z = -1$ in both $H_0(z)$ and $F_0(z)$. When f_0 with binomial coefficients gives a spline, stability and smoothness may require a longer h_0 than would be needed for perfect reconstruction.

A third important quantity is the *coding gain*, to give the expected compression for inputs that have Markov correlation 0.95 between neighboring pixels. None of those measures is totally consistent with human visual perception. Therefore filters are generally chosen for good looks on well–known images.

We were able to compare the FBI 9/7 with the binary 9/7 on a "boats" image [1].

25:1 compression			50:1 compression	
FBI	binary		FBI	binary
40.55	40.14	MEAN SQUARE ERROR	84.55	85.92
32.05	32.10	PEAK SIGNAL/NOISE RATIO	28.86	28.79
43.71	45.07	MAXIMUM ERROR	68.81	81.99
OK	better	PERCEPTUAL QUALITY	OK	better

Note the slight advantage of the FBI, objectively. Note the equally slight advantage of the binary filters, subjectively. We believe that the binary property and the interpolation property may be significantly useful in applications.

Acknowledgement I would like to add a less conventional tribute to Ron Mitchell by thanking him for teaching my sons to play "football".

References

1. Gilbert Strang and Truong Nguyen, *Wavelets and Filter Banks*, Wellesley–Cambridge Press (1996).

2. Albert Cohen and Ingrid Daubechies, A stability criterion for biorthogonal wavelet bases and their related subband coding schemes, Duke Math. J., 68:313–335 (1992).
3. Lars Villemoes, Wavelet analysis of refinement equations, SIAM J. Math. Anal., 25:1433–1466 (1994).
4. Gilbert Strang, Eigenvalues of $(\downarrow 2)H$ and convergence of the cascade algorithm, IEEE Trans. on Signal Processing 44 (1996) 233-238.
5. Wim Sweldens, The lifting scheme: a custom–design construction of bior thonormal wavelets, Appl. Comput. Harm. Anal., (To appear).

Signal Analysis And Synthesis With 1-D Quasi-Continuous Wavelet Transform

Stéphane H. Maes *

IBM, T. J. Watson Research Center,
Human Language Technologies, Acoustic Processing Department,
Room J2-N52, P.O. Box 704, Yorktown Heights, NY 10598, USA.
Phone: +1-(914)-784-6376
Fax: +1-(914)-784-6534
e-mail: smaes@watson.ibm.com

Summary. The wavelet transform is a widely used time-frequency tool for signal processing. Because continuous versions are time-consuming, restricted to some types of wavelets or too approximate, the use of wavelets in signal processing is usually limited to discrete-time critically sampled transforms. This paper formalizes the quasi-continuous wavelet transform. A fast quasi-continuous algorithm for analysis is proposed. It is valid for almost any generating analysis wavelet. Thereafter, a synthesis algorithm is described with filter bank implementations. It has the peculiarity to give the same weight to each point of the time-scale plane as required for selective reconstructions of portions of this plane.

Keywords: quasi-continuous wavelet transform, filter design, fast analysis algorithms, fast synthesis algorithms.

1. Introduction

This paper is part of a global program which attempts to provide solutions to some of the drawbacks of the continuous *WT* (*CWT*) [20, 21, 19, 16, 17, 15, 8]: it introduces the concept of quasi-continuous wavelet transform (*QCWT*) and presents fast algorithms for implementation and inversion.

2. The wavelet transform

It is useful to briefly present the notations for the *WT* which are used in this paper. The *WT* of a signal $f(t)$ using a generating analysis wavelet $g(t)$ is defined as:

$$\mathcal{W}T_f(a,b,g) = < g_{(a,b)} | f > \qquad (2.1)$$

* This work was performed when Dr. S. Maes was a Research Assistant of the Belgian National Fund for Scientific Research (*FNRS*). The author was affiliated to *CAIP* Center, Rutgers University, Piscataway, NJ 08855, USA and Unités *TELE* and *FYMA*, Université Catholique de Louvain, B-1348 Louvain-la-Neuve, Belgium. This work was done in collaboration with Prof. I. Daubechies, Princeton University.

where,

$$g_{(a,b)}(t) = \frac{1}{\sqrt{a}} g(\frac{t-b}{a}) \qquad (2.2)$$

$a \in I\!R_+ \setminus \{0\}$ is the scale or dilation parameter and $b \in I\!R$ is the translation parameter[1]. a is often related to the inverse of the frequency while b can be considered as the time variable. $< .|. >$ is the classical functional inner product. The WT at a given scale level a appears as the projectio n of $f(t)$ over the set of translated versions of $g_{(a,0)}(t)$[2].

Any signal $f(t)$ in a satisfying functional Hilbert space $\mathcal{H}$ can be reconstructed from its WT, using a generating analysis wavelet $h(t)$. This operation is called *resolution of the identity*,

$$\forall f(t) \in \mathcal{H} \ : \ f(t) = \frac{1}{C_{g,h}} \int_{-\infty}^{+\infty} \int_{0}^{+\infty} h_{(a,b)}(t) \ \frac{db \ da}{a^2} \ WT_f(a,b,g) \quad (2.4)$$

and it is subject to the following *admissibility condition* [3]:

$$0 < \left| C_{g,h} = \int_0^{+\infty} \frac{\hat{h}(\omega)\hat{g}^*(\omega) \ d\omega}{\omega} = \int_0^{+\infty} \frac{\hat{h}(-\omega)\hat{g}^*(- \ omega) \ d\omega}{\omega} \right| < +\infty \quad (2.5)$$

Note that these equations are written for the continuous biorthogonal case: the generating analysis wavelet $g(t)$ can be different from the generating synthesis wavelet [12].

3. Definition of the lquasi-continuousl wavelet transform

The $QCWT$ is a discrete time transform with no downsampling along the time axis and the possibility to choose any sampling grid along the scale axis. Actually, a very useful sampling grid for analysis purposes is defined by

---

[1] In the literature, another convention is also often used where $a \in I\!R \setminus \{0\}$.

[2] The projection of $f(t)$ on the space spanned by the set $\{e_k(t)\}_k$ is defined as: $\{P_{\{e_k(t)\}}\{f(t)\}\}_k \triangleq \{< e_k|f >\}_k$. When the term "projection" is used instead of orthogonal projection, the reader should always use this definition. It is the series of lengths of the orthogonal projections of $f(t)$ on $e_k(t)$, multiplied by the length of $e_k(t)$. The absence of normalization by the length of $e_k(t)$ allows to select different normalizations for the wavelet transform . For example, some authors use another normalization for $g_{(a,b)}(t)$:

$$\breve{WT}_f(a,b,g) = \frac{1}{\sqrt{a}} < g_{(a,b)}|f > \qquad (2.3)$$

[3] $\hat{g}(\omega)$ stands for the Fourier transform of $g(t)$. It is normalized as: $\hat{g}(\omega) \triangleq \int_{-\infty}^{+\infty} g(t) \frac{e^{-j\omega t}}{\sqrt{2\pi}} \ dt$

$$a = a_0^{-m - \frac{\epsilon}{n_{so}+1}}, \quad a_0 > 0, a_0 \neq 1, \quad m \in \mathbb{Z}, \quad \epsilon = 0, \cdots, n_{so} \quad (3.1)$$

where $n_{so} \in I\!N \setminus \{0\}$.

In the *QCWT* case, a_0 is usually chosen equal to 2. In this particular case, the different levels, numbered by the parameter m, are arranged in octaves and the sub-levels are arranged in fractions of octaves and called sub-octaves or sub-subbands. For other choices of the value of a_0, the same denomination is maintained. However these choices are of very little interest, as the sampling grid (3.1) should always give enough resolution in scale.

If $a_0 \neq 2$, the generating analysis wavelets are usually not dilated version of a same generating analysis wavelet. Hence, it creates an additional difficulty of interpretation of the time-scale plane, unless it can be transformed in a representation independent of the generating analysis wavelet, as discussed in [17, 15]. The fast *QCWT* proposed in this paper can easily be extended to any choice of a_0.

4. A filter bank-based fast quasi-continuous *WT* algorithm

This section is devoted to our new algorithm for fast computation of the *WT*. It has the advantage to be a natural extension of wavelet subband coding schemes [28, 1, 30], with an "à trous" flavor [29, 27, 26]. It is implemen table for any generating analysis wavelet [4].

Thereafter, the coefficients of the wavelet transform, associated to the first octave and its sub-octaves, are computed with a bandpass filter bank applied to this projected signal. This projected signal is also re-projected on the set *scaling function* at the next octave level by lowpass filtering[5]. Then, the algorithm is iterated with dilated filters.

The approach is very similar to subband coding schemes. The use of the intermediate *scaling functions* facilitates the design of the filters.

In order to limit the size of the filters and guarantee stability, it is not po ssible to chose any dominant frequency for the generating analysis wavelet. Hence, to cover completely the bandwidth of the signal, it is often mandatory to add an upper-octave interpolation, where the *QCWT* is computed as if the signal was sampled on $\frac{1}{2}\mathbb{Z}$ instead of $\mathbb{Z}$. This is an important difference from the continuous *WT*. It results from the fact that *QCWT* start at one level and use dilated analysis wavelets but no compressed analysis wavelets[6].

---

[4] In what follows, the generating analy sis wavelet is denoted by $\psi(t)$.

[5] The lowpass filter defines the generating *scali ng function*.

[6] In other words, the dilation factors take values larger than a_0 but never smaller values. The up per-octave interpolation is an exception to this statement.

4.1 Details of the algorithm

In a first phase, the signal is projected on the intermediate set of functions called *scaling functions*. A priori, any set of functions which have the properties of *scaling functions*, as described in [23, 5], can be c hosen. In practice, their choices depend essentially on the generating analysis wavelet.

In digital signal processing, a discrete signal is supposed to result from sampling a bandlimited continuous signal with a sampling frequency larger than the associated Shannon frequency. In order to guarantee that the *WT* is computed for the same continuous function as the one which has produced the inp ut sample series, a Whittaker interpolation is performed on the signal before its projection on the set of *scaling functions* [13].

Suppose that the initial level of decomposition of the associated discrete time *WT* is V_0 and that the continuous signal is bandlimited to the interval $[-\pi, \pi]$, after normalization of the frequency.

The interpolation is performed with the filter of impulse response $\beta_\phi(k)$

$$\beta_\phi(k) \; = < sinc(t)|\phi(t-k) > \tag{4.1}$$

Knowing the sequence of samples associated to the signal $f(t)$, the projection of $f(t)$ on the set of translated versions of the generating *scaling function* is obtained as output of the interpolating fi lter $\beta_\phi^*(k)$. This results from the development

$$< f(t)|\phi(t-n) > = \sum_m f(m) < sinc(t-m)|\phi(t-n) > \tag{4.2}$$

or

$$< f(t)|\phi(t-n) > = \sum_m f(m)\beta_\phi(n-m) \tag{4.3}$$

Although derived independently, this result is similar to Shensa's algorithm [26, 27].

The projection on the *scaling function* at the initial dilation level is denoted $\{P_{\phi(t-k)}f\}(k)$. It is obtained by filtering the di screte signal[7]:

$$\{P_{\phi(t-k)}f\}(k) \;\stackrel{\Delta}{=}\; < \phi(t-k)|f(t) > = \;(f \;*\; \beta_\phi^*)(k) \tag{4.4}$$

t is a mute parameter of integration.

Filtering with $\beta_\phi(k)$ produces the exact projection of the initial signal into the subspace generated by all the translations, with a translation step which is a multiple of the sampling period[8], of th e *scaling functions* at the initial dilation level.

Whittaker interpolation is applied on $\phi(t)$, which can be chosen with a compact support or at least a fast decay [5, 4, 7]. Therefore, the interpolation

---

[7] The reader should notice the definition of projection th at will be repeatedly used in the rest of this paper. It is in agreement with the preceding convention.

[8] The sampling period is assumed to be 1 in this paper.

followed by the projection on the set of *scaling functions* can be approximated, up to any accuracy, with short *FIR* filters. The fast decay of the *scaling function*, accelerates the decay of the coefficients of $\beta_\phi(k)$. It is important that the reader well understands that the $sinc(t)$ interpolation is introduced only to compute the correct projection and not because of any behavior of the coefficients of interpolation . Indeed, the decay of Whittaker interpolation coefficients is rather poor. However, the glob al decay of the interpolating filter is governed by the regularity of the *scaling function* [18].

In this paper, the *scaling function* is generated by

$$\hat{G}_0(\omega) = \frac{(1+cos(\omega))(cos(\frac{2\pi}{3}) - cos(\omega))(cos(\frac{3\pi}{4}) - cos(\omega))}{2(cos(\frac{2\pi}{3}) - 1)(cos(\frac{3\pi}{4}) - 1)} \tag{4.5}$$

Note that the *scaling function* is compact as $\hat{G}_0(\omega)$ is a *FIR* filter.

Once $\{P_{\phi(t-k)}f\}(k)$ is known, the projection of the signal on the set of translated wavelets at the same level is obtained by filtering with the filter of impulse response $g_1^*(k)$. The $g_1(k)$ are the coefficients of decomposition of the generating wavelet into the set of translated *scaling functions* at the same level of dilation,

$$< \phi(t-l)|\psi(t) > = \sum_k g_1(k) < \phi(t-l)|\phi(t-k) > = g_1(l) < \phi(t)|\phi(t) > \tag{4.6}$$

The generating *scaling function* is supposed normalized.

By analogy with (4.4), the projection of the signal on the set of translated wavelets at the same level is denoted by

$$\{P_{\psi(t-k)}f\}(k) = < \psi(t-k)|f(t) > = (\{P_{\phi(t-.)}f\} * g_1^*)(k) \tag{4.7}$$

The dot denotes the mute parameter of the convolution.

In order to shorten the filters required by our algorithm, we impose as much regularity as possible on the $\phi(t)$ and $\psi(t)$. Usually, the generating analysis wavelet is quite regular. In such a case, $\hat{G}_1(\omega)$ can easily be approximated with a short *FIR* filter. If the regularity of $\psi(t)$ is not satisfying enough, a solution consists into increasing the regularity of $\phi(t)$. The reader should note that this is the main argument to justify our use of *scaling functions* [18, 17].

The following Morlet wavelet is considered as generating analysis wavelet [11, 5],

$$\psi(t) = e^{\frac{-t^2}{16}}(e^{j\,0.69\,\pi\,t} - 4.89098\;10^{-4}) \tag{4.8}$$

The choice of a complex generating analysis wavelet is motivated in [17, 20, 21, 8], where a *WT* phase anal ysis is required. Because $\psi(t)$ is complex, the resulting filter banks are also complex. Real cases are handled similarly.

To actually reduce the number of non-negligible coefficients, the generating analysis wavelet must be finely tuned, by dilation, until the minimum amount of non-negligible coefficients is achieved while still giving an acceptable approximation. (4.8) is the result of such a fine tuning.

However, this does not completely justify the choice (4.8), which is rather a compromise, for the filter (4.5), between the minimum number of non-zero coefficients and the requirements of having a good coverage of the upper part of the bandwidth by the generating analysis wavelet. In the absence of this constraint, which can perfectly be relaxed provided that an upper-octave interpolation is performed as described later, the number of non-negligible coefficients can be reduced even more.

Any additional sub-subband, located between the current level of dilation and the next octave, is obtained by defining additional filters $g_{\epsilon\,+\,1}(k)$ applied to the same coefficients of projection of the signal as the filter $g_1(k)$: $\left\{P_{\phi(t-k)}f\right\}(k)$.

The filters $g_{\epsilon\,+\,1}(k)$ are designed by

$$\begin{aligned} < \phi(t-l)|\frac{1}{\alpha(\epsilon)}\psi(\frac{t}{\alpha(\epsilon)}) > &= \sum_k g_{\epsilon\,+\,1}(k) < \phi(t-l)|\phi(t-k) > \\ &= g_{\epsilon\,+\,1}(l) < \phi(t)|\phi(t) > \\ \alpha(\epsilon) = a_0^{\frac{\epsilon}{n_{so}\,+\,1}} &, \quad n_{so} \in \mathbb{N},\ \epsilon = 1,\ 2,\ \cdots,\ n_{so} + 1 \end{aligned}$$

where n_{so} is the number of sub-levels between two consecutive octave levels, $\epsilon = 0$ is associated to $g_1(k)$ and $\epsilon = n_{so} + 1$ is associated to the projection on the next octave level.

This choice for the sub-octave wavelets is independently proposed in [26]. However, just as for the discrete dilation operator based fast *WT* algorithm [10], the sub-octaves are obtained with an "octave-by-octave" approach: a similar "à trous" algorithm is applied on each of the $n_{so} + 1$ generating wavelets. This approach has one major drawback: the numerical accuracy which can be achieved is limited by coarser approximations of the filters.

The projection of the signal on the set of translated wavelets at the same level is denoted by

$$\left\{P_{\frac{1}{\alpha(\epsilon)}\psi(\frac{t-k}{\alpha(\epsilon)})}f\right\}(k) = (\{P_{\phi(t-.)}f\} * g^*_{\epsilon\,+\,1})(k) \qquad (4.9)$$

Note that convention (2.3) is used: the *WT* computed by this algorithms is normalized with wavelets multiplied by $\frac{1}{a}$ instead of $\frac{1}{\sqrt{a}}$ [17, 16]. This is more appropriate for the speech analysis application which motivates this work [17, 20, 21, 19, 8]. If convention (2.1) is preferred, each subband must be accordingly renormal ized. In other words, each filter $g_{\epsilon\,+\,1}(k)$ must be accordingly re-normalized.

The next octave level, $\left\{P_{\frac{1}{a_0}\phi(\frac{t-k}{a_0})}f\right\}(k)$ is obtained with the filter $g_0(k)$ applied to the coefficients of projection of the signal into the *scaling functions* at the current dilation level $(\left\{P_{\phi(t-k)}f\right\}(k))$,

$$\left\{P_{\frac{1}{a_0}\phi(\frac{t-k}{a_0})}f\right\}(k) = (\{P_{\phi(t-.)}f\} * g^*_0)(k) \qquad (4.10)$$

The projection on the set of wavelets at this new octave level can be obtained by two different ways,

$$\left\{P_{\frac{1}{a_0}\psi(\frac{t-k}{a_0})}f\right\}(k) = (\left\{P_{\frac{1}{a_0}\phi(\frac{t-\cdot}{a_0})}f\right\} * g_1^*(\frac{\cdot}{2}))(k) \tag{4.11}$$

or,

$$\left\{P_{\frac{1}{a_0}\psi(\frac{t-k}{a_0})}f\right\}(k) = (\{P_{\phi(t-\cdot)}f\} \ * \ g^*_{n_{so}\,+\,2})(k) \tag{4.12}$$

As $\hat{\psi}(a_0\omega)$ is better covered by $\hat{\phi}(\omega)$ than by $\hat{\phi}(a_0\omega)$, its approximation is more accurate when developed in the set of translated of $\phi(t)$ than of $\frac{1}{a_0}\phi(\frac{t}{a_0})$. Therefore, at the price of a longer filter, the numerical accuracy is improved by choosing approach (4.12).

The next sub-octave levels are obtained, by application of the discrete filter bank $g^*_{\epsilon\,+\,1}(\frac{k}{a_0})$ to $\left\{P_{\frac{1}{a_0}\phi(\frac{t-k}{a_0})}f\right\}(k)$.

Thereafter, all the following octaves and sub-octaves are obtained by iteration of dilated versions of the filter bank

$$g_{\epsilon\,+\,1}(k), \quad \epsilon \neq 0, \quad \epsilon = -1,\, 1,\, 2,\, \cdots,\, n_{so} + 1 \tag{4.13}$$

This part is similar to the "à trous" approach.

4.2 The upper-octave interpolation

It has been explained in the previous section that a compromise must exist between the coverage, by the generating analysis wavelet, of the high frequency content of the signal and the length of the filter bank $g^*_{\epsilon\,+\,1}(k)$.

However, even if any filter length is acceptable, part of the high frequency content can not be validly covered with the generating analysis wavelet. Unfortunately, for analysis it is often important to correctly represent the high frequencies.

The solution consists in the implementation of an *upper-octave interpolation*: the signal is interpolated and projected on the set of translated version of $2\phi(2t)$, with a Whittaker interpolation on the half-integers, followed by application of the contracted filter bank

$$g_{\epsilon\,+\,1}(2k), \quad \epsilon \neq -1, \quad \epsilon \neq n_{so} + 1 \tag{4.14}$$

Thereafter, the signal is downsampled, to keep only the integer samples, and to remain on the same sampling grid as all the other levels and sub-levels.

In detail, the signal is interpolated and projected on the upper-octave *scaling functions*, with a filter $\tilde{\beta}_\phi(k)$,

$$\tilde{\beta}_\phi(k) \ = \ < sinc(t)|a_0\phi(a_0(t-k)) > \tag{4.15}$$

$$< f(t)|a_0\phi(a_0(t-n))) > \ = \ \sum_m f(m)\tilde{\beta}_\phi(n-m) \tag{4.16}$$

$$\left\{P_{a_0\phi(a_0(t-k))}f\right\}(k) = (f * \tilde{\beta}^*_{\phi})(k) \tag{4.17}$$

$$\begin{aligned}\left\{P_{\frac{a_0}{\alpha(\epsilon)}\psi(\frac{a_0(t-k)}{\alpha(\epsilon)})}f\right\}(k) &= (\left\{P_{a_0\phi(a_0(t-.))}f\right\} * g^*_{\epsilon+1}(a_0.))(k) \\ \alpha(\epsilon) &= a_0^{\frac{\epsilon}{n_{so}+1}},\ n_{so} \in I\!N,\ \epsilon = 1, 2, \cdots, n_{so}\end{aligned}$$

The complete fast $QCWT$ algorithm is schematically depicted in figure 4.1.

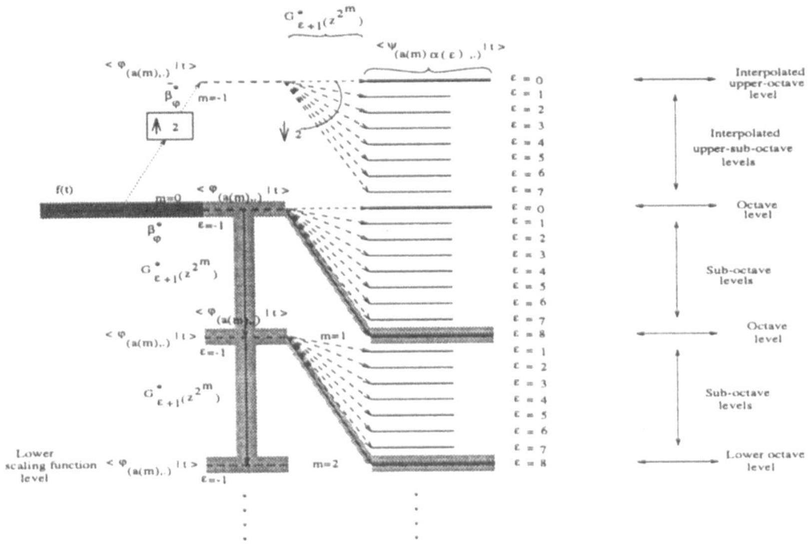

Fig. 4.1. Fast analysis quasi-continuous wavelet transform algorithm. $n_{so} = 7$, $a_0 = 2$. For the sake of simplicity, $n_{oct} = 2$. The light-gray shaded parts indicates the associated "à trous" algorithm. The dark gray shaded part indicates the additional Whittaker interpolation which can be associated to any subband coding scheme.

4.3 Performances

The complexity of the algorithm is:

$$\mathcal{O}_{comp}(N((n_{oct} + d_{ui})(n_{so} + 1) + 1)) \tag{4.18}$$

where N is the length of the input signal, n_{oct} is the number of octave blocks (octave and sub-octaves) used and n_{so} is the number of sub-octaves between two successive octaves. d_{ui} is equal to 2 if there is an upper level interpolation and it is equal to 1 otherwise.

Because of the iterative filter bank approach, the number of operations is kept to a minimum. The *QCWT* design is now replaced by a filter bank design, which is a problem far more easy to optimize.

Furthermore, because of the filter bank concept, this approach is well suited for hardware (VLSI) and parallel or pipe-lined implementations. In case of a fully parallel implementation, it loses its *"à trous"* properties.

Also, because of the "à trous" flavor of the algorithm, it can be implemented with an interleaved subband coding "à trous" algorithm (*ISCT A*) approach [26, 17, 18].

5. The fair synthesis *QCWT* algorithm

Depending on the application, different strategies can be developed to implement a *reconstruction of the identity*. In some cases, it is only required to reconstruct the signal with the *WT* coefficients, independently of their origins in the time-scale plane. In other cases, some regions of the time-scale plane are selected and only the associated portions of the signal are desired. This strategy is called the Selective Fusion Algorithm (*SFA*) [17, 20, 21, 19, 16].

Different alternatives are presented in [17, 18].

This algorithm is based on an approach similar to subband coding synthesis schemes. Equal weights are given to each of the levels and sub-levels.

The design of the filters results from the structure of the following recurrent equations.

Let

$$
\begin{aligned}
< \phi(t) \quad | \quad \breve{\mathcal{D}}_{n_{so}+3}\phi(t - l) > &\triangleq < \phi(t)|\phi(t - l) > \\
&- \sum_k h_{n_{so}+3}(k - l) \; < \phi(t)|\frac{1}{a_0}\phi(\frac{t - k}{a_0}) >
\end{aligned}
\tag{5.1}
$$

where $h_{n_{so}+3}(n)$ minimizes,

$$
| < \phi(t)|\phi(t) - \sum_k h_{n_{so}+3}(k) \, \frac{1}{a_0}\phi(\frac{t - k}{a_0}) > |
\tag{5.2}
$$

(5.2) represents the orthogonal projection on $\phi(t)$ on the space s panned by $\{\frac{1}{a_0}\phi(\frac{t-k}{a_0})\}_k$, which is an over-complete set [6]. Therefore, $\{h_{n_{so}+3}(k)\}_k$ is chosen to minimize the amount of non-negligibl e coefficients. This can easily be implemented with a matching pursuit strategy with $\{\frac{1}{a_0}\phi(\frac{t-k}{a_0})\}_k$ as dictionary [22, 24, 25].

Similarly,

$$
\begin{aligned}
< \phi(t) \quad | \quad \breve{\mathcal{D}}_{\epsilon+1}\phi(t - l) > &\triangleq < \phi(t)|\breve{\mathcal{D}}_{\epsilon+2}\phi(t - l) > \\
&- \sum_k h_{\epsilon+2}(k - l) \; < \phi(t)|\frac{1}{\alpha(\epsilon)}\psi(\frac{t - k}{\alpha(\epsilon)}) > \\
\alpha(\epsilon) &= a_0^{\frac{\epsilon}{n_{so}+1}}, \; n_{so} \in I\!N, \; \epsilon = 0, 1, \cdots, n_{so} + 1
\end{aligned}
\tag{5.3}
$$

where $h_{\epsilon+2}(n)$ minimizes,

$$| < \phi(t)|\breve{\mathcal{D}}_{\epsilon+2}\phi(t) - \sum_k h_{\epsilon+2}(k) \frac{1}{\alpha(\epsilon)}\psi(\frac{t-k}{\alpha(\epsilon)}) > | \tag{5.4}$$

Once again, we want to minimize the amount of non-negligible coefficients. Now, we can use either matching pursuit or when $\epsilon \leq n_{so}$, we can seed an optimization procedure for $\{h_{\epsilon+2}(k)\}_k$ using $\{h_{\epsilon+3}(k)\}_k$.

The same formulas can be extended on a half-integer grid for the upper-octave. In which case, $\breve{\mathcal{D}}_1$ is obtained from the upper-octave as illustrated in figure 5.1. $\breve{\mathcal{D}}_1$ is not applied on the lower octave.

The operators $\breve{\mathcal{D}}_\eta$ with $\eta = 1, 2, \cdots, n_{so} + 3$ fairly share the extracted information used to reconstruct $\{P_{\phi(t-k)}f\}(k)$ among the different sets of coherent states associated to the octave which means the scale levels which hav e been obtained in the fast $QCWT$ by direct application of the filter bank $g_{\epsilon+1}(k)$.

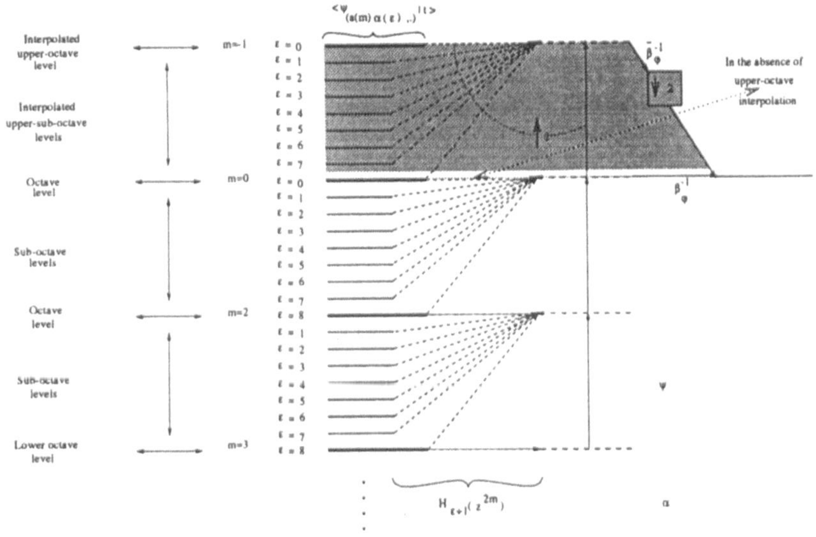

Fig. 5.1. Fast synthesis quasi-continuous wavelet transform algorithm, b ased on the "fair strategy". The gray shaded parts is a sequence of operation which is implemented only if an upper-octave interpolation is used. $n_{so} = 7$, $a_0 = 2$. For the sake of simplicity, $n_{oct} = 2$.

The filter banks used for synthesis are defined by h_η with $\eta = 1, 2, 3, \cdots, n_{so} + 3$.

The application of the filters $h_{n_{so}+3}(k)$ and $h_{\epsilon+1}(k)$ respectively to $\left\{P_{\frac{1}{a_0}\phi(\frac{t-k}{a_0})}f\right\}(k)$ and $\left\{P_{\frac{1}{\alpha(\epsilon)}\psi(\frac{t-k}{\alpha(\epsilon)})}f\right\}(k)$ followed by summation of the resulting sequences produces the sequence $\{P_{\phi(t-k)}f\}(k)$ But in fact, the sequence $\left\{P_{\frac{1}{a_0}\phi(\frac{t-k}{a_0})}f\right\}(k)$ results from applying the same process with the same filters dilated by a_0. And the iteration is initiated at the lowest level, by the knowledge of the projection of the signal over the set of the translated versions of the dilated *scaling function*, which often is set to zero in an analysis context. Usually, the consequence is merely a loss of the average value of the signal. But of course, nothing prevents the storage of these coefficients.

For the last non-interpolated octave, if no upper-octave interpolation is used, the filter $h_1(k)$ has also to be applied to $\{P_{\psi(t-k)}f\}(k)$ In case of an upper-octave interpolation, $h_1(k)$ is not used but rather replaced by repeating the same process on the u pper-octave but on a half-integer sampling grid.

The last step results, if no upper-interpolation is used, from equation (4.2). The filter $(\beta_\phi)^{-1}(k)$ is applied to $\{P_{\phi(t-k)}f\}(k)$ to restore the initial continuous input:

$$\sum_{k\in\mathbb{Z}} \frac{\{P_{\phi(t-k)}f\}(k)}{<\phi(t-k)|\,phi(t-k)>}\varphi(t-k) \tag{5.5}$$

where $\varphi(t)$ is the result of bandlimiting $\phi(t)$ to $[-\pi,\pi]$. It is in that sense that notation $(\beta_\phi)^{-1}(k)$ must be understood.

With an upper-octave interpolation, the signal is reconstructed with a similar strategy applied on the upper-octave , which replaces the contribution of $\breve{\mathcal{D}}_1$.

The fair synthesis *QCWT* algorithm is summarized in figure 5.1.

The computational complexity of the synthesis algorithm is almost the same as for the analysis process,

$$\mathcal{O}_{comp}(N((n_{oct}\ +\ d_{ui})(n_{so}\ +\ 2)\ +d_{lsf})) \tag{5.6}$$

with the same notations as for equation (4.18) and d_{lsf} equal to 1 if the lower level of *scaling function* level is taken into account for the synthesis, 0 otherwise.

Knowing the synthesis filter bank, the signal can be reconstructed with the synthesis *ISCTA* [9].

Actually, this strategy encompasses the discrete time *WT* which is invertible [17]. The discrete time *WT* is particular limit case of the combination of the fast *QCWT* algorithm and the fair synthesis *QCWT* algorithm.

[9] If the generating synthesis wavelet is different from the generating analysis wavelet, the same process is directly applied on the synthesis wavelets and, at the end, the final sequence is correctly renormalize d by the ratio of the constants of admissibility.

6. Conclusion

This paper proposes different filter bank-based algorithms for *QCWT* analysis and synthesis of the resulting time-scale plane. It is shown that the design of the filter bank is systematic for any generating analysis or generating wavelet. The use of intermediate *scaling functions* is the main design improvement proposed by our approach.

The example which is presented also provides filters for complex subband coding. Indeed the method proposed for analysis as well as synthesis can be used in the particular case of critically sampled discrete time *WT*.

The solution proposed for reconstruction offers the possibility to selectively synthesize the time-scale plane.

Different examples of analysis and reconstruction using the "fair synthesis" strategy can be found in [20, 21, 19, 17, 16, 8]. In these papers, the interest of complex filtering is illustrated for extraction of phase-derived features. The reader will also find an introduction to the concepts of synchrosqueezing measures which provide representations essentially independent of the generating analysis wavelet.

This paper provides the tools necessary to design and implement on-line *CWT*. The algorithm are also compatible with hybrid *WT* [15], however, it is not designed for easy on-line tuning or adaptation to a signal[10].

Acknowledgments

The author's pre-eminent thanks to his research advisor, Professor I. Daubechies, for her help, encouragement and guidance during this research. Special thanks to Professor R. Mammone and Professor J. L. Flanagan for their comments and suggestions and for welcoming the author in the speech recognition team at *CAIP*, Rutgers University, where most of this research was performed. Eventually the author wants to give thanks, many thanks, to his thesis advisors, Professor J-P. Antoine and Professor P. Delogne.

The author is also grateful to the members of the departments *TELE* and em FYMA at *UCL*, *CAIP* at Rutgers University, the speech research de partment at *AT&T* Bell Laboratories and the department of human language technolog ies at *IBM* T. J. Watson Research Center for their continuous assistance and friendship.

[10] On-line tuning or adaptation seems challenging because all the design process would have to be repeated on-line. However, it is probably feasible to adapt the algorithm accordingly. This subject is definitively worth investigating.

References

1. A. N. Akansu and R. A. Haddad. *Multiresolution signal decomposition: transforms, subbands, and wavelets.* Academic Press, 1992.
2. Ch. K. Chui, editor. *Wavelet analysis and its applications: a tutorial in wavelet theory and applications.* Academic Press, 1992.
3. J.M. Combes, A. Grossmann, and Ph. Tchamitchian, editors. *Wavelets, time-frequency methods and phase space.* Springer-Verlag, 1989. Proc. of the Int. Conf., Marseille, France, December 14-18, 1987.
4. I. Daubechies. Orthonormal bases of compactly supported wavelets. *Com. Pure Appl. Math.*, 41:pp. 909–996, 1988.
5. I. Daubechies. *Ten lectures on wavelets.* Number 61 in CBMS-NSF Regional Conference Series in Applied Mathematics. SIAM, Philadelphia, PA, 1990.
6. I. Daubechies. The wavelet transform, time-frequency localization and signal analysis. *IEEE Trans. Info. Theor.*, 36(5):pp. 961–1005, 1990.
7. I. Daubechies. Orthonormal bases of compactly supported wavelets II: variations on a theme. *SIAM J. Math. Anal.*, 24:pp. 499–519, 1993.
8. I. Daubechies and S. Maes. Wavelets in medicine and biology. In A. Alroubi and M. Unser, editors, *A nonlinear squeezing of the continuous wavelet analysis based on auditory nerve models.* CRC Press, April 1996.
9. P. Delogne. *Signaux de télécommunication, théorie des fonctions aléatoires.* Lecture notes: ELEC 2880, Université Catholique de Louvain, Louvain-la-Neuve, Belgium, 1988.
10. P. Dutilleux. An implementation of the "algorithme à trous" to compute the wavelet transform. In Combes et al. [3]. Proc. of the Int. Conf., Marseille, France, December 14-18, 1987.
11. A. Grossmann, R. Kronland-Martinet, and J. Morlet. Reading and understanding continuous wavelet transforms. In Combes et al. [3]. Proc. of the Int. Conf., Marseille, France, December 14-18, 1987.
12. M. Holschneider. General inversion formulas for wavelet transforms. *J. Math. Phys.*, 34(9):pp. 4190–4198, September 1993.
13. A. Laloux and P. Delogne. *Traitement numérique du signal.* Lecture notes: ELEC2900, Université Catholique de Louvain, Louvain-la-Neuve, Belgium, 1987.
14. P. G. Lemarié, editor. *Les ondelettes en 1989.* Lecture notes in mathematics - volume 1438. Springer-Verlag, 1990.
15. S. Maes. *The generalized hybrid wavelet transform yields a new reading of subband coding decompositions.* PhD thesis, Université Catholique de Louvain, Louvain-la-Neuve, Belgium, June 1994. (Annex PhD Dissertation).
16. S. Maes. The synchrosqueezed plane representation, a new time frequency representation applied to speaker identification. Technical report, CAIP Center - Rutgers University, The State Univeristy of New Je rsey, New Brunswick, NJ, June 1994.
17. S. Maes. *The wavelet tranform in signal processing, with application to the extraction of the speech modulation model features.* PhD thesis, Université Catholique de Louvain, Louvain-la-Neuve, Belgium, June 1994.
18. S. Maes. Fast quasi-continuous wavelet algorithms for analysis and synthesis of 1-D signals. *preprint submitted to SIAM J. Applied Math.*, April 1995.
19. S. Maes. The synchrosqueezed representation yields a new reading of the wavelet transform. In H. H. Szu, editor, *Proc. SPIE 1995 on OE/Aerospace Sensing and Dual Use Photonics - Wavelet Applications for Dual Use - Session*

on Acoustic and Signal Processing, Wavelet Applications II, volume 2491, pages 532–559, Orlando, FL, April 1995. Part I.
20. S. Maes. The wavelet-derived synchrosqueezed plane representation yields a new time-frequency analysis of 1-D signals, with application to speech. *preprint submitted to IEEE Trans. Signal Processing*, April 1995.
21. S. Maes. The wavelet-derived synchrosqueezed plane representation yields new front-ends for automatic speech recognition. *preprint submitted to IEEE Trans. Speech and Audio Processing*, April 1995.
22. S. G. Mallat. Computation with adapted time-frequency atoms. Viewgraphs presented at the conference Wavelet and Applications, Toulous e, June 08-13, 1 992.
23. S. G. Mallat. Multiresolution approximation and wavelets. *Trans. Am. Math. Soc.*, 315:pp. 69–88, 1989.
24. S. G. Mallat and Z. Zhang. Local time/frequency multilayer orthogonal transforms. *preprint*, 1992.
25. S. G. Mallat and Z. Zhang. Matching pursuits with time-frequency dictionaries. *IEEE Trans. Sig. Proc.*, 41(12):pp. 3397–3415, 1993.
26. O. Rioul and P. Duhamel. Fast algorithms for discrete and continuous wavelet transforms. *IEEE Trans. on Inf. Theory*, 38(2 Part II):pp. 569–586, 1992.
27. M. Shensa. Affine wavelets: wedding the Atrous and the Mallat algorithms. *IEEE Trans. Ac. Speech. Proc.*, 40(10):pp. 2464–2482, October 1992.
28. P. P. Vaidyanathan. *Multirate systems and filter banks*. Prentice-Hall, Englewood Cliffs, NJ, 1993.
29. M. Vetterli. Wavelet and signal processing, tutorial session II. In *Time-frequency and time-scale analysis*, Proc. IEEE-SP, Victoria, Canada, 1992. The IEEE Signal Processing Society.
30. M. Vetterli and J. Kovačević. *Wavelets and subband coding*. Prentice Hall, Englewood Cliffs, NJ, 1995.

Adaptive Wavelet Collocation for Nonlinear BVPs

Silvia Bertoluzza and Paola Pietra

Istituto di Analisi Numerica del C.N.R., Pavia (Italy)

Abstract

An adaptive collocation method based on interpolating wavelets for solving nonlinear boundary value problems is introduced. The method is tested on an Euler/Poisson system arising in semiconductor modelling.

Key-Words:

Interpolating Wavelets, Nonlinear BVP, Collocation, Semiconductors

1 INTRODUCTION

One of the difficulties that one encounters in the application of wavelet based adaptive schemes to the numerical solution of partial differential equations is evaluating the action of nonlinear terms.

If the nonlinearity is of multilinear type (as uu_x in Burgers equation) then one can apply the results of [DM], which allow to take such terms into account in a Galerkin scheme. When the nonlinearity is more complex, one has to pass through the physical space, by means of some approximate procedure consisting in evaluating the integrals of nonlinear expressions through some quadrature formula.

Both approaches are based on computing some fundamental quantities (either the integrals of nonlinear expressions or their numerical approximation) on an uniform fine discretization. The need to work (even only at certain stages) on an uniform fine mesh, is a severe drawback in view of an application of such schemes to realistic problems.

In this paper we propose a way to overcome such a difficulty. It is based on two different ingredients: first, a collocation scheme, which makes use of the nodal values at collocation points and allows to avoid the computation of integrals, and, second, the choice of a so-called *interpolating wavelet basis*, for which the degrees of freedom are the nodal values themselves. We wish to remark that these two ingredients could be used independently. Though the use of interpolating wavelets in the framework of collocation schemes is quite natural [BN2], other bases might be used with such an approach. On the other hand, interpolating wavelets can be used within a Galerkin discretization. Both approaches allow us to work on the adapted space at all stages of the procedure.

We test the proposed method on a one-dimensional Euler-Poisson system for semiconductor devices.

2 THE BASIS

Let θ be the *Deslaurier-Dubuc fundamental function* of order $N = 2L - 1$ ([Do,DD]). θ may be defined as the auto-correlation function of the Daubechies compactly supported orthonormal scaling function ϕ_L of order L ([D1,BS]). We remark that θ is supported in $[-N, N]$ and that, as a trivial consequence of the orthonormality of the system $\{\phi_L(x-n),\ n \in \mathbb{Z}\}$, the function θ verifies the *interpolation property*: $\theta(n) = \delta_{n0}$, for $n \in \mathbb{Z}$. The sequence $\{\widetilde{V}_j,\ j \in \mathbb{Z}\}$ of the spaces $\widetilde{V}_j = span < \theta(2^j x - n),\ n \in \mathbb{Z} > \subset L^2(\mathbb{R})$, forms a multiresolution analysis as defined in [Me]. A multiresolution on the interval is constructed by applying to the spaces $\widetilde{V}_j$ the ideas of [CDV,Do]. This results in an increasing sequence $\{V_j,\ j > j_0 = [\log_2 N]\}$ ([$\cdot$] stands for the integer part) of closed subspaces of $L^2(0,1)$. Each space V_j has a basis of the form $\{\theta_{jn}\}$ where for $n = N, \ldots, 2^j - N$ (interior basis functions), $\theta_{jn} = \theta(2^j x - n)$, while the basis functions which cross the boundaries are corrected in order to retain the property of exact interpolation of polynomials: $\forall p$ polynomial of degree lower or equal than N we have $p(x) = \sum_{n=0}^{2^j} p(n)\theta_{jn}(x)$, in $[0,1]$. The boundary corrections are performed in such a way that the interpolation property of the basis functions is maintained: for $n, k = 0, \ldots, 2^j$ we have $\theta_{jn}(k2^{-j}) = \delta_{nk}$. It is then easy to define an interpolation operator $I_j : C^0(0,1) \longrightarrow V_j$

$$I_j f = \sum_{n=0}^{2^j} f(x_{jn})\theta_{jn}, \qquad x_{jn} = n2^{-j}. \tag{1}$$

The spaces V_j satisfy a Strang-Fix condition of order N, which implies that the following approximation property holds ([BN1]): if $f \in H^s(0,1)$, $1/2 < s \leq N+1$ then $\forall r \leq s$, $0 \leq r \leq R$ we have

$$\|f - I_j f\|_{r,(0,1)} \leq C 2^{-j(s-r)} \|f\|_{r,(0,1)}. \tag{2}$$

(Here R is the regularity of the function θ, *i.e.*, θ is Hölder continuous of order R).

We can now introduce the complement spaces $W_j = span < \theta_{j+1,2n+1}$, $n = 0, \ldots, 2^j - 1 > \subset V_{j+1}$. We have $V_{j+1} = V_j \oplus W_j$. For analogy with orthonormal wavelets we introduce the notations

$$\psi_{jn} = \theta_{j+1,2n+1}, \qquad y_{jn} = x_{j+1,2n+1}. \tag{3}$$

Such a multiscale decomposition has been extensively investigated in [Do], where several properties which will play an important role in the following have been stated. It is possible to define an *interpolating wavelet transform* mapping any continuous function f into the sequence of coefficients $IWT(f) = \{\{\beta_{j_0n}\}, \{\alpha_{j_0n}\}, \{\alpha_{j_0+1,n}\}, \ldots\}$, defined as

$$\beta_{j_0n} = f(x_{j_0n}), \qquad \alpha_{jn} = f(y_{jn}) - I_j f(y_{jn}). \tag{4}$$

Any function $f \in C^0$ can be reconstructed from its transform by means of $f = \sum_{n=0}^{2^j_0} \beta_{j_0n}\theta_{j_0n} + \sum_{j \geq j_0} \sum_{n=0}^{2^j} \alpha_{jn}\psi_{jn}$.

Formula (4) clearly shows that the wavelet coefficients α_{jk} measure the lack of approximation of f by $I_j f$. In fact, a stronger result holds. It is possible to prove ([Do,D1]that the result of characterization of Besov spaces that holds for orthonormal wavelets, carries on to interpolating wavelets on $(0,1)$, provided that the Besov space considered is embedded in $C^0(0,1)$. In particular in dimension one this holds for the Sobolev spaces $H^s(0,1)$ provided $s > 1/2$.

As in the usual framework of biorthogonal wavelets we have a *fast wavelet transform* algorithm, acting on functions in V_j. Given the coordinates of a fuction $f \in V_j$ with respect to the *scaling functions* basis $\{\theta_{jn},\ n = 0, \ldots, 2^j\}$ the FWT provides the coordinates of f with respect to the wavelet basis $\{\theta_{j_0n},\ n = 0, \ldots, 2^{j_0}\} \cup_{m=j_0}^{j-1} \{\psi_{mn},\ n = 1, \ldots, 2^m\}$. However, unlike in the usual biorthogonal case, the coordinates with respect to the scaling function basis are nothing but the values of f at the dyadic points.

Moreover, the matrix realizing such a change of basis turns out to be lower triangular. This plays an important role in the efficient implementation of the method we are going to propose.

3 THE METHOD

Let us consider an equation of the following form

$$\begin{cases} \mathcal{A}u = 0 & \text{in } (0,1) \\ u(0) = a_0, \quad u(1) = a_1 \end{cases} \tag{5}$$

where $\mathcal{A}$ is a nonlinear differential operator. We assume that $\mathcal{A}$ is elliptic, with a possibly small ellipticity constant. In order to solve (5) we reduce it to a sequence of linear continuous problems of the form

$$\begin{cases} \mathcal{A}_k(u^{k+1}, u^k) = 0 & \text{in } (0,1) \\ u^{k+1}(0) = a_0, \quad u^{k+1}(1) = a_1 \end{cases} \tag{6}$$

with u^0 initial guess. This can be done, for instance, with the aid of a pseudo-temporal iteration method. This consists in considering u as the steady state limit of an auxiliary evolution problem of the form

$$\begin{cases} u_t + Au = 0 & \text{in } (0,1) \\ u(0,t) = a_0, \quad u(1,t) = a_1 & \forall t \\ u(x,0) = u^0(x) \end{cases} \tag{7}$$

Such problem is then discretized in time by a one-step semi-implicit scheme, resulting in a sequence of problems of the form (6).

Each linear problem is solved by a collocation method on a non-uniform grid. This consists in the following. Let G_k

$$G_k = \{x_{j_0 n}\} \cup \{y_{jn}, \quad (j,n) \in \Lambda_k\} \tag{8}$$

(with $\Lambda_k \subset \{(j,k), \quad j \geq j_0, \quad 1 \leq k \leq 2^j\}$) be a of dyadic points, composed by a "coarsest uniform grid" with stepsize 2^{-j_0} (which will remain untouched all along the iterative procedure), plus other points y_{jn}, which a priori might be unevenly distributed, and given the space generated by the corresponding basis functions

$$U_k = V_{j_0} \oplus span < \psi_{jn}, \quad (j,n) \in \Lambda_k > \tag{9}$$

we look for $u^k \in U_k$ verifying the equation exactly at the grid points:

$$\begin{cases} u^k(0) = a_0, \quad u^k(1) = a_1 \\ A_k(u^k, u^{k-1})(x_{j_0 n}) = 0 & \text{for } 0 \leq n \leq 2^{j_0} \\ A_k(u^k, u^{k-1})(y_{jn}) = 0 & \text{for } (j,n) \text{ in } \Lambda_k \end{cases} \tag{10}$$

This can be rewritten in the form of a linear system in which both the matrix and the righthand side may depend on u^{k-1}.

Assembling the linear system implies the evaluation if some nonlinear expressions involving u^{k-1}. We remark that, as a result of using a collocation technique (together with the particular form taken by the basis functions) such a task can be performed in a very efficient way. In fact, unlike what happens with other wavelet based techniques, thanks to the particular shape of the FWT, it is possible to perform all the computations on the nonuniform space U_k, using only the values of u^{k-1} at the points of G_k. Moreover, the use of collocation allows us to treat general nonlinearities, as we will see in the following section.

Following the ideas of [MPR] the choice of the grid G_k is driven by the size of wavelet coefficient of u_{k-1}.

Given $u^{k-1} = \sum_n \beta_{j_0 n}^{k-1} \theta_{j_0 n} + \sum_{(j,n) \in \Lambda_{k-1}} \alpha_{jn}^{k-1} \psi_{jn}$ we define the new grid G_k by removing those points y_{jn} if the corresponding coefficient α_{jn} is "small", and adding some points in the neighboorhood of those y_{jn} corresponding to "big" coefficients α_{jn}. More precisely we fix two tolerances $\epsilon_r < \epsilon_a$, and define the new index set

$$\Lambda_k = \{(j,n),\ |\alpha_{jn}^{k-1}| > \epsilon_r\} \cup \{(j+l, 2^l n + i), l = 0,1,\ i = 0,1,\ |\alpha_{jn}^{k-1}| > \epsilon_a\}$$

and then build up G_k and U_k according to (8) and (9).

4 NUMERICAL TESTS

The method described in the previous section is tested on a hydrodynamic model for a unipolar semiconductor device. The problem consists of an Euler-type system for a gas of electrons, coupled with a Poisson equation for the electric potential. We refer to [MSR] for a description of the hydrodynamic models. The problem is a good test for the proposed method because on one hand the non linearity s not of multilinear type, and on the other hand the solution can exhibit very sharp profiles.

We consider the stationary, isentropic, isothermal model in the one dimensional case. In this case the current density J is constant and for current-driven devices J can be prescribed. Then, after introducing a consistent viscous regularization (see [G]), the resulting problem is the following

Given $J = const$, find n, V such that

$$\left(\frac{J^2}{n} + n\right)_x - V_x n = -\delta n_{xx} \qquad \text{in } (0,b) \tag{11}$$

$$-V_{xx} = n - 1 \qquad \text{in } (0,b) \tag{12}$$

$$n(0) = n(b) = \bar{n} \quad V(b) = 0, \quad V_x(0) = (1 - (J/\bar{n})^2) n_x(0)/\bar{n}, \tag{13}$$

where n is the electron density and V is the electric potential. We recall that the electron velocity is defined as $u = J/n$ and the sound speed is $c = 1$, so that a subsonic flow is characterized by $n > J$ and a supersonic flow by $n < J$, (for $J > 0$). It is well known ([AMPS], [G]) that, in the $\delta = 0$ limit, the transition from a subsonic region to a supersonic region is smooth, while the transition from a supersonic to a subsonic region might develop shocks or boundary jumps.

To solve the system, we use pseudotemporal iterations in equation (11) with respect to the variable n. We point out that such a procedure is a way of introducing an iterative scheme, but it does not correspond to a discretization of the transient hydrodynamic problem. By discretizing the resulting system by a one-step semi-implicit scheme we end up with the following sequence of linear problems:

$$n^{k+1} - \delta \Delta t n_{xx}^{k+1} - \Delta t \left(1 - \left(\frac{J}{n^k}\right)^2\right) n_x^{k+1} = n^k - \Delta t n^k V_x^k \tag{14}$$

$$-V_{xx}^{k+1} = n^{k+1} - 1 \tag{15}$$

+ boundary conditions.

In iterating (14-15) we use continuation in the parameter δ, which allows much larger time-steps and fewer spatial degrees of freedom at the beginning of the iterative procedure, when we are far from the true solution. Since we are only interested in the steady-state solution and not in the proper evolution of the equation, we can also use fewer remeshing procedures during the iterative loop. In other words, we fix a parameter $NPAR > 1$ and we remesh once every $NPAR$ iterations according to the strategy described in the previous section.

Figure 1: delta=.0001 - b=1 - J=2 - d.o.f.=113

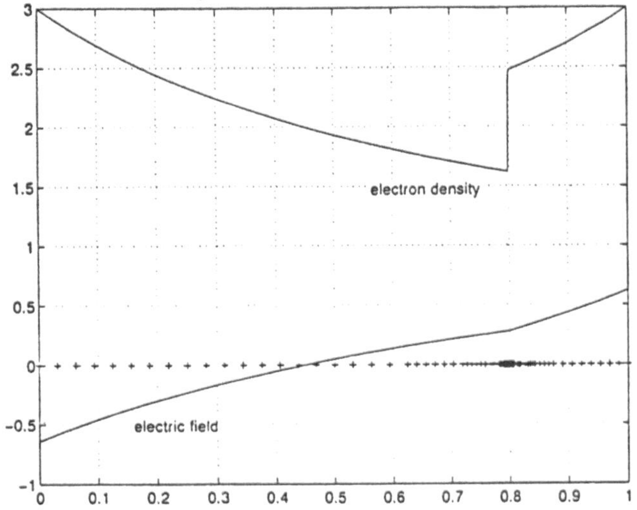

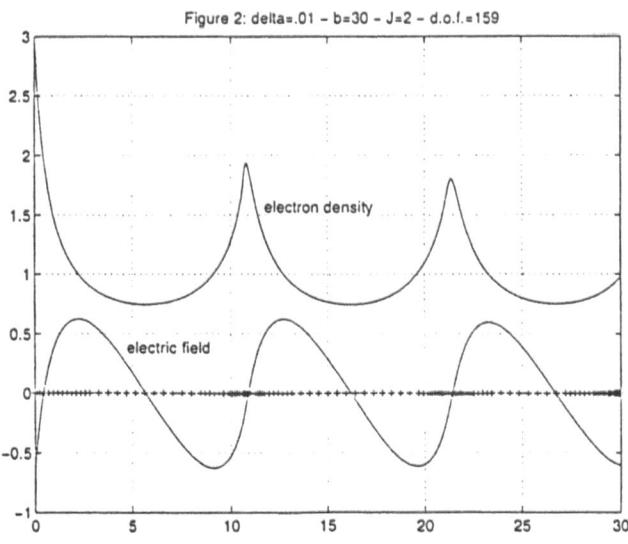

Figure 2: delta=.01 - b=30 - J=2 - d.o.f.=159

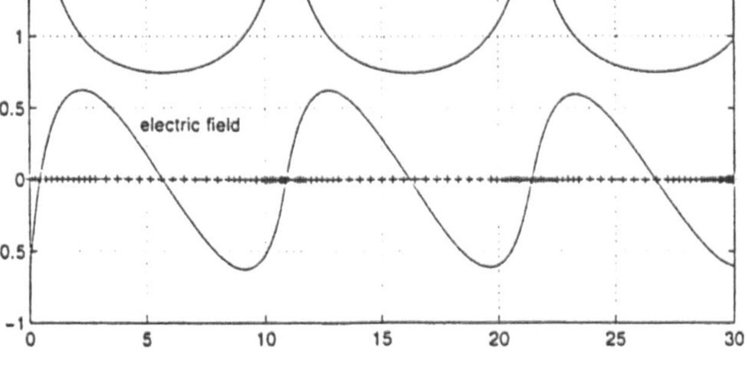

Figs.1-2 show possible structures of the density n and the electric field $E = -V_x$ for a fixed current ($J = 2$) and fixed boundary values ($\bar{n} = 3$). Different choices of the interval length provide different behaviours of the solution. For $b = 1$ the transition from the supersonic flow to the subsonic flow occurs with a shock (see Fig.1). Fig.2 exhibits supersonic oscillations and a boundary layer ($b = 30$). The good performances of the remeshing strategy are clearly illustrated by the adapted grid displayed in the pictures.

References

[AMPS] U.Ascher, P.Markowich, P.Pietra and C.Schmeiser, A phase plane analysis of transonic solutions for the hydrodynamic semiconductor model, *Math.Mod.-Meth. in the Appl.Sci.*, **1** (1991)

[BMR] S. Bertoluzza, Y. Maday and J.C. Ravel, A dinamically adaptive wavelet method for solving partial differential equations, *Computer Methods in Applied Mechanics and Engineering*, **116** (1994)

[BN1] S. Bertoluzza and G. Naldi, Some Remarks on Wavelet Interpolation, *Matematica Aplicada e Computacional*, **13**, n. 1, (1994)

[BN2] S. Bertoluzza and G. Naldi, A Wavelet Collocation Method for the Numerical Solution of Partial Differential Equations, to appear in *ACHA* (1996)

[BS] G. Beylkin, and N. Saito, Multiresolution Representation using the auto-correlation functions of compactly supported wavelets, in *Progress in wavelet Analysis and Applications, Proceedings of the International Conference on Wavelets and Applications*, Toulouse, 1992.

[D1] W. Dahmen, *Stability of multiscale transformations*, RWTH Aachen, Preprin

[DM] W. Dahmen, and C. A. Micchelli, Using the refinement equation for evaluating integrals of wavelets, *SIAM J. Numer. Anal.* **30** (2) (1993)t 1994.

[DM] W. Dahmen, and C. A. Micchelli, Using the refinement equation for evaluating integrals of wavelets, *SIAM J. Numer. Anal.* **30** (2) (1993)

[D2] I. Daubechies, Orthonormal Bases of Compactly Supported Wavelets, *Comm. on Pure and Appl. Math* **XLI** (1988)

[Do] D. Donoho, Interpolating Wavelet Transform, Department of Statistics, Stanford University, 1992, preprint.

[G] I.M.Gamba, Stationary transonic solutions on a one-dimensional hydrodynamic model for semiconductors, *Comm. P.D.E.*, **17** (1992)

[M] Y. Meyer, *Ondelettes et Operateurs I - Ondelettes*, Hermann, Paris, 1990

[MPR] Y. Maday, V. Perrier, and J.C. Ravel, Adaptivité dynamique sur bases d:ondelettes pour l'approximation d'équations aux derivées partielles, *C. R. Acad. Sci Paris*

Multiwavelets and Two-scale Similarity Transform

Vasily Strela

Department of Mathematics, MIT, Cambridge MA 02139, strela@math.mit.edu

Summary. An important object in wavelet theory is the *scaling function* $\phi(t)$, satisfying a *dilation equation* $\phi(t) = \sum C_k\phi(t-k)$. Properties of a scaling function are closely related with the properties of the *symbol* or *mask* $P(\omega) = \sum C_k e^{-i\omega k}$. The approximation order provided by $\phi(t)$ is the number of zeros of $P(\omega)$ at $\omega = \pi$, or in other words the number of factors $(1+e^{-i\omega})$ in $P(\omega)$. In the case of *multiwavelets* $P(\omega)$ becomes a *matrix trigonometric polynomial.* The factors $(1+e^{-i\omega})$ are replaced by a matrix factorization of $P(\omega)$, which defines the approximation order of the *multiscaling function.* This matrix factorization is based on the *two-scale similarity transform* (TST) which connects matrix polynomials $P(\omega)$ and $Q(\omega)$ by the relation $Q(\omega) = M(2\omega)P(\omega)M^{-1}(\omega)$. In this paper we study properties of the TST and show how it is connected with the theory of multiwavelets. This approach leads us to new results on regularity, symmetry and orthogonality of multi-scaling functions and opens an easy way to their construction.

Key words: approximation order, symmetry, orthogonality, regularity, multiscaling functions, multiwavelets.

1. Introduction

This note studies the *multi-scaling functions* that solve a *matrix dilation (refinement) equation*:

$$\phi(t) = \sum_k C_k\phi(2t-k). \tag{1.1}$$

Here $\phi(t) = [\phi_0(t)\dots\phi_{n-1}(t)]^T$ is a multi-scaling function and C_k are n by n matrices with constant elements. If $n = 1$, C_k are real numbers and $\phi(t) = \phi_0(t)$. In this case we say that (1.1) is a *scalar* dilation (refinement) equation.

Solutions of equation (1.1) have been extensively studied in recent years [3, 4, 5, 13, 7, 6, 8]. It turned out that it is more convenient to work with the Fourier transform $\widehat{\phi}(\omega)$ of the scaling function than with $\phi(t)$ itself. By taking the transform of (1.1), one can see that $\widehat{\phi}(\omega)$ satisfies the *dilation equation in frequency domain*:

$$\widehat{\phi}(\omega) = P(\omega/2)\widehat{\phi}(\omega/2). \tag{1.2}$$

Here $P(\omega) = \frac{1}{2}\sum e^{-ik\omega}C_k$ is the *two-scale symbol* corresponding to $\phi(t)$. In the scalar case $P(\omega)$ is a trigonometric polynomial. In the multi case $P(\omega)$

becomes a matrix trigonometric polynomial — a trigonometric polynomial with matrix coefficients.

Properties of $P(\omega)$ play a crucial role in the whole theory of scalar wavelets. In this paper we generalize some of these properties from the scalar to the multi case. In order to do it we need some new tools, particularly the *two-scale similarity transform* (TST).

The outline of the paper is as follows.

In Section 2. we define TST and show how it changes the eigenvalues and eigenvectors of a matrix.

Section 3. is devoted to factorizations of the symbol $P(\omega)$. Recently G. Plonka gave an example of such factorization [8]. We generalize her result and find all possible factorizations based on TST.

In Section 4. we show how a factorization of $P(\omega)$ is connected with the smoothness of the multi-scaling function $\phi(t)$. We also generalize some basic results on regularity from scalar to multi case. Similar results were obtained independently in [2, 10].

Section 5. shows how a non-orthogonal multi-scaling function can be orthogonalized using TST.

Finally, in Section 6. we establish a condition of symmetry of a multi-scaling function and find the TST which preserves symmetry. Some similar results were independently obtained in [1].

Everywhere in the text i stands for imaginary unit, $\widehat{(\cdot)}$ denotes a fourier image of $(\cdot)$. Hat over a vector or a matrix means that we take Fourier transform of each component of this vector or matrix. $(D^k\cdot)(x)$ denotes derivative of order k of $(\cdot)$ taken at point x. Derivative of a vector or a matrix means that all components of this vector or matrix are differentiated corresponding number of times. Except otherwise stated, all vectors have length n and all matrices have size n by n. Generally, $M(\omega), P(\omega), Q(\omega), \ldots$ are $n \times n$ matrices with components depending on real variable ω. We assume that those components have a sufficient number of derivatives and are 2π periodic.

All proofs and details can be found in [14]

2. Two-scale Similarity Transform

$Q(\omega)$ is a two-scale similarity transform (TST) of $P(\omega)$ if

$$Q(\omega) = M(2\omega)P(\omega)M^{-1}(\omega). \tag{2.1}$$

$\widetilde{P}(\omega)$ is an inverse two-scale similarity transform (ITST) of $P(\omega)$ if

$$\widetilde{P}(\omega) = M^{-1}(2\omega)P(\omega)M(\omega). \tag{2.2}$$

$M(\omega)$ is the transform matrix.

In this section we study how a TST changes the eigenvalues and eigenvectors of a matrix. The first result is obvious but useful.

Lemma 2.1. *Suppose $M(\omega)$ is invertible for all ω. Then TST (and the ITST) of $P(\omega)$ do not change the eigenvalues of $P(0)$.*

The situation changes when $M(0)$ is degenerate.

Lemma 2.2. *Suppose $M(\omega)$ is invertible for all ω except $\omega = 0$, $M(0)$ has a simple eigenvalue $\lambda_M(0) = 0$ corresponding to an eigenvector $r(0)$ and left eigenvector $l(0)$:*

$$M(0)r(0) = 0, \qquad l^T(0)M(0) = 0.$$

Moreover, $P(0)$ shares the eigenvector $r(0)$ with $M(0)$:

$$P(0)r(0) = \lambda_P(0)r(0),$$

and $\lambda_P(0)$ is a simple eigenvalue of $P(0)$. Then TST (2.1) preserves all eigenvalues of $P(0)$ except $\lambda_P(0)$. This eigenvalue changes to $2\lambda_P(0)$ and has the left eigenvector $l(0)$:

$$l^T(0)Q(0) = 2\lambda_P(0)l^T(0).$$

3. Factorizations of the Refinement Mask

Let us start with an example.

Example 3.1. Suppose that a scalar trigonometric polynomial $p(\omega)$ has m zeros at $\omega = \pi$ and $p(0) = 1$:

$$p(\omega) = \left(\frac{1+e^{-i\omega}}{2}\right)^m p_0(\omega), \qquad p_0(0) = 1.$$

Then the scaling function $\phi(t)$ corresponding to $p(\omega)$, $\widehat{\phi}(\omega) = p(\omega/2)\widehat{\phi}(\omega/2)$ has approximation order m [4]. Consider an ITST of $p(\omega)$

$$\widetilde{p}(\omega) = M^{-1}(2\omega)p(\omega)M(\omega) = \frac{M(\omega)}{M(2\omega)}p(\omega).$$

(We are dealing with the scalar polynomials and they commute!) Clearly there are two principally different cases $M(0) \neq 0$ and $M(0) = 0$.

If $M(\omega) \neq 0$, then

$$\widetilde{p}(\omega) = \left(\frac{1+e^{-i\omega}}{2}\right)^m \frac{M(\omega)}{M(2\omega)}p_0(\omega), \quad \widetilde{p}(0) = p_0(0) = 1,$$

and $\widetilde{p}(\omega)$ has the same approximation order as $p(\omega)$. Mention that the scaling function corresponding to $\widetilde{p}(\omega)$ is $\widehat{\psi}(\omega) = M(\omega)\widehat{\phi}(\omega)$.

If $M(\omega) \neq 0$ for all $\omega \neq 0$ but $M(0)$ has a simple zero at $\omega = 0$, then $M(\omega) = (1 - e^{-i\omega})m_0(\omega)$, $\quad m_0(0) \neq 0$ and

$$\tilde{p}(\omega) = 2M^{-1}(2\omega)p(\omega)M(\omega) = 2\frac{(1-e^{-i\omega})m_0(\omega)}{(1-e^{-2i\omega})m_0(2\omega)}p(\omega)$$
$$= \frac{1}{2^{m-1}}(1+e^{-i\omega})^{m-1}\frac{m_0(\omega)}{m_0(2\omega)}p_0(\omega)$$

which means that $\tilde{p}(\omega)$ has approximation order $m-1$.

Example 3.1 suggests that in the matrix case, the approximation order of $\tilde{P}(\omega) = 2M^{-1}(2\omega)P(\omega)M(\omega)$ with some singular $M(\omega)$ is less by one then the approximation order of $P(\omega)$, and on the contrary a TST with nonsingular transform matrix preserves the approximation order of $P(\omega)$. In this section we study how the TST and ITST influence the approximation order in the matrix case. This enables us to find a whole class of factorizations of the symbol $P(\omega)$. Our results generalize those achieved by Plonka in [8].

Suppose we are given a dilation equation in frequency domain (1.2). In [8, 7] was proved that a scaling function $\phi(t)$ has m-th approximation order (i.e. all polynomials $1, t, \ldots, t^{m-1}$ can be exactly represented as a linear combination of translates $\phi(t-n)$) if and only if there exist starting vectors $u_0^j \in R^n$; $u_0^0 \neq 0$ $(j = 0, \ldots, m-1)$ such that for $k = 0, \ldots, m-1$

$$\sum_{j=0}^{k} \binom{k}{j} (u_0^j)^T (2i)^{j-k} (D^{k-j}P)(0) = 2^{-k}(u_0^k)^T \tag{3.1}$$

$$\sum_{j=0}^{k} \binom{k}{j} (u_0^j)^T (2i)^{j-k} (D^{k-j}P)(\pi) = 0 \tag{3.2}$$

We say that $P(\omega)$ provides approximation order m if relations (3.1),(3.2) are satisfied.

Theorem 3.1. *If the transform matrix $M(\omega)$ is 2π periodic and invertible for all ω then both TST $Q(\omega) = M(2\omega)P(\omega)M^{-1}(\omega)$, and ITST $\tilde{P}(\omega) = M^{-1}(2\omega)P(\omega)M(\omega)$ have the same approximation order as $P(\omega)$. The starting vectors for $Q(\omega)$ and $\tilde{P}(\omega)$ are*

$$(q_0^k)^T = \sum_{l=0}^{k} \binom{k}{l} i^{l-k} (u_0^l)^T (D^{k-l}M^{-1})(0)$$

and

$$(\tilde{u}_0^k)^T = \sum_{l=0}^{k} \binom{k}{l} i^{l-k} (u_0^l)^T (D^{k-l}M)(0)$$

respectively.

As was expected, the situation changes in the case of singular $M(0)$.

Theorem 3.2. *Suppose $P(\omega)$ has approximation order $m \geq 1$ and for $0 \leq k \leq m-1$ (3.1),(3.2) are satisfied with starting vectors $u_0^0, \dots u_0^{m-1}$. Let $M(\omega)$ be 2π periodic and invertible for all $\omega \neq 0$. Assume also that $M(0)$ has a simple eigenvalue $\lambda_M(0) = 0$ corresponding to the left eigenvector u_0^0 and $\left.\frac{d\lambda_M(\omega)}{d\omega}\right|_{\omega=0} \neq 0$.*

Then

$$\widetilde{P}(\omega) = 2M^{-1}(2\omega)P(\omega)M(\omega) \tag{3.3}$$

provides approximation order at least $m-1$ and for $0 \leq k \leq m-2$ conditions (3.1),(3.2) are satisfied with starting vectors $\widetilde{u}_0^0, \dots \widetilde{u}_0^{m-2}$,

$$(\widetilde{u}_0^k)^T = \frac{1}{k+1}\sum_{j=0}^{k+1} \binom{k+1}{j} i^{j-k-1}(u_0^j)^T(D^{k+1-j}M)(0), \quad \widetilde{u}_0^0 \neq 0.$$

Corollary *If $P(\omega)$ has approximation order m, then it can be factored:*

$$P(\omega) = \frac{1}{2^m}M_m(2\omega)\dots M_1(2\omega)P_0(\omega)M_1^{-1}(\omega)\dots M_m^{-1}(\omega), \tag{3.4}$$

where $M_i(\omega)$ is invertible for all $\omega \neq 0$, $M_i(0)$ has a simple eigenvalue $\lambda_{M_i}(0) = 0$, corresponding to the left eigenvector $l_i^T(0)$.

Note 3.1. In the scalar case, (3.4) becomes the famous factorization of symbol $p(\omega)$ [4]

$$p(\omega) = \frac{1}{2^m}(1-e^{-2i\omega})^m p_0(\omega)(1-e^{-i\omega})^{-m} = \frac{1}{2^m}(1+e^{-i\omega})^m p_0(\omega) \tag{3.5}$$

which ensures that $p(\omega)$ has approximation order m.

Note 3.2. For the construction of multi-scaling functions with given approximation order is crusial the statement reverse to the statement of the Theorem 3.2, i.e. that the factorization (3.4) impliest the approximation order m for $P(\omega)$. This statement is proved in [9]. Also, there are presented severel new examples and an algorithm for the construction of multi-scaling functions.

4. Estimate of Regularity of Multiwavelets

In this section we study how a TST of symbol $P(\omega)$ influences the corresponding scaling function $\phi(t)$. This leads us to an estimate of regularuty of $\phi(t)$

Let $\phi(t)$ be a refinable vector such that its Fourier transform satisfies a dilation equation in frequency domain (1.2) with symbol $P(\omega)$. It is easy to see that $P(0)$ must have an eigenvalue 1 with eigenvector $\widehat{\phi}(0)$. We assume that this eigenvalue is simple and absolut values of all other eigenvalues of $P(0)$ are less then 1 ([6] shows that this actually guarantees existence and uniqueness of asolution of the dilation equation in the scence of tempered distributions).

What happens with (1.2) if we apply a TST to $P(\omega)$? Let us start with a scalar example.

Example 4.1. In Example 3.1 we mentioned that a scalar TST with invertible $M(\omega)$ changes the scaling function from $\widehat{\phi}(\omega)$ to $\widehat{\psi}(\omega) = M(\omega)\widehat{\phi}(\omega)$. On the other hand in [11] was shown that the multiplication of a symbol $\widetilde{p}(\omega)$ by $\frac{1}{2}(1-e^{-2i\omega})/(1-e^{-i\omega}) = (1+e^{-i\omega})/2$ produces a factor $\frac{1-e^{-i\omega}}{\omega}$ in $\widehat{\psi}(\omega)$. This means that if $p(\omega) = \frac{1}{2^m}(1+e^{-i\omega})^m p_0(\omega)$ than $\widehat{\phi}(\omega) = \frac{1}{\omega^m}(1-e^{-i\omega})^m \widehat{\phi}_0(\omega)$ and each order of approximation "adds" one more derivative to $\phi(t)$.

Now we are ready to start with the multi case.

Assume that the transform matrix $M(\omega)$ is invertible for all ω. Apply a TST with $M(\omega)$ to $P(\omega)$. Then (1.2) still holds , but for other scaling vector $\widehat{\psi}(\omega) = M(\omega)\widehat{\phi}(\omega)$ and other symbol $Q(\omega)$. Really, from (1.2) follows that

$$M(\omega)\widehat{\phi}(\omega) = M(\omega)P(\omega/2)M^{-1}(\omega/2)M(\omega/2)\widehat{\phi}(\omega/2),$$

or in other words,

$$\widehat{\psi}(\omega) = Q(\omega/2)\widehat{\psi}(\omega/2).$$

(Absolutely analogously to the scalar case!) By Lemma 2.1, $Q(0)$ has a simple eigenvalue 1 with nonzero eigenvector $\widehat{\psi}(0) = M(0)\widehat{\phi}(0)$.

As one could expect, the situation is different when $M(0)$ is not invertible.

Lemma 4.1. *Suppose $\widehat{\phi}(\omega)$ satisfies (1.2), $M(\omega)$ is invertible for all $\omega \neq 0$ and $M(0)$ has eigenvector $\widehat{\phi}(0)$ corresponding to a simple eigenvalue $\lambda_M(0) = 0$, such that $\left.\frac{d\lambda_M(\omega)}{d\omega}\right|_{\omega=0} \neq 0$. Then*

$$\widehat{\psi}(\omega) = \frac{1}{\omega}M(\omega)\widehat{\phi}(\omega)$$

is the solution of the dilation equation $\widehat{\psi}(\omega) = Q(\omega/2)\widehat{\psi}(\omega/2)$, where $Q(\omega) = \frac{1}{2}M(2\omega)P(\omega)M^{-1}(\omega)$.

In the previous section we proved that if a refinement mask $P(\omega)$ provides approximation order m, then it can be factored in the form (3.4). Lemma 4.1 then suggests that Fourier transform of the scaling function $\widehat{\phi}(\omega)$ corresponding to $P(\omega)$ has form

$$\widehat{\phi}(\omega) = \frac{1}{\omega^m}M_m(\omega)\dots M_1(\omega)\widehat{\psi}(\omega), \tag{4.1}$$

where $\widehat{\psi}(\omega)$ is a scaling function corresponding to $P_0(\omega)$. Or in other words $\phi(t)$ has "m more derivatives than $\psi(t)$".

Theorem 4.1 summarizes all said above and gives an estimate of regularity of $\phi(t)$, analogous to corresponding estimates in the scalar case [4]. Similar results were obtained independently in [2, 10] using different techniques.

Let us clear up the notation. We say that $\phi(t) \in C^\alpha$ if for all components $\phi_i(t)$, $\int |\phi_i(\omega)|(1+|\omega|)^\alpha d\omega < \infty$. Following [6], by $||\cdot||$ we understand such matrix norm that $||P_0(0)|| = 1$. Such a norm can always be found as long as $P_0(0)$ has nondegenerate eigenvalue 1 and absolut values of all its others are less than 1.

Theorem 4.1. *Let $P_0(\omega)$ be defined by the relation:*

$$P_0(\omega) = 2^m M_1^{-1}(2\omega)\dots M_m^{-1}(2\omega)P(\omega)M_m(\omega)\dots M_1(\omega)$$

Suppose $P(\omega)$ has approximation order m, $P(0)$ has a simple eigenvalue 1 and absolut values of all its other eigenvalues are less then 2^{-m+1}. Then $\phi(t) \in C^\alpha$ if $\log_2 R = \log_2(\sup_{\omega \in R} ||P_0(\omega)||) < m - \alpha - 1$.

Example 4.2. Geronomo-Hardin-Massopust symbol [5]

$$P(\omega) = \frac{1}{20}\begin{bmatrix} 6+6e^{-i\omega} & 8\sqrt{2} \\ (-1+9e^{-i\omega}+9e^{-2i\omega}-e^{-3i\omega})/\sqrt{2} & -3+10e^{-i\omega}-3e^{-2i\omega} \end{bmatrix},$$

can be factorized in the following form [8]:

$$P(\omega) = \tfrac{1}{40} \begin{bmatrix} \sqrt{2}/2 & -\sqrt{2}/2 \\ -e^{-2i\omega} & 1 \end{bmatrix}\begin{bmatrix} 1/2 & -1/2 \\ -e^{-2i\omega}/2 & 1/2 \end{bmatrix} \times$$
$$\begin{bmatrix} 10 & 0 \\ -1+20e^{-i\omega}-e^{-2i\omega} & -4-4e^{-i\omega} \end{bmatrix} \times$$
$$\begin{bmatrix} 1/2 & -1/2 \\ -e^{-i\omega}/2 & 1/2 \end{bmatrix}^{-1}\begin{bmatrix} \sqrt{2}/2 & -\sqrt{2}/2 \\ -e^{-i\omega} & 1 \end{bmatrix}^{-1}.$$

One can see that $m = 2$, $\sup_{\omega \in R} ||P_0(\omega)|| = 1$, so $\alpha = 1$ and by Theorem 4.1 this scaling functions must be continuous.

5. TST and Orthogonality of Multi-Scaling Functions

The next question is how a TST influences the orthogonality of a multi-scaling function.

Let $\phi(t)$ be refinable with a refinament mask $P(\omega)$. We say that $\phi(t)$ is an orthogonal multi-scaling function if all components $\phi_i(t)$ of $\phi(t)$ are orthogonal to the integer shifts of themselves and each other:

$$\int \phi(t)\phi^*(t-n)dt = \delta(n)I. \tag{5.1}$$

Suppose that $\phi(t)$ is not orthogonal but the integer translates of its components $\phi_i(t-n)$, $\quad 0 \le i \le n-1$ are linearly independent and the cascade algorithm converges in L^2. *Is there any TST which makes it orthogonal?* The answer is *yes*. The corresponding transform matrix $M(\omega)$ and the orthogonalized multi-scaling function $\psi(t)$ can be found using Algorithm 1.

Algorithm 1

1. *Starting from* $\phi(t)$ *compute the autocorrelation matrix* $A(\omega) = \int \widehat{\phi}(\omega)\widehat{\phi}^*(\omega)d\omega$.
2. *Find the square root* $\widetilde{M}(\omega)$ *of* $A(\omega)$: $A(\omega) = \widetilde{M}(\omega)\widetilde{M}^*(\omega)$. $\widetilde{M}(\omega)$ *exists and is invertible for all* ω.
3. *Compute the Fourier transform* $\widehat{\psi}(\omega)$ *of the orthogonolized scaling function:*
$$\widehat{\psi}(\omega) = \widetilde{M}^{-1}(\omega)\widehat{\phi}(\omega) = M(\omega)\widehat{\phi}(\omega).$$
It satisfies a dilation equation
$$\widehat{\psi}(\omega) = Q(\omega/2)\widehat{\psi}(\omega/2) = M(\omega)\widehat{\psi}(\omega/2)M^{-1}(\omega/2)\widehat{\psi}(\omega/2).$$
4. *Compute* $\psi(t)$ *using inverse Fourier transform.*

Note 5.1. As $M(\omega)$ is invertible for all ω, $\psi(t)$ has the same order of approximation as $\phi(t)$.

Note 5.2. Generally, $M(\omega) = \widetilde{M}^{-1}(\omega)$ is a matrix trigonometric polynomial with infinite number of terms, which means that $\psi(t)$ has infinite support.

One can see that the results of this section are absolutely consistent with the corresponding results in the scalar case.

6. Symmetry of Multi-Scaling Functions

In this section we are going to show how a TST can be implemented to get a condition of the symmetry of a multi-scaling function. We also find the TST which preserves symmetry.

Suppose a scalar function $f(t)$ has finite support and is symmetric about point $T/2$, then $\widehat{f}(\omega) = e^{-i\omega T}\widehat{f}(-\omega)$. Analogously, $f(t)$ is antisymmetric about $T/2$ if and only if $\widehat{f}(\omega) = -e^{-i\omega T}\widehat{f}(-\omega)$. If $\phi(t)$ is a vector of n functions then all its componenets are symmetric or antisymmetric if and only if

$$\widehat{\phi}(\omega) = E(\omega)\widehat{\phi}(-\omega), \quad E(\omega) = \text{diag}(\pm e^{-i2T_0\omega}, \dots, \pm e^{-i2T_{n-1}\omega}). \tag{6.1}$$

Here T_i is the point of symmetry of $\phi_i(t)$ and $+$ is chosen if $\phi_i(t)$ is symmetric. Mention that $E(\omega)$ is invertible for all ω. We say that a vector $\phi(t)$ is symmetric if $\widehat{\phi}(\omega)$ satisfies (6.1) with some $E(\omega)$.

Theorem 6.1. *Let $\widehat{\phi}(\omega)$ satisfy a dilation equation $\widehat{\phi}(\omega) = P(\omega/2)\widehat{\phi}(\omega/2)$, then $\phi(t)$ is symmetric if*

$$P(\omega) = E(2\omega)P(-\omega)E^{-1}(\omega), \tag{6.2}$$

where $E(\omega)$ is defined by equation (6.1).

Now let us find the TSTs which preserve symmetry of a multi-scaling function.

Lemma 6.1. *If $M(\omega)$ is such that $M(\omega) = E(\omega)M(-\omega)\widetilde{E}^{-1}(\omega)$, $E(\omega) = diag(\pm e^{-i2T_0\omega}, \ldots, \pm e^{-i2T_{n-1}\omega})$, $\widetilde{E}(\omega) = diag(\pm e^{-i2\widetilde{T}_0\omega}, \ldots, \pm e^{-i2\widetilde{T}_{n-1}\omega})$ and the pionts of symmetry of the components $\phi_i(t)$ of $\phi(t)$ are $\widetilde{T}_i$, $0 \leq i \leq n-1$, then TST with $M(\omega)$ preserves symmetry of this multi-scaling function, and the points of symmetry of components $\psi_i(t) = M(\omega)\phi(\omega)$ are T_i $0 \leq i \leq n-1$.*

References

1. C. K. Chui and J. Lian, *A study of orthonormal multi-wavelets*, preprint, 1995.
2. A. Cohen, I. Daubechies and G. Plonka, *Regularity of Refinable Function Vectors*, preprint, 1995.
3. A. Cavaretta, W. Dahmen and C. A. Micchelli, *Stationary Subdivision*, Memoirs Amer. Math. Soc. **453**, 1-186, 1991.
4. I. Daubechies, *Ten Lectures on Wavelets*, SIAM, Philadelphia, 1992.
5. J. Geronimo, D. Hardin, and P. R. Massopust, *Fractal functions and wavelet expansions based on several functions*, J. Approx. Th., **78** (1994), pp. 373-401.
6. C. Heil and D. Colella, *Matrix refinement equations: existence and uniqueness*, preprint, 1994.
7. C. Heil, G. Strang, and V. Strela, *Approximation by translates of refinable functions*, Numerische Mathematik, 1996.
8. G. Plonka, *Approximation order provided by refinable function vectors*, Constructive Approximation, 1996.
9. G. Plonka and V. Strela, *Construction of Multi-Scaling Functions with Approximation and Symmetry*, preprint, Rostock Univ., 1995.
10. Z. Shen, *Refinable function vectors*, preprint, 1995.
11. G. Strang, *Eigenvalues of $(\downarrow 2)H$ and Convergence of the Cascade Algorithm*, IEEE Trans. Sig. Proc. 1996.
12. G. Strang, *The cascade algorithm for the dilation equation*, preprint, 1994.
13. G. Strang and V. Strela, *Short wavelets and matrix dilation equations*, IEEE Trans. on SP. vol. 43. pp 108-115, 1995.
14. V. Strela, *Multiwavelets: regularity, orthogonality and symmetry via two-scale similarity transform*, Studies in Appl. Math. (1996).

Image Segmentation

TEXTURE SEGMENTATION BY VARIATIONAL METHODS

C. Ballester and M. González

Dpt. Matemàtiques i Informàtica. Univ. de les Illes Balears, 07071-Palma de Mallorca, Baleares, Spain. E-mail: dmicbn0@ps.uib.es, dmimgh0@ps.uib.es

Summary. We present a novel texture segmentation algorithm which is affine invariant. We prove that the texton densities (affine invariant orientation channels) can be computed using affine invariant intrinsic neighborhoods and affine invariant intrinsic orientation matrices. We discuss several possibilities for the definition of the channels and give comparative experimental results where an affine invariant Mumford-Shah type energy functional is used to compute the multichannel affine invariant segmentation. We prove that the method is able to retrieve faithfully the texture regions and to recover the shape from texture information in images where several textures are present.
Key words. Texture segmentation, affine invariance, variational method, dominant local orientation, Zernike moments, shape from texture.

1. Introduction

Texture is an important characteristic in order to analyse images. Texture analysis has been used in classification tasks concerning 2D images in general and segmentation of aerial or medical images in particular. Texture segmentation has been attempted in many different ways. Most of them follow the same general strategy: a process of feature extraction followed by a segmentation. The main purpose of feature extraction is to map differences in spatial structures, either stochastic or geometric, into difference values in feature space. Then segmentation methods analyse the feature space in order to extract homogeneous regions.

Feature extraction has been effectuated by several methods ([9], [8], [22], [23], [15], [13], [19], [4]). There are non-local methods which are able to take into account textures with repetition frequencies ranging from low to high frequencies while other methods are restricted to the size of a prescribed neighborhood. The textural structures which can be described within a neighborhood are naturally restricted to those which can be observed in the neighborhood. Hence, features based on measurements on a neighborhood of fixed size have poor discrimination power when applied to textural structures not observable within the size of the neighborhood, because of the wrong scale selected. In general, this size information is not available. The psychophysical experiments indicate that there exists frequency and orientation selectivity in the human visual system, giving a hint of how this scale problem can be solved [6].

Using these ideas as a guide, we shall generate an oriented texture field, also called a representation of dominant local orientation, computed on an adaptative intrinsic elliptic neighborhood at each point which captures the texture pattern. It represents, in some sense, a particular lenght scale related to the texture pattern from which we construct the feature channels for a segmentation algorithm (this construction agrees with the texton theory of Julesz for texture discrimination [10]). Indeed, we propose in the next two sections several affine invariant Mumford-Shah type energy functionals based on the second moment matrix (see [3], [18], [7]), dominant local orientation and Zernike moments as channels for an affine invariant functional. Also, we shall use in Section 4 the affine invariant analogue of the second moment matrix, introduced in Section 2, to obtain information about shape from texture. The introduction of affine invariance as a requirement for texture analysis goes beyond what is known of the human performance and also beyond the psychophysical theories. Indeed, the texton doctrine is basically euclidean, but the texture discrimination must be, in many aspects, affine invariant.

2. Texture segmentation by a multichannel affine invariant method.

The affine invariant analogue of Mumford-Shah functional for grey level image segmentation was introduced in [1]. Based on it, we propose a multichannel model for texture segmentation. We associate to the given image $g : \Omega \to \mathcal{R}$ a function G with values in a feature space, i.e., let $G : \Omega \to E$ be a bounded measurable function with values in a Hilbert space E representing the feature space. In this case the functional can be written

$$E_{af}(u, B) = \int_{\Omega} \|u - G\|_E^2 dx + \lambda ATV(B), \tag{2.1}$$

where u is an approximating vector function equal to the mean value of G on each connected component of $\Omega \setminus B$ and $ATV(B)$ is the *affine total variation* of B ([1], [2]). Let us recall that the ATV was not an arbitrary functional. It was shown in [1] to be the only positive functional, up to a scaling factor, associating to each pair of Jordan curves a quantity which is geometric, affine invariant, biadditive and continuous (in the $W^{1,1}$ topology of the space of parametric curves).

Let us recall that, based on the Mumford-Shah functional [17], multichannel models for texture segmentation were proposed in [12] and in [14] using, respectively, a wavelet decomposition of g and multiscale curvature densities at each position and orientation as texture features (see also [13]).

As observed in [1], the ATV functional is affine invariant. Moreover, the feature vectors G we are going to propose are affine invariant in the following sense:

If $g, h : \Omega \to \mathcal{R}_+$ are images such that $h(x) = g(Ax)$, A being any linear map with $\det A \neq 0$, and $G, H : \Omega \to E$ are the vector features associated to g, h, respectively, then there exists an isometry U of E depending on g, A, h such that

$$H(x) = UG(Ax).$$

Then, we shall recover the affine invariance of functional 2.1.

Concerning the existence of minimizers for the functional 2.1 we have the following theorem which can be proved along the lines of [1].

Theorem 2.1. *The functional E_{af} attains its infimun.*

Now, we briefly explain the construction of some affine invariant feature vectors G. At each point x we find an intrinsic elliptic Δ-neighborhood $\varepsilon(x)$, determined by the texture pattern (see below for the definition), in which the histogram is stable. This neighborhood $\varepsilon(x)$ represents in some sense, a particular length scale related to the texture pattern and to the concept of $\Delta-$neighborhood of Julesz ([10]). Using this neighborhood, we compute the matrix

$$Q_g(x) = \frac{1}{|\varepsilon(x)|} \int_{\varepsilon(x)} \nabla g(y) \nabla g(y)^t dy, \tag{2.2}$$

in order to obtain an estimate of the dominant local orientation at x. We define

$$E_{1af}(Q,B) = \int_\Omega tr\left(Q_g^{-1/2} Q_g(x) Q_g^{-1/2} - Q\right)^2 dx + \lambda ATV(B), \tag{2.3}$$

where

$$Q_g = \frac{1}{|\Omega|} \int_\Omega Q_g(x) dx,$$

and

$$E_{2af}(w,B) = \int_\Omega \|Tw_g(x) - w\|^2 dx + \lambda ATV(B), \tag{2.4}$$

where $w_g(x)$ is the eigenvector of $Q_g^{-1/2} Q_g(x) Q_g^{-1/2}$ associated to its largest eigenvalue and $Tw_g(x) = Re^{2i\theta}$ if $w_g(x) = Re^{i\theta}$. The main role of Q_g is to reinforce affine invariance. Also, it plays a role as a normalization factor with respect to the magnitude of the gradient.

In Figure 2.1 we display a numerical experiment using the functionals above. We have used a region growing algorithm similar to the one described in [12], [1].

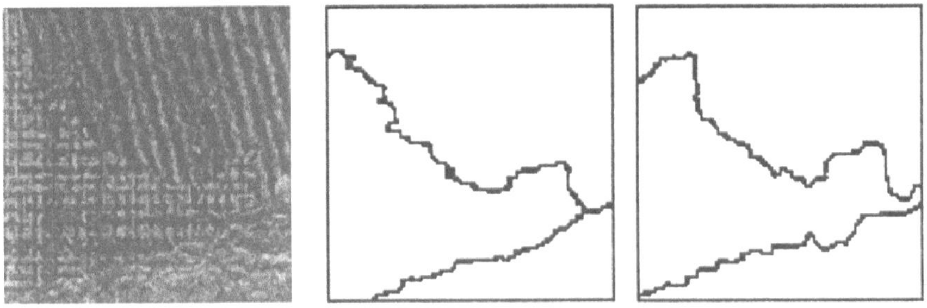

Fig. 2.1. Left: original image. Center: segmentation boundaries obtained from E_{1af}. Right: segmentation boundaries obtained from E_{2af}.

On the other hand, we shall use the matrix $Q_g^{-1/2} Q_g(x) Q_g^{-1/2}$ to obtain information about shape from texture, in the same sense that in [24],[3], where they used the second moment matrix.

2.1 The intrinsic elliptic Δ-neighborhood.

At each point x, we try to find an elliptic neighborhood $\varepsilon(x)$ determined by the texture pattern in which the histogram is stable. This neighborhood $\varepsilon(x)$ represents in some sense, a particular length scale related to the texture pattern. The definition of this intrinsic elliptic Δ-neighborhood is as follows:

Definition 2.1. *Let $g : \Omega \to \mathcal{R}_+$ be a given image. Suppose that we have for each point a neighborhood $\varepsilon_g(x)$, a matrix $Q_g(x)$ and a real number $\rho_g(x)$ such that*

$$Q_g(x) = \frac{1}{|\varepsilon_g(x)|} \int_{\varepsilon_g(x)} \nabla g(y) \nabla g(y)^t dy, \tag{2.5}$$

$$\varepsilon_g(x) = \{y \in \mathcal{R}^2 : \langle Q_g(x)(y-x), y-x\rangle \leq \rho_g(x)\}. \tag{2.6}$$

Then $\varepsilon_g(x)$ is called an intrinsic elliptic $\Delta-$neighborhood of the image g at point x, where $\rho_g(x)$ has been selected large enough so that the histogram of g in $\varepsilon_g(x)$ is stable.

This definition generalizes in an affine framework the Julesz definition of $\Delta-$neighborhood, which is circular in the original definition.

Let us, for convenience, use also the notation $\varepsilon_g(x; \rho_g(x))$ to refer to $\varepsilon_g(x)$. We notice that if A is any linear map in $\mathcal{R}^2$ with $\det A \neq 0$ and $h : A^{-1}\Omega \to \mathcal{R}_+$, $h(x) = g(Ax)$, then we can define an intrinsic elliptic $\Delta-$neighborhood for h by $\rho_h(x)$, $\varepsilon_h(x; \rho_h(x))$ and $M_h(x)$ associated with h by

$$Q_h(x) = A^t Q_g(Ax)A \tag{2.7}$$

$$\varepsilon_h(x, \rho_h(x)) = A^{-1}\varepsilon_g(Ax, \rho_g(Ax)) \tag{2.8}$$

$$\rho_h(x) = \rho_g(Ax). \tag{2.9}$$

Following Julesz, we assume the existence of an intrinsic elliptic $\Delta-$neighborhood for g at any x. With this on hand and with the notion of histogram stability ([2]), we can prove the following result:

Proposition 2.1. *Both functionals E_{1af} and E_{2af} are affine invariant.*

Moreover, we propose a set of features based on Zernike moments computed on the intrinsic elliptic Δ-neighborhood $\varepsilon(x)$, as a feature space for texture discrimination.

3. Features based on Zernike moments.

Moments and functions of moments have since long been used as features in a number of applications to pattern recognition ([20], [21]). Our purpose is to use the method of moments to compute locally at the intrinsic elliptic Δ-neighborhood of each point, a set of features which will be used as channels in a Mumford-Shah type segmentation functional for texture discrimination.

As above let $g : \Omega \to \mathcal{R}_+$ be a given image. At each point $x \in \Omega$, we consider the elliptic neighborhood $\varepsilon_g(x, \rho_g(x))$ defined in terms of the matrix Q_g in 2. by

$$\varepsilon_g(x, \rho_g(x)) := \{y \in \Omega : < Q_g(y-x), y-x > \leq \rho_g(x)\}.$$

For simplicity we write $N_g(x) = \rho_g(x)^{-1}Q_g$. Then

$$\varepsilon_g(x, \rho_g(x)) = \{y \in \Omega : < N_g(x)^{1/2}(y-x), N_g(x)^{1/2}(y-x) > \leq 1\}.$$

Since

$$\varepsilon_g(x, \rho_g(x)) = x + N_g(x)^{-1/2}D(0,1),$$

where $D(0,1) = \{z \in \mathcal{R}^2 : ||z|| \leq 1\}$, we can exploit this identity to define the following moments:

$$m_{nl}(g,x) = \int_{||z||\leq 1} V_{nl}(z)g(x + N_g^{-1/2}(x)z)dz, \tag{3.1}$$

where $z = (z_1, z_2)$, $x = (x_1, x_2) \in \Omega$ and $V_{nl}(z)$ are the Zernike polynomials (see [20], [21] for a detailed account of these). We use the moments m_{nl} to define the following functional

$$E_{Zaf}(\{m_{nl}\}, B) = \sum_{n,l} \int_{\Omega} \|m_{nl}(x) - m_{nl}\|^2 dx + \lambda ATV(B), \tag{3.2}$$

The normalization with the matrix $N_g(x)$ is needed to use the Zernike moments on the intrinsic neighborhood of each point and it gives for free the affine invariance of the energy functional. Figure 3.1 shows a numerical result obtained from the texture segmentation algorithm when the previous functional is applied.

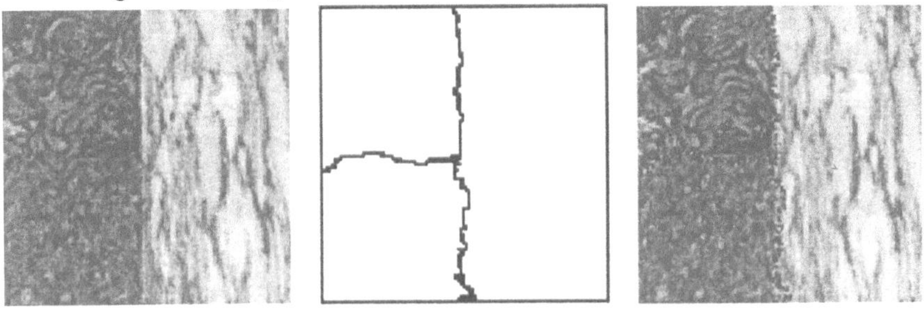

Fig. 3.1. Left: Original image. Center: segmentation boundaries obtained from E_{Zaf}. Right: segmentation superposed original image.

4. Applications to Shape from texture.

The image of a slanted textured surface contains cues that can be used to estimate the shape and orientation of the surface. As pointed by Witkin [24], the image of an slanted circle is an ellipse and the orientation and eccentricity of the ellipse indicate the magnitude and direction of the slant.. The projection on a plane of two pathes on a surface with the same (o similar) texture will be different depending on their position and orientation in the surface. The two important parameters measuring this distorsion projection effect are the *slant* and *tilt*. As we can see in [7], these parameters can be recovered from the distortion tangent map. It can be shown ([2]) that, if we made an assumption about the surface reflectance pattern, then we can estimate the distorsion map (*slant* and *tilt* parameters) from the second moment matrix $Q_g(x)$ computed on the image plane. More especifically, we can compute the slant and tilt parameters (denoted by σ and τ, respectively) by

$$\cos\sigma = \sqrt{\frac{\lambda_2}{\lambda_1}} \quad \text{and} \quad t = (\cos\tau, \sin\tau),$$

where λ_1 and λ_2 are the maximum and minimum eigenvalues of the matrix $Q_g(x)$ and t is the unit eigenvector associated to λ_1.

Following this strategy, given and image g, we use the functional E_{1af} in order to compute the slant and tilt parameters of each region with a homogeneous surface shape. Then, the steps of the algorithm are:

- First: Minimize the functional E_{1af}.

- Second: use the matrix function $Q_g^{1/2} Q_O Q_g^{1/2}$ to compute the slant (σ_O) and tilt (τ_O) at each region O obtained in the First Step, where Q_O is the affine invariant analogue of the second moment matrix associated to the region O.
- Finally, estimate the local surface normal, given by

$$n_O = (\sin\sigma_O \cos\tau_O, \sin\sigma_O \sin\tau_O, -\cos\sigma_O)$$

We present in Figure 4.1 an experiment for an image of a 3D-shape made of a same texture with different orientations planes.

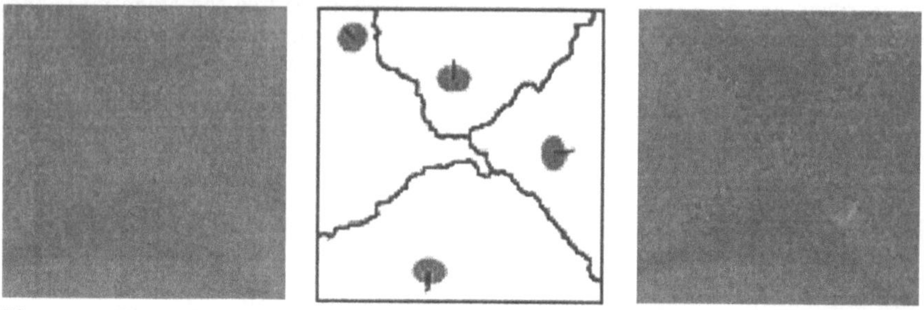

Fig. 4.1. Shape from texture information. Left: original image. Center: segmentation boundaries with the estimated surface orientations. Right: segmentation superposed original image.

5. Summary and discussion.

We propose an affine invariant model to, both extract affine invariant texture features from images of natural scenes and to find the boundaries between the different textures in the scene. We take an intrinsic approach to solve the usual problem in texture discrimination of choosing an appropiate window in which one computes the texture features, by incorporating what we call the intrinsic elliptic Δ-neighborhood determined by the texture pattern. From our affine intrinsic neighborhoods we compute affine intrinsic orientation matrices giving an estimate of the dominat local orientation and set of moments. These features are used in a multichannel analogue to the affine invariant energy functional introduced in [1] for grey level image segmentation. The models we studied are able to discriminate and segment a wide range of natural textures. We also showed that the affine intrinsic matrix captures the essential information of surface shape in the scene, which is in agreement with the work in [7], where they use the second moment matrix.

References

1. C. Ballester, V. Caselles, and M. González. Affine invariant segmentation by variational methods. To appear in SIAM J. Appl. Math.
2. C. Ballester and M. González. Affine invariant texture segmentation and shape from texture by variational methods. Preprint, 1995.

3. J. Bigun, G. H. Granlund, and J. Wiklund. Multidimensional orientation estimation with applications to texture analysis and optical flow. *IEEE Trans. Patt. Anal. Machine Intell.*, 13(8):775–790, 1991.
4. R. R. Coifman and N. Saito. Selection of best bases for classification and regression. *Internat. Congress on Ind. and Applied Mathematics*, 1995.
5. R. W. Connors and C. A. Harlow. A theoretical comparison of texture algorithms. *IEEE Trans. Patt. Anal. Machine Intell.*, 2(3):204–222, 1980.
6. F. H. Crick, D. C. Marr, and T. Poggio. An information processing approach to understanding the visual cortex. A.I. Memo 557, MIT, April 1980.
7. J. Garding and T. Lindeberg. Direct computation of shape cues by multiscale retinotopic processing. Report from comput. vision and active perception (CVAP), Dept. of Numer. Anal. and Comp. Science., Royal Inst. of Technology. Stockholm, Sweden, 1992.
8. S. Geman and D. Geman. Stochastic relaxation, Gibbs distributions, and the Bayesian restoration of images. *IEEE Trans. Patt. Anal. Machine Intell.*, PAMI-6(6):721–741, Nov. 1984.
9. R. M. Haralick, K. Shanmugam, and I. Dinstein. Textural features for image classification. *IEEE Trans. System Man Cybernet.*, 3(1):610–621, 1973.
10. B. Julesz. Textons gradients: the texton theory revisited. *Biol. Cyber.*, 54:245–251, 1986.
11. M. Kass and A. Witkin. Analyzing oriented patterns. *Comput. Vision Graphics Image Process.*, 37:362–385, 1987.
12. G. Koepfler, J. M. Morel, and C. Lopez. A multiscale algorithm for image segmentation by variational methods. To appear in SIAM J. on Numerical Analysis.
13. T. S. Lee, D. Mumford, and A. Yuille. *Texture Segmentation by Minimizing Vector Valued Energy Functionals: The Coupled-Membrane Model*, volume 588 of *Lecture Notes in Computer Science*, pages 165–173. Springer-Verlag, 1992.
14. C. Lopez and J. M. Morel. Axiomatisation of shape analysis and application to texture hyperdiscrimination. Proceedings on the Trento Conference of Surface Tension and Movement by Mean Curvature. De Grayter Publishers. Berlin.
15. J. Malik and J. Perona. Preattentive texture discrimination with early vision mechanisms. *Journ. of the Opt. Society of America A*, 7:923–932, 1991.
16. J. M. Morel and S. Solimini. *Variational Methods for image segmentation*. Birkhauser Verlag, 1994.
17. D. Mumford and J. Shah. Optimal approximations by piecewise smooth functions and variational problems. *Communications on Pure and Applied Mathematics*, XLII:577–685, 1989.
18. A. R. Rao and B. G. Schunk. Computing oriented texture fields. *CVGIP: Graphical Models and Image Processing*, 53(2):157–185, 1991.
19. T. R. Reed and J. M. Hans du Buf. A review of recent texture segmentation and feature extraction techiques. *CVGIP: Image Understanding*, 57(3):359–372, 1993.
20. M. R. Teague. Image analysis via the general theory of moments. *J. Opt. Soc. Amer.*, 70:920–930, Aug. 1980.
21. C. H. Teh and R. T. Chin. On image analysis by the methods of moments. *IEEE Trans. on Pattern Analysis and Machine Intell.*, 10(4):496–513, July 1988.
22. M. Unser. Local linear transforms for texture measurements. *Signal Processing*, 11:61–67, 1986.
23. H. Voorhees and T. Poggio. Computing texture boundaries in images. *Nature*, 333:364–367, 1988.
24. A. P. Witkin. Recovering surface shape and orientation from texture. *Artificial Intelligence*, 17:17–45, 1981.

Affine Invariant Gradient Flows*

Peter J. Olver[1], Guillermo Sapiro[2], and Allen Tannenbaum[1]

[1] University of Minnesota, Minneapolis, MN 55455
[2] HP Labs., 1501 Page Mill Rd., Palo Alto, CA 94304

Summary. An affine invariant metric allowing one to compute affine invariant gradient descent flows is first presented in this work. This means that given an affine invariant energy, we compute based on this metric the flow that minimizes this energy as fast as possible and in an affine invariant way. Two examples are then presented. The first one shows that the affine flow minimizing the area enclosed by a planar curve is given by the affine geometric heat flow. We then extend this energy to derive affine invariant active contours for invariant image segmentation.

1. Introduction

A number of problems in image processing and computer vision are approached via energy minimization techniques. The minimizer can be obtained for example via gradient descent flows, which are flows minimizing this energy as fast as possible according to certain metric. In a number of cases, as for example in object recognition, the energy being minimized is invariant to certain transformation group, and the solution is expected to be invariant as well (see for example [2]). In order to obtain a gradient flow which is also invariant, an invariant metric needs to be defined.

In this work, we present an affine invariant metric which will help us to define affine invariant gradient descent flows for affine invariant energies. Two applications of this gradient flow are then provided. The first one shows that the affine invariant gradient flow corresponding to the area enclosed by a planar curve is given by the affine geometric heat flow introduced in [19, 20] (see also [1]). We then extend this area and present affine invariant active contours, extending the results in [5, 11]. In this case, active contours (introduced by Terzopoulos *et al.* [10, 22]) are given by an affine invariant weighted distance [16]. The affine gradient flow minimizing this distance converges to the objects boundaries, obtaining affine invariant detection.

2. Planar curve evolution

The theory of planar curve evolution has been considered in a large variety of fields. See [18] and references in there for pointers to some of the relevant

* This work was partially supported by NSF ECS 91-22106, DMS 92-04192 and DMS 95-00931, AFOSR F49620-94-1-00S8DEF, ARO DAAH04-94-G-0054 and DAAH04-93-G-0332.

literature. Formally, let $C(p,t) : S^1 \times [0,\tau) \to \mathbb{R}^2$ be a family of smooth embedded closed curves in the plane (boundaries of planar shapes). Assume that this family of curves evolves according to ($C(p,0) = C_0(p)$)

$$\frac{\partial C(p,t)}{\partial t} = \frac{\partial^2 C(p,t)}{\partial v^2} = \kappa(p,t)\mathcal{N}(p,t). \tag{2.1}$$

Here $v(p) = \int_0^p \| C_p \| \, dp$ is the *Euclidean arc-length* ($\| C_v \| \equiv 1$), $\kappa = [C_v \times C_{vv}]$ the *Euclidean curvature*, and $\mathcal{N}$ the *inward unit normal*. The flow given by (2.1) is called the *Euclidean shortening flow*, since the curve perimeter shrinks as fast as possible when the curve evolves according to it [9]. Gage and Hamilton [8] and Grayson [9] proved that any embedded curve in the plane converges to a round point via the flow given in (2.1). The non-linear flow (2.1) is also called the *Euclidean geometric heat flow*. It has been utilized for the definition of a geometric, Euclidean invariant, multiscale representation of planar shapes [1, 12].

Recently, we introduced a new curve evolution equation, the *affine geometric heat flow* [19, 20]:

$$\frac{\partial C(p,t)}{\partial t} = \frac{\partial^2 C(p,t)}{\partial s^2}, \tag{2.2}$$

where $s(p) = \int_0^p [C_p \times C_{pp}]^{1/3} dp$, is the *affine arc-length* ($[C_s \times C_{ss}] \equiv 1$) [3], and C_{ss} is the *affine normal*. This evolution is the affine analogue of equation (2.1), and its solution space is affine invariant. Since the affine normal C_{ss} exists just for non-inflection points, we formulated the natural extension of the flow (2.3) for non-convex initial curves in [20, 21]:

$$\frac{\partial C(p,t)}{\partial t} = \begin{cases} 0, & p \text{ an inflection point,} \\ C_{ss}(p,t), & \text{otherwise.} \end{cases} \tag{2.3}$$

The flow (2.3) defines a geometric, affine invariant, multiscale representation of planar shapes [20]. The curve first becomes convex, as in the Euclidean case, and after that it converges into an ellipse [19].

We should also add that in [21], we give a general method for writing down invariant flows with respect to any Lie group action on $\mathbb{R}^2$. This was formalized and extended to $\mathbb{R}^n$, together with uniqueness results, in [14, 15]. Results for the projective group were recently also reported in [7].

3. Affine invariant curve metric

Let $C = C(p,t)$ be a smooth family of closed curves where t parametrizes the family and p the given curve, say $0 \le p \le 1$. Consider the length functional $L(t) := \int_0^1 \| C_p \| \, dp$, and its first variation $L'(t) = -\int_0^{L(t)} \langle \frac{\partial C}{\partial t}, \kappa\mathcal{N} \rangle \, dv$. The goal now is to compute from this derivation, the flow minimizing $L(t)$ as fast as possible, in an Euclidean invariant way. For this, we need to define an Euclidean invariant metric. In the standard way, we can define a norm $\| \cdot \|_{euc}$ on the (Fréchet) space of twice-differentiable closed curves in the plane. Indeed, the (curve) norm is given by the length

$$\| C \|_{euc} := \int_0^1 \| C_p \| \, dp = \int_0^1 < C_p, C_p >^{1/2} dp = \int_0^L dv = L,$$

of the curve C. Thus the direction in which $L(t)$ is decreasing most rapidly is when C satisfies the gradient flow $C_t = \kappa \mathcal{N}$, as previously pointed out [8, 9].

We extend now the results above, that is the definition of a curve norm, for the affine group. We use in analogy to the case above the minimization of affine length and area. Since affine geometry is defined only for convex curves [3], we will initially have to restrict ourselves to the (Fréchet) space of differentiable convex closed curves in the plane. Being $L_{aff} := \oint ds$ the *affine length* [3], we define

$$\| C \|_{aff} := \int_0^1 \| C(p) \|_a \, dp = \int_0^{L_{aff}} \| C(s) \|_a \, ds,$$

where $\| C(p) \|_a := [C(p) \times C_p(p)]$. Observe that $\| C_s \|_a = [C_s, C_{ss}] = 1$, $\| C_{ss} \|_a = [C_{ss}, C_{sss}] = \mu$, where μ is the *affine curvature*. This makes the affine norm $\| \cdot \|_{aff}$ consistent with the properties of the Euclidean norm on curves relative to the Euclidean arc-length dv ($\| C_v \| = 1$, $\| C_{vv} \| = \kappa$).

Note that the area enclosed by C is just

$$A = \frac{1}{2} \int_0^1 \| C(p) \|_a \, dp = \frac{1}{2} \int_0^1 [C, C_p] \, dp = \frac{1}{2} \| C \|_{aff} . \qquad (3.1)$$

A straightforward computation reveals that the first variation of the area is $A'(t) = -\int_0^{L_{aff}(t)} [C_t, C_s] \, ds$. Therefore the affine invariant gradient flow which will decrease the area as quickly as possible relative to $\| \cdot \|_{aff}$ is exactly $C_t = C_{ss}$, which, modulo tangential terms, is equivalent to $C_t = \kappa^{1/3} \mathcal{N}$ [19], precisely the affine invariant heat equation studied in [1, 19, 20]. Note that based on the Euclidean norm, the flow minimizing the area is $C_t = \mathcal{N}$. Both $\mathcal{N}$ and C_{ss} are normal vectors, each one in its corresponding group. Computations similar to the above show that the affine invariant flow minimizing the affine length is given by $C_t = \mu C_{ss}$ [19].

4. Affine invariant active contours

The goal now is to extend the works in [5, 11] to affine invariant detection. The developments in [5, 11] are strongly based on the original energy-based snakes introduced by Terzopoulos *at al.* [10, 22] as well as the curve evolution ones in [4, 13]. We refer the interested readers to the mentioned papers for details and relations between the models.

It is important to note that after affine edges are computed locally based on the scale-space or affine gradient derived in [16], [1] affine invariant fitting can be performed (see [23] and references therein). In this work, the affine invariant integration is done by means of active contours.

4.1 Deformable models based on curve shortening

Assume in the 2D case that the deforming curve C is given as a level-set of a function $u : \mathbb{R}^2 \to \mathbb{R}$. Then, we can represent the deformation of C via the deformation of u. The 2D segmentation is obtained via the evolution equation [4, 13] $(u(0,x) = u_0(x))$

$$\frac{\partial u}{\partial t} = \phi \parallel \nabla u \parallel \operatorname{div}\left(\frac{\nabla u}{\parallel \nabla u \parallel}\right) + \nu\phi \parallel \nabla u \parallel \qquad (t,x) \in [0,\infty[\times\mathbb{R}^2 \quad (4.1)$$

where the stopping term typically has the form $\phi = (1 + \parallel \nabla\hat{\Phi} \parallel^m)^{-1}$, $m = 1$ or 2, and $\hat{\Phi}$ is a regularized version of the original image Φ. Using the fact that $\operatorname{div}\left(\frac{\nabla u}{\|\nabla u\|}\right) = \kappa$, where κ is the Euclidean curvature of the level-sets C of u, equation (4.1) can be written in the form $u_t = \phi \cdot (\nu + \kappa) \parallel \nabla u \parallel$. The flow $u_t = (\nu + \kappa) \parallel \nabla u \parallel$, means that the the (smooth) level-set C of u we are considering is evolving according to $C_t = (\nu + \kappa)\mathcal{N}$, where $\mathcal{N}$ is the inward normal to the curve. This equation was first proposed in [17], were extensive numerical research on it was performed. It was introduced in computer vision in [12], where deep research on its importance for shape analysis is presented.

The motion $C_t = \kappa\mathcal{N}$, is the *Euclidean heat flow* presented before, very well know for its excellent geometric smoothing properties [8, 9]. The constant velocity $\nu\mathcal{N}$, acts as the balloon force in [6] and is related to classical mathematical morphology. If $\nu > 0$, this velocity pushes the curve inwards and it is crucial in the model in order to allow convex initial curves to become non-convex. The external force is given by ϕ, which is supposed to prevent the propagating curve from penetrating into the objects in the image.

This curve evolution model given by (4.1) automatically handles different topologies, allowing the detection any number of objects in the image, without knowing their exact number. This is achieved with the help of the numerical algorithm developed by Osher and Sethian [17].

4.2 Euclidean geodesic active contours

We present now the geodesic active contours derived in [5, 11]. These models are based on the models in [4, 10, 13, 22], as well as the concepts of shortening and gradient flows described in Section 3. In [5], the model is

[1] A local computation of edges is one of the ingredients of active contours schemes.

derived from the principle of least action in physics, showing the mathematical relation between energy and curve evolution based snakes. In [11], the model is derived immediately from curve shortening, and is compared to similar flows in continuum mechanics. One of the basic ideas is to change the ordinary Euclidean arc-length function $dv = \| C_p \| dp$ along a curve $C(p)$ by multiplying by a conformal factor $\phi(x, y) > 0$, which is assumed to be a positive, differentiable function. The resulting *conformal Euclidean metric* on $\mathbb{R}^2$ is given by $\phi\, dxdy$, and its associated arc length element is $dv_\phi = \phi\, dv = \phi \| C_p \| dp$. As in ordinary curve shortening, we want to compute the corresponding gradient flow for the modified length functional $L_\phi(t) := \int_0^L \phi\, dv = \int_0^1 \| C_p \| \phi\, dp$. Taking the derivative and integrating by parts, we find that $-L'_\phi(t) = \int_0^{L_\phi(t)} \langle C_t, \phi\kappa\mathcal{N} - (\nabla\phi \cdot \mathcal{N})\mathcal{N}\rangle\, dv$ [5, 11], which based on the Euclidean metric in Section 3, means that the (gradient) direction in which the L_ϕ perimeter is shrinking as fast as possible is given by $\frac{\partial C}{\partial t} = \phi\kappa\mathcal{N} - (\nabla\phi \cdot \mathcal{N})\mathcal{N}$. As long as the flow remains regular, we will get convergence to a closed geodesic in the plane relative to the conformal Euclidean metric $\phi\, dxdy$. Regularity may be deduced from the classical curve shortening case.

As in [4, 13], we may add an inflationary constant of the form $\nu\phi\mathcal{N}$, and embedd the flow as a level-set. In the context of image processing, we take ϕ to be a stopping term depending on the image. The new gradient term $\nabla\phi$ directs the curve towards the boundary of the objects, increasing attraction to them. Existence, uniqueness and stability results for the gradient active contour model above were studied in [5, 11]. See [5, 11] for details, examples, and relations with other active contours schemes.

4.3 Affine invariant geodesic active contours

We can now formulate the functionals that will be used to define the affine invariant snakes. Assume that $\phi_{aff} = \phi(w_{aff})$ is an affine invariant stopping term, based on the affine invariant edge detectors in [16], in analogy with the Euclidean case developed in previous Section. Therefore, ϕ_{aff} behaves as the weight ϕ in L_ϕ, being now affine invariant. As in the Euclidean case, we regard ϕ_{aff} as an affine invariant conformal factor, and replace the affine arc length element ds by a conformal counterpart $ds_{\phi_{aff}} = \phi_{aff}\, ds$ to obtain the first possible functional for the affine active contours

$$L_{\phi_{aff}} := \int_0^{L_{aff}(t)} \phi_{aff}\, ds. \tag{4.2}$$

The obvious next step is to compute the gradient flow corresponding to $L_{\phi_{aff}}$ in order to produce the affine invariant model. [2] Unfortunately, as we will

[2] Using the connection $ds = \kappa^{1/3} dv$ [19], $L_{\phi_{aff}} = \int_0^{L(t)} \phi_{aff}\, \kappa^{1/3}\, dv$.

see, this will lead to an impractically complicated geometric contour model which involves four spatial derivatives.

The snake model which we will use comes from another (special) affine invariant, namely *area*, cf. (3.1). Following the results in Section 3, we define the conformal area functional to be

$$A_{\phi_{aff}} := \int_0^1 [C, C_p]\phi_{aff}\, dp = \int_0^{L_{aff}(t)} [C, C_s]\phi_{aff}\, ds.$$

Using the definition $Y^\perp := (-y_2, y_1)^T$ where $(y_1, y_2)^T \in \mathbb{R}^2$, we obtain

Lemma 4.1.

$$\frac{dL_{\phi_{aff}}(t)}{dt} = -\int_0^{L_{aff}(t)} [C_t, (\nabla\phi_{aff})^\perp]ds + \int_0^{L_a(t)} \phi_{aff}\,\mu[C_t, C_s]ds. \quad (4.3)$$

$$\frac{dA_{\phi_{aff}}(t)}{dt} = -\int_0^{L_{aff}(t)} [C_t, (\phi_{aff}\, C_s + \frac{1}{2}[C, (\nabla\phi)^\perp C_s])]ds. \quad (4.4)$$

The affine invariance of the resulting variational derivatives follows from a general result governing invariant variational problems having volume preserving symmetry groups proved in [15].

We now consider the corresponding gradient flows computed with respect to $\|\cdot\|_{aff}$. First, the flow corresponding to the functional $L_{\phi_{aff}}$ is

$$C_t = \{(\nabla\phi_{aff})^\perp + \phi_{aff}\,\mu C_s\}_s = ((\nabla\phi_{aff})^\perp)_s + (\phi_{aff}\,\mu)_s C_s + \phi_{aff}\,\mu C_{ss}.$$

As before, we ignore the tangential components, which do not affect the geometry of the evolving curve, and so obtain the following possible model for geometric affine invariant active contours:

$$C_t = \phi_{aff}\,\mu\kappa^{1/3}\mathcal{N} + \langle((\nabla\phi_{aff})^\perp)_s, \mathcal{N}\rangle\mathcal{N}. \quad (4.5)$$

The geometric interpretation of the affine gradient flow (4.5) minimizing $L\phi_{aff}$ is analogue to the one of the corresponding Euclidean geodesic active contours [16]. This flow involves μ which makes it difficult to implement.

The gradient flow coming from the first variation of the modified area functional on the other hand is much simpler:

$$C_t = (\phi_{aff}\, C_s + \frac{1}{2}[C, (\nabla\phi_{aff})^\perp]C_s)_s. \quad (4.6)$$

Ignoring tangential terms, this flow is equivalent to

$$C_t = \phi_{aff}\,\kappa^{1/3}\mathcal{N} + \frac{1}{2}\langle C, \nabla\phi_{aff}\rangle\kappa^{1/3}\mathcal{N}. \quad (4.7)$$

Notice that both models (4.5) and (4.7) were derived for *convex curves*, even though the flow (4.7) makes sense in the non-convex case. Formal results regarding existence of (4.7) can be derived following [1, 4, 5, 11].

The figure below illustrates simulations of these active contour models (the implementation is as in [5, 11, 13], based on the level-sets formulation [17]).

References

1. L. Alvarez, F. Guichard, P. L. Lions, and J. M. Morel, *Arch. Rational Mechanics* **123**, 1993.
2. C. Ballester, V. Caselles, and M. Gonzalez, "Affine invariant segmentation by variational method," *Technical Report, U. of Illes Balears*, 1994.
3. W. Blaschke, *Vorlesungen über Differentialgeometrie II,* Verlag Von Julius Springer, Berlin, 1923.
4. V. Caselles, F. Catte, T. Coll, F. Dibos, *Num. Mathematik* **66**, pp. 1-31, 1993.
5. V. Caselles, R. Kimmel, and G. Sapiro, " Geodesic active contours," *International Journal of Computer Vision*, to appear. Also in *Proc. ICCV*, Cambridge, MA, June 1995.
6. L. D. Cohen, *CVGIP: IU* **53**, pp. 211-218, 1991.
7. O. Faugeras, *C. Rendus de l'Acad. des Sciences de Paris* **317**, pp. 565-570, 1993.
8. M. Gage and R. S. Hamilton, *J. Differential Geometry* **23**, pp. 69-96, 1986.
9. M. Grayson, *J. Differential Geometry* **26**, pp. 285-314, 1987.
10. M. Kass, A. Witkin, and D. Terzopoulos, *International Journal of Computer Vision* **1**, pp. 321-331, 1988.
11. S. Kichenassamy, A. Kumar, P. J. Olver, A. Tannenbaum, and A. Yezzi, "Conformal curvature flows: From phase transitions to active vision," to appear *Archive for Rational Mechanics and Analysis*. Also in *Proc. ICCV*, June 1995.
12. B. B. Kimia, A. Tannenbaum, and S. W. Zucker, *International Journal of Computer Vision* **15**, pp. 189-224, 1995.
13. R. Malladi, J. A. Sethian and B. C. Vemuri, *IEEE Trans. on PAMI* **17**, pp. 158-175, 1995.
14. P. J. Olver, G. Sapiro, and A. Tannenbaum, in [18].
15. [–] "Invariant geometric evolutions of surfaces and volumetric smoothing," *SIAM J. of Appl. Math.*, to appear.
16. [–] P. J. Olver, G. Sapiro, and A. Tannenbaum, "Affine invariant edge maps and active contours," *Geometry Center Technical Report* **90**, University of Minnesota, October 1995.
17. S. J. Osher and J. A. Sethian, *Journal of Computational Physics* **79**, pp. 12-49, 1988.
18. B. Romeny (Ed.), *Geometry Driven Diffusion in Computer Vision*, Kluwer, 1994.
19. G. Sapiro and A. Tannenbaum, *Journal of Functional Analysis* **119:1**, pp. 79-120, 1994.
20. [–] *International Journal of Computer Vision* **11:1**, pp. 25-44, 1993.
21. [–] *Indiana University Mathematics Journal* **42:3**, 1993.
22. D. Terzopoulos, A. Witkin, and M. Kass, *AI* **36**, pp. 91-123, 1988.
23. I. Weiss, *International Journal of Computer Vision*, pp. 207-231, 1993.

Convex Variational Segmentation of Multi-Channel Images

C. Schnörr

Universität Hamburg, FB Informatik, AB KOGS,
Vogt-Kölln-Str. 30, D-22527 Hamburg, Germany

Summary. A variational approach to nonlinear smoothing and segmentation of multi-channel images is presented. The corresponding functional is strictly convex and combines a standard first-order smoothness measure with a total variation based "discontinuity"-measure. The gradient dynamics perform a coupled nonlinear diffusion process that yields a segmentation of given images into smooth patches and transition regions. Compared to related variational models, the advantages of our model are: Uniqueness of the solution, continuous dependence on both data and parameters, consistent numerical approximation with the FEM-method. The approach involves two global parameters that determine the degree of smoothing and the sensivity to signal structures. Criteria to control these two parameters are given. Applications of our approach to the segmentation of color images and image motion vector fields are reported.

Keywords: computer vision, variational segmentation, nonlinear diffusion

1. Introduction

Image segmentation and feature extraction are basic tasks of low-level computational vision. These tasks require the use of local processes with different properties (detection/localization vs. smoothing) at the same time. Consequently, to compute a representation of images in terms of edges, corners and smooth patches, models have to be devised that combine these processes in a suitable way. Based on a comprehensive review of past research, Morel and Solimini [1] suggested variational approaches as proper models and advocated to consider the approach of Mumford and Shah [2] as a generic model for image segmentation. This approach directly addresses the problem to compute a cartoon-like segmentation of a given image in terms of smooth patches divided by (at least) rectifiable curves. Despite considerable mathematical progress with respect to the existence and regularity of functions minimizing the Mumford-Shah functional (see the survey [1]), sound algorithms have been reported only for a simplified version of the approach concerning piecewiese constant functions [3]. Besides the existence of many local minima, the major obstacle with respect to the evaluation of the Mumford-Shah functional appears to be the lack of consistent dicretization schemes (cf. [4]).

In the present paper, we consider a variation of the Mumford-Shah approach which mathematically is much more simpler [8]. Our variational model has always a unique solution $u(x)$ which can consistently be approximated with the standard FEM-method in the sense that $\|u - u_h\| \to 0$, where u_h is an approximate solution and the discretization parameter h goes to zero. Furthermore, the solution $u(x)$ continuously depends on both the data and the parameters involved. Corresponding perturbations therefore can only cause *smooth* changes of the computed representation, and control schemes with respect to parameter values can be devised. This is a decisive advantage over traditional methods of low-level computer vision (see the review in [1]).

A related variational approach (for scalar images) has been proposed by Nordström [5]. To our knowledge, an existence proof of minimizing solutions is not known. Concerning multi-channel images, many other attractive ideas for nonlinear diffusion approaches have been presented on an experimental basis in [6, 7].

2. Convex variational segmentation

2.1 Variational model

For given image data $g : \mathbb{R}^d \supset \Omega \to \mathbb{R}$, we consider the following minimization problem:

$$J(u) = \inf_{v \in H^1(\Omega)} J(v) , \qquad J(v) = \frac{1}{2} \int_\Omega \Big\{ (v-g)^2 + \lambda(|\nabla v|) \Big\} dx , \tag{2.1}$$

where

$$\lambda(t) = \begin{cases} \lambda_h^2 t^2 & , 0 \le t \le c_\rho , \\ \lambda_l^2 t^2 + 2(\lambda_h^2 - \lambda_l^2) c_\rho t - (\lambda_h^2 - \lambda_l^2) c_\rho^2 & , 0 < c_\rho \le t , \end{cases} \tag{2.2}$$

with parameters $\lambda_h, \lambda_l, c_\rho > 0$. The functional (2.1) is strictly convex, so that any given image data $g \in L^2(\Omega)$ is mapped to a unique minimizing function $u \in H^1(\Omega)$.

To interpret the cost functional (2.1), we drop in (2.2) the terms involving the small constant λ_l and use the notation

$$\Omega_r = \{x \in \Omega \, : \, |\nabla v(x)| < c_\rho\} , \quad \Omega_t = \{x \in \Omega \, : \, |\nabla v(x)| > c_\rho\} , \tag{2.3}$$

and

$$E_\alpha = \Omega_t \cap \{x : v(x) > \alpha\} . \tag{2.4}$$

Denoting further with $P(E_\alpha, \Omega_t)$ the perimeter of E_α in Ω_t, the cost functional J in (2.1) can be written as follows:

$$2J(v) = \int_\Omega (v-g)^2 dx + \lambda_h^2 \int_{\Omega_r} |\nabla v|^2 dx + 2\lambda_h^2 c_\rho \int_{\mathbb{R}} P(E_\alpha, \Omega_t) d\alpha - (\lambda_h c_\rho)^2 |\Omega_t| . \tag{2.5}$$

The first two terms in (2.5) are well-known, and the remaining terms define a "discontinuity"-measure in terms of the length of the level curves of $v(x)$

within Ω_t. Thus, depending on the parameters λ_h and c_ρ, a partition of the underlying domain Ω is obtained into regions with smooth patches Ω_r and regions with signal transitions Ω_t. Note that the measure with respect to Ω_t involves the total variation of $v(x)$ which has proven to be useful in the context of image restoration [9]. Accordingly, the variational approach (2.5) may be viewed as adaptively switching between smoothing of image data and restoration of signal transitions.

2.2 Scale and direction selective smoothing

We investigate the smoothing process encoded by the cost-functional (2.1). The unique minimizing function $u(x)$ may be interpreted as the weak steady-state solution to the nonlinear diffusion equation

$$\frac{dv}{dt} = \nabla \cdot \Big(\rho(|\nabla v|)\nabla v\Big) + g - v \,, \tag{2.6}$$

with Neumann boundary condition and $\rho(t) = \lambda'(t)/(2t) \;,\; t \geq 0$. First, we note that a constant diffusion coefficient $\rho(\cdot) = c$ yields the isotropic smoothing term: $\nabla \cdot (\rho(|\nabla v|)\nabla v) = c\Delta v$. By contrast, a diffusion coefficient as defined in (2.6) results in the anisotropic smoothing term:

$$\nabla \cdot \Big(\rho(|\nabla v|)\nabla v\Big) = \Big(\rho'(|\nabla v|)|\nabla v| + \rho(|\nabla v|)\Big)\left(\frac{d^2}{de_1^2}v\right) + \rho(|\nabla v|)\left(\frac{d^2}{de_2^2}v\right) \,, \tag{2.7}$$

where e_1, e_2 denote the directions parallel and perpendicular to the gradient of v. Computing the corresponding coefficient functions in (2.7), we obtain the following scale and direction selective smoothing characteristics of (2.7):

The scale varies between λ_h and $\lambda_l \ll \lambda_h$, and that like a step function along the direction of the gradient of v, and $\sim |\nabla v|^{-1}$ along the direction perpendicular to the gradient. Thus, a certain amount of smoothing remains *along* edges, which is a consequence of the third term in (2.5). Decrease of scale only happens at locations with significant structures (as specified by the parameter c_ρ) and automatically adapts to the spatially varying directions e_1, e_2.

2.3 Control of parameter values

To determine the parameter λ_h, we adopt the usual representation of image data g as an unknown function u and additive noise n, $g = u + n$, and assume the noise level to be known: $\int_\Omega n^2 dx = c$. Next, we introduce the notation $\alpha = 1/\lambda_h^2$ and modify the functional J in (2.1): $\tilde{J}(v) = F(v) + \alpha G(v)$, where

$$F(v) = \frac{1}{2\lambda_h^2}\int_\Omega \lambda(|\nabla v|)dx \,, \quad G(v) = \frac{1}{2}\int_\Omega (v-g)^2 dx \,. \tag{2.8}$$

Exploiting the known noise-level as a constraint we arrive at the minimization problem: $F(u) = \inf_{v \in \mathcal{M}_c} F(v)$; $\mathcal{M}_c = \{v \in \mathcal{H} : G(v) = c\}$, $\mathcal{H} = H^1(\Omega)$. A solution $u(x)$ is determined as the steady-state solution $v(t = \infty)$ to the following differential equation on $\mathcal{M}_c$ ($\mathcal{R} : \mathcal{H}^* \to \mathcal{H}$ denotes the Riesz-mapping):

$$\frac{dv(t)}{dt} = -\mathcal{R}F'(v) + \frac{\langle F'(v), \mathcal{R}G'(v)\rangle}{\langle G'(v), \mathcal{R}G'(v)\rangle}\mathcal{R}G'(v) , \quad v(0) \in \mathcal{M}_c . \qquad (2.9)$$

$u(x)$ is unique within the intersection of $\mathcal{M}_c$ and a ball around $v(0)$ (see [10]. For λ_h we obtain:

$$\lambda_h(u) = \alpha^{-1/2}(u) , \quad \alpha(u) = -\frac{\langle F'(u), \mathcal{R}G'(u)\rangle}{\langle G'(u), \mathcal{R}G'(u)\rangle} .$$

To determine the parameter c_ρ, we propose to pre-specify the size of the area (but not the location, of course) of a subset in the image where significant signal variations are expected: $|\Omega_t| = c$. For fixed λ_h, $|\Omega_t|$ monotonically increases for decreasing c_ρ. Therefore, it is not difficult to specify a control strategy to achieve $|\Omega_t| = c$ (see [10]).

3. Application to multi-channel images

3.1 Segmentation of color images

The generalization of problem (2.1) to the case of multi-channel images $g : \mathbb{R}^d \supset \Omega \to \mathbb{R}^n$ is straightforward:

$$J(u) = \inf_{v \in \mathcal{H}} J(v) , \quad J(v) = \frac{1}{2}\int_\Omega \Big\{|v - g|^2 + \lambda(||Dv||)\Big\}dx , \qquad (3.1)$$

with $\mathcal{H} = \left(H^1(\Omega)\right)^n$, $\lambda(\cdot)$ as in (2.2), $Dv = (\nabla v_1, \ldots, \nabla v_n)$, $(Du, Dv) = \mathrm{tr}(Du^t Dv)$, and $||\cdot||$ denoting the Frobenius norm. The minimizing vector field u is determined by the variational equation $\langle J'(u), v\rangle = 0$, $\forall v \in \mathcal{H}$, where

$$\begin{aligned} \langle J'(u), v\rangle &=: \langle A(u) - f, v\rangle & (3.2)\\ &= \int_\Omega \Big\{u \cdot v + \frac{\lambda'(||Du||)}{2||Du||}(Du, Dv)\Big\}dx - \int_\Omega g \cdot v dx . & (3.3)\end{aligned}$$

Using $\rho(t) = \lambda'(t)/(2t)$, $t \geq 0$ and the inequality $\big(\rho(x)x - \rho(y)y\big) \cdot (x - y) \geq c_1|x - y|^2$, $\forall x, y \in \mathcal{R}^{nd}$, we conclude that the operator A defined above is strongly monotone:

$$\begin{aligned} &\langle A(u) - A(v), u - v\rangle = & (3.4)\\ &\int_\Omega \Big\{|u - v|^2 + \Big(\rho(||Du||)Du - \rho(||Dv||)Dv, Du - Dv\Big)\Big\}dx & (3.5)\\ &\geq c_2||u - v||^2_{\mathcal{H}} . & (3.6)\end{aligned}$$

Hence the minimizer u exists and is the unique (weak) steady-state solution to the following system of coupled nonlinear diffusion equations (H_i denotes the Hessian of $v_i(x)$):

$$\frac{dv_i}{dt} = \rho'(\|Dv\|)\nabla v_i \cdot \frac{1}{\|Dv\|}\sum_{j=1}^{n} H_j \nabla v_j + \rho(\|Dv\|)\Delta v_i - (v_i - g_i) , \quad (3.7)$$

$i = 1, \ldots, n$. Comparison with eqn. (2.6) shows how the smoothing process with respect to image channel $v_i(x)$ interacts with those of the other image channels.

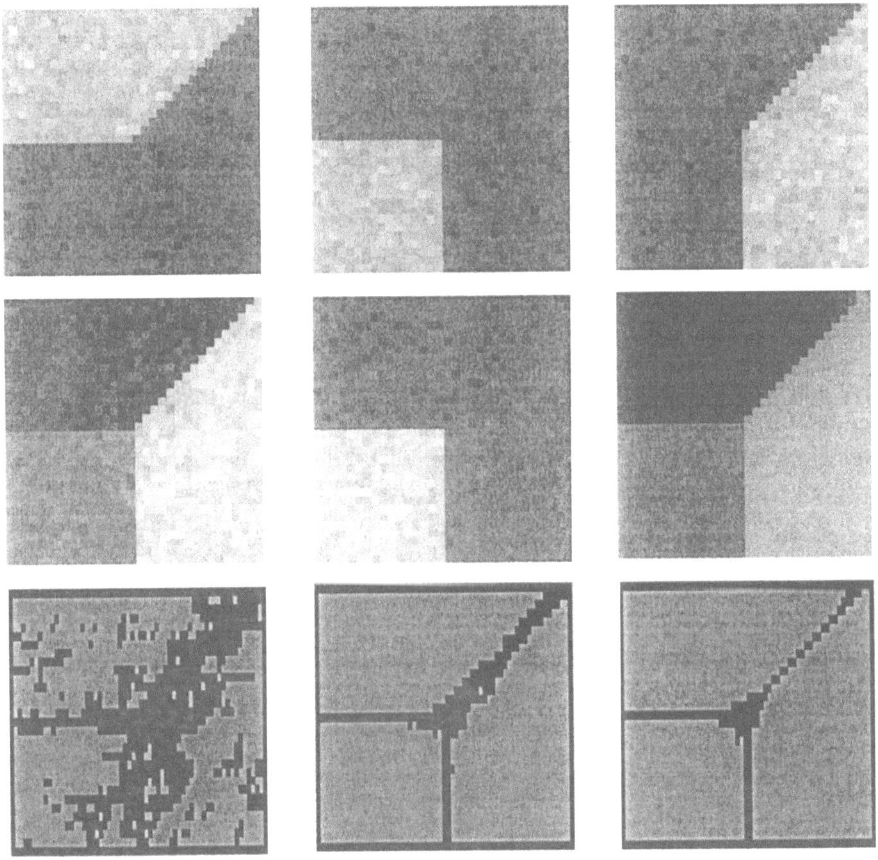

Fig. 3.1. Top: Noisy red, green and blue input images. Middle row: The two extracted feature channels (left and center, with the mean shifted for better visibility), and the first channel of the nonlinearly smoothed feature vector image ($\lambda_h = 4.0, c_\rho = 3.0$). Bottom, left to right: Segmentations obtained with decreasing sensivity to signal transitions ($\lambda_h = 4.0; c_\rho = 0.3, 1.0, 3.0$). The set Ω_t is marked with black.

In the case of color images we have $g : \mathbb{R}^2 \supset \Omega \to \mathbb{R}^3$, where the functions g_1, g_2, g_3 denote the red, green and blue image channel, respectively. The approach (3.1) is applied to the functions $\mathbb{R}^2 \ni \tilde{g}(x) := W^t g(x)$, $W = (w_1, w_2)$, where the feature vectors w_1, w_2 denote the eigenvectors corresponding to the largest eigenvalues of the sample-covariance matrix $E\Big\{(g - E\{g\})(g - E\{g\})^t\Big\}$. w_1 and w_2 compress in an unsupervised way the maximal amount of information provided by the sum of the within-cluster scatter matrix and the scatter matrix of the cluster means.

The essential properties of the approach are demonstrated in Figure 3.1. Note that the red, green and blue channel have been chosen such that a greyvalue-channel would merely "see" the superimposed noise signal. Furthermore, observe how the feature extraction stage compresses the three input channels and provides two complementary images. The interaction of the corresponding two nonlinear smoothing processes yields a correct segmentation.

3.2 Application to image motion vector fields

Suppose that a camera moves relative to its environment. Then the corresponding motion vector field is defined by the velocities $(\dot{x}_1, \dot{x}_2)^t$ of projected scene-points within the image plane. Now suppose that an estimate $\big(g_1(x), g_2(x)\big)^t$ of the motion field, often called optical flow field, has been extracted from the image sequence by some technique. Approach (3.1) may be applied to obtain a segmentation of this vector field. This task is important for separating moving objects from the background, for example. Numerical examples have been reported in [11].

As an alternative to the approach (3.1), we propose to minimize the following functional:

$$J(v) = \int_\Omega \Big\{ |v - g|^2 + \frac{1}{2}\Big(\lambda_d\big(div(v)\big) + \lambda_r\big(rot(v)\big) + \lambda_s\big(|sh(v)|\big)\Big)\Big\} dx \; , \tag{3.8}$$

where

$$div(v) = v_{1,x1} + v_{2,x2} \; , \quad rot(v) = v_{2,x1} - v_{1,x2} \; , \tag{3.9}$$

$$sh(v) = (v_{2,x2} - v_{1,x1}, \; v_{1,x2} + v_{2,x1})^t \; . \tag{3.10}$$

The functions $\lambda_d(\cdot), \lambda_r(\cdot), \lambda_s(\cdot)$ are defined according to (2.2). In contrast to (3.1), the functional (3.8) involves measures of quantities that are more directly related to kinematical behaviours of moving objects (divergence $div(v)$ may signal an approaching object, for example). Using the identity

$$\frac{1}{2}\Big(div^2(v) + rot^2(v) + |sh(v)|^2\Big) = (Dv, Dv) \; , \tag{3.11}$$

it is not difficult to show that $J(\cdot)$ in (3.8) is strictly convex.

The essential properties of the approach are demonstrated in Figure 3.2. Note that the nonlinear smoothing process encoded by (3.8) does not necessarily produce piecewise constant vector fields but is able to adapt to the locally dominant flow structure in terms of the vector gradient components divergence, vorticity and shear. Note also that the motion boundary is preserved.

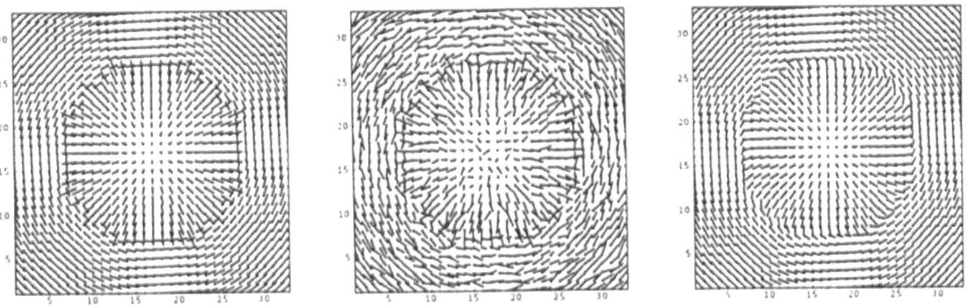

Fig. 3.2. A vector field (left), contaminated with noise (middle), and the unique minimizer of the functional in (3.8) (right).

References

1. J.-M. Morel and S. Solimini. *Variational Methods in Image Segmentation.* Birkhäuser, Boston, 1995.
2. D. Mumford and J. Shah. Optimal approximations by piecewise smooth functions and associated variational problems. *Comm. Pure Appl. Math.*, 42:577–685, 1989.
3. G. Koepfler, C. Lopez, and J. Morel. A multiscale algorithm for image segmentation by variational method. *SIAM J. Numer. Anal.*, 31(1):282–299, 1994.
4. S.R. Kulkarni, S.K. Mitter, T.J. Richardson, and J.N. Tsitsiklis. Local versus nonlocal computation of length of digitized curves. *IEEE Trans. Patt. Anal. Mach. Intell.*, 16(7):711–718, 1994.
5. N. Nordström. Biased anisotropic diffusion - a unified regularization and diffusion approach to edge detection. *Image and Vis. Comp.*, 8(4):318–327, 1990.
6. M. Proesmans, E. Pauwels, and L. van Gool. Coupled geometry-driven diffusion equations for low-level vision. In [12].
7. R. Whitaker and G. Gerig. Vector-Valued Diffusion. In [12].
8. C. Schnörr. Unique reconstruction of piecewise smooth images by minimizing strictly convex non-quadratic functionals. *J. of Math. Imag. Vision*, 4:189–198, 1994.
9. L.I. Rudin, S. Osher, and E. Fatemi. Nonlinear total variation based noise removal algorithms. *Physica D*, 60:259–268, 1992.
10. C. Schnörr. Representation of images by a convex variational diffusion approach. Technical Report FBI-HH-M-256/96, Universität Hamburg, FB Informatik, Germany, Feb. 1996.
11. C. Schnörr. Segmentation of visual motion by minimizing convex non-quadratic functionals. In *12th Int. Conf. on Pattern Recognition*, Jerusalem, Israel, Oct 9-13 1994.
12. Bart M. ter Haar Romeny, editor. *Geometry-Driven Diffusion in Computer Vision*, Dordrecht, The Netherlands, 1994. Kluwer Academic Publishers.

Segmentation d'images texturées par l'analyse anisotropique basée sur la transformation de Radon

Mo DAI

Equipe Image – Institut de Géodynamique
Université Michel de Montaigne – Bordeaux III
Avenue des Facultés, 33405 Talence Cedex, FRANCE
E-mail: dai@igd.u-bordeaux.fr

Résumé Cette communication présente une nouvelle approche de segmentation d'images texturées. Une image peut être reconstruite par la transformation de Radon inverse à partir de ses projections. En autre terme, l'image peut être modélisée par sa transformée de Radon inverse si la reconstruction est parfaite. L'analyse sur ce modèle montre que la contribution d'une projection pour la reconstruction est fortement liée à sa variation. Donc, on introduit une courbe de variance en fonction de l'angle projectif qui illustre l'importance de projections pour la reconstruction de l'image. C'est aussi cette courbe qui indique les quantités d'informations de l'image conservées dans les projections. Ensuite, un filtre multidirectionnel permettant de calculer les variances de projection pour un voisinage centré à un pixel est proposé pour la segmentation d'images constituées des régions remplies par des textures anisotropes. Les résultats expérimentaux donnés à la fin de la communication montrent la performance du filtrage.

Mots-clés Texture, Anisotropie, Transformation de Radon, Projection, Variance, Segmentation d'images, Filtrage multidirectionnel.

1. Introduction

La caractérisation texturale des objets est un élément fondamental dans le traitement d'images. Le concept de texture est lié à la perception visuelle. La texture se décrit en terme linguistique tel que la rugosité, le contraste, la finesse, la régularité, etc. La texture est une mesure quantitative pour décrire les contenus d'une région. On peut la considérer comme une structure spatiale constituée par l'organisation de primitives (motifs de base) ayant chacune un aspect aléatoire[1]. Au niveau microscopique, nous ne nous intéressons qu'à une primitive. A la perception visuelle, nos impressions sont plutôt macroscopiques, c'est-à-dire une organisation homogène. Cette homogénéité se traduit formellement par le concept "invariance par translation", c'est-à-dire l'observation d'une texture laisse la même impression visuelle quelle que soit la partie de la texture observée. Parfois l'anisotropie est une propriété très importante de textures, tant pour des images de microscope, sismiques que médicales. En fait une texture est anisotrope si dans laquelle l'arrangement de primitives s'oriente grossièrement le long d'un ou de plusieurs sens vus par les observateurs. La figure 1 donne deux exemples de textures anisotropes.

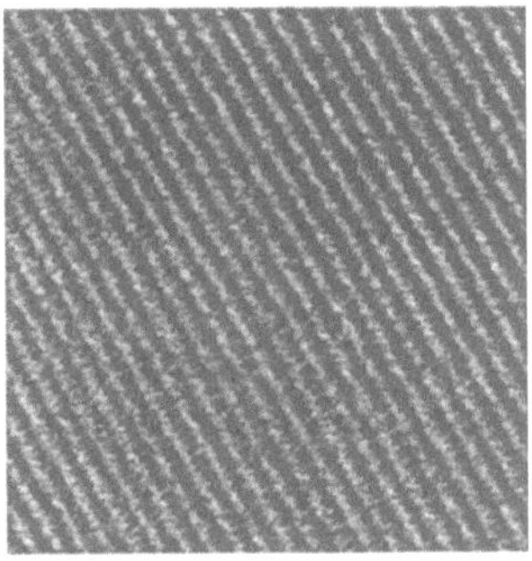

(a) (b)

Fig. 1 - Deux textures anisotropes (résolution des imges : 256x256)

En général deux approches principales ont été envisagées pour l'analyse de textures[2][3]: les méthodes dites statistiques qui sont fondées sur l'étude des fréquences spatiales d'une image, et les méthodes structurelles fondées elles, sur l'étude des primitives existant dans l'image. Depuis quelques années, un nombre de nouvelles méthodes sont proposées et basées sur la modélisation de textures, par exemple, le modèle fractal et multicannal. C'est ce dernier qui permet de réaliser l'analyse de textures anisotropes et de segmenter des images comportant des régions couvertes par les textures anisotropes[4].

Dans cette communication nous présentons une autre méthode qui peut être utilisée également pour analyser et segmenter les textures anisotropes et qui est basée sur une modélisation de la transformée de Radon. Dans un premier temps nous pouvons élaborer une nouvelle représentation de la texture. Il s'agit donc de transformer l'image de la texture dans le domaine de projections et de mesurer la variation de chaque projection. Ainsi nous pouvons obtenir une courbe, c'est-à-dire une représentation monodimensionnelle, permettant d'extraire l'information la plus importante sur l'anisotropie de la texture. A partir de cette caractérisation, nous pouvons réaliser une segmentation de l'image texturée par son anisotropie.

Après cette première partie introductive la suite de l'article sera composée de 4 parties. La partie 2 rappelle rapidement la définition de la transformation de Radon à 2D. La partie 3 présente la caractérisation d'une texture par la courbe de variance de projection. La partie 4 nous amènera à appliquer à la segmentation des images texturées à notre étude. Enfin dans la partie 5 nous conclurons en discutant les résultats obtenus, et nous présenterons quelques idées pour la suite des travaux.

2. Transformation de Radon

Soit une image $f(x,y)$ dont la transformée de Fourier s'exprime comme $F(u,v)$, la transformée de Radon de $f(x,y)$ peut être considérée comme une famille de projections. Chacune des projections est donc définie par la somme de $f(x,y)$ le long d'une ligne inclinée d'un angle ϕ par rapport à l'axe X et à une distance t de l'origine de XoY[5][6]. Mathématiquement, la transformée de Radon de f s'écrit comme

$$\Lambda_\phi f = p_f(\phi,t) = \sum_y \sum_x f(x,y)\delta(x\sin\phi - y\cos\phi - t) \qquad (1)$$

où $\Lambda_\phi f$ est la transformée de Radon de f en fonction de l'angle projectif ϕ.

Inversement, la donnée des projections pour tous les angles ($0 - \pi$) permet de déterminer f. La reconstruction de f à partir de l'ensemble des projections, c'est-à-dire, la transformée de Radon inverse s'exprime sous la forme:

$$\hat{f}(x,y) = \frac{1}{2\pi} \sum_{\phi=0}^{\pi} V_\phi p_f \qquad (2)$$

où $\hat{f}(x,y)$ est l'image reconstruite. On utilise ici le symbole $\hat{f}$ afin de la différencier de l'image originale f.

Pour un ϕ fixé, le terme $V_\phi p_f$ peut être interprété comme une rétro-projection, c'est-à-dire, une fonction de deux variables définie à partir une fonction d'une seule variable. Cette rétroprojection est en fait la convolution de $p_f(\phi,t)$ avec un filtre $r(t)$.

$$V_\phi p_f = p_f(\phi,t) * r(t) \qquad (3)$$

avec $F\,[r(t)] = |\omega| = (u^2+v^2)^{1/2}$ ($F\,[\bullet]$ étant l'opérateur de la transformation de Fourier).

Par conséquent, La transformée de Radon inverse est obtenue en deux étapes. D'abord, chaque projection $p_f(\phi,t)$ est filtrée par un filtre monodimensionnel dont la réponse de fréquence est $|\omega|$. Le résultat $V_\phi p_f$ est la rétroprojection. La reconstruction de f est alors un demi de la moyenne des rétroprojections intégrées sur ϕ.

Maintenant, on considère une image uniforme f dans laquelle l'intensité est égale à A partout. Donc, la reconstruction de cette image à partir de ses projections

$$\hat{f}(x,y) = \frac{1}{2\pi} \sum_{\phi=0}^{\pi} V_\phi p_f = \frac{1}{2\pi} \sum_{\phi=0}^{\pi} V_\phi(n_t A) \qquad (4)$$

où n_t est le nombre de pixels situés sur la même ligne projective à la distance t de l'origine de XoY. Si la reconstruction est parfaite, c'est-à-dire, $\hat{f} = f$, alors il est bien évident que

$$\frac{1}{2\pi} \sum_{\phi=0}^{\pi} V_\phi(n_t A) = A \qquad (5)$$

Pour une image quelconque f dont l'intensité moyenne est M_f, elle peut être considérée comme la somme de M_f et de la variation d'intensité par rapport à M_f pour chaque pixel. Ainsi, la formule de transformation de Radon inverse (2) peut se récrire comme la suivante:

$$\hat{f}(x,y) = \frac{1}{2\pi}\sum_{\phi=0}^{\pi} V_\phi p_f = M_f + \frac{1}{2\pi}\sum_{\phi=0}^{\pi} [\, V_\phi p_f - V_\phi (n_t M_f)\,]$$
$$= M_f + \frac{1}{2\pi}\sum_{\phi=0}^{\pi} r(t) * [\, p_f(\phi,t) - n_t M_f]$$
$$= M_f + \frac{1}{2\pi}\sum_{\phi=0}^{\pi} r(t) * n_t [\frac{p_f(\phi,t)}{n_t} - M_f] \qquad (6)$$

On note $\overline{\Lambda}_\phi f = p_f(\phi,t) \,/\, n_t$. Ceci dit qu'on peut redéfinir la projection par la moyenne de $f(x,y)$ au lieu de prendre la somme le long d'une ligne orientée dans un angle ϕ. La figure 2 montre la transformée de Radon de la figure 1(a) et celle de la figure 1(b).

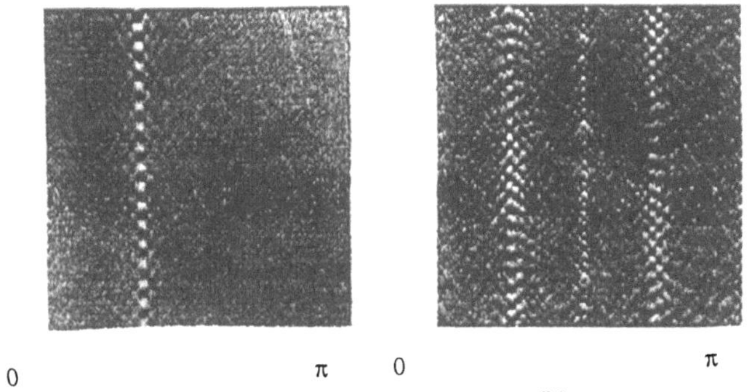

(a) (b)

Fig. 2 - Transformées dc Randon dc Fig.1

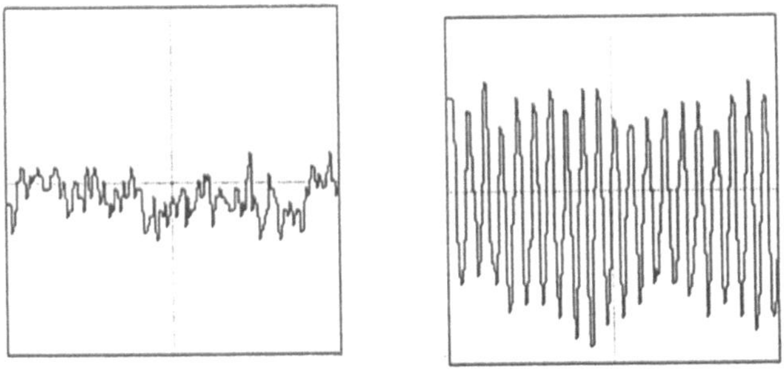

(a) (b)

Fig. 3 - Profils de Fig.2(b)
(a) colonne 29; (b) colonne 46.

Pour simplifier le problème, on considère n_t comme une constante N pour toutes les t, alors la reconstruction de f à partir de $\overline{\Lambda}_\phi f$

$$\hat{f}(x,y) = M_f + \frac{N}{2\pi}\sum_{\phi=0}^{\pi} V_\phi (\overline{\Lambda}_\phi f - M_f) \qquad (7)$$

Observant l'équation ci-dessus(7), il est clair que la contribution de $\overline{\Lambda}_\phi f$ pour la reconstruction de f dépend de sa variation par rapport à la moyenne des intensités de f. Plus la variation est élevée, plus la contribution pour la reconstruction est importante. Selon ce principe, la reconstruction peut être effectuée à partir des projections partielles dont les variations sont les plus importantes. Cela peut servir pour la compression des images[7]. Les profils montrés dans la figure 3 correspondent aux colonnes 29 et 46 de la figure 2 (b). Respectivement, la colonne 46 contient beaucoup plus d'information sur l'image originale (la figure 1 (b)) et la colonne 29 est beaucoup moins informative.

3. Courbe de variance de projection

Une description convenable permettant d'illustrer l'importance à la conservation des informations dans une image pour chacune de ses projections peut être intervenue. C'est la courbe de variance de projection (CVP) en fonction de l'angle projectif ϕ. Pour un ϕ fixé, la variance de la projection $\overline{\Lambda}_\phi f$ est calculée par

$$\mathrm{var}(\overline{\Lambda}_\phi f) = \frac{1}{T}\sum_t \left[\frac{p_f(\phi,t)}{N} - M_f\right]^2 = \frac{1}{TN}\sum_t p_f(\phi,t)^2 - M_f^2 \quad (8)$$

T étant le nombre de lignes projectives. La figure 4 donne la CVP de la figure 1(a) et celle de la figure 1(b).

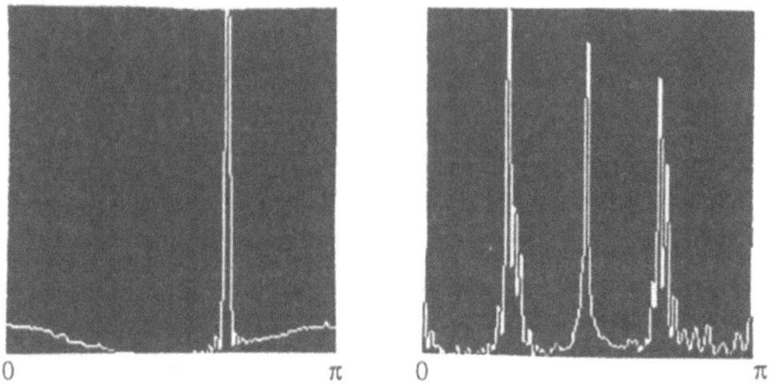

(a) (b)

Fig. 4 - CVPs de Fig.1

Observant les courbes dans Fig. 4, on trouve que respectivement il y a 1 pic sur la courbe de la figure 1(a) et 4 pics sur la courbe de la figure 1(b) et que les abscisses des pics correspondent aux orientations principales de primitives dans les textures. Cela signifie que les projections suivant les orientations principales d'une texture anisotrope contiennent la majorité d'informations sur la structure de la texture. Cela n'est pas seulement significatif pour la reconstruction et la compression des images, mais peut-être aussi nous offre une approche pour l'analyse des images texturées.

4. Segmentation d'images texturées

La segmentation est un traitement de bas-niveau qui consiste à créer une partition de l'image f en sous-ensemble R_i, appelés régions[8]. Une telle région est un ensemble connexe de pixels ayant des propriétés communes parmi lesquelles la texture est un attribut très important.

Comme on a décrit dans la section III, la CVP peut être utilisée pour mesurer l'anisotropie d'une texture. De plus, en utilisant le même principe on peut effectuer une segmentation d'images texturées. Suppose maintenant qu'une image $f(x,y)$ peut être partitionnée en K régions texturées R_1, R_2, ..., R_K et que chaque région texturée comporte une orientation privilégiée ϕ_i (i =1, 2, ..., K). Un filtre spatial $P_{\phi,\lambda}$ permettant d'estimer la variance de la projection $\overline{\Lambda}_\phi f$ pour une sous-image $f(m,n)$ centrée au pixel (x,y) est défini par

$$P_{\phi,\lambda}[(f(x,y)] = \mathrm{var}\,[\overline{\Lambda}_\phi f\,(m,n)] \tag{9}$$

avec $|(m-x)\cos\phi + (n-y)\sin\phi| \leq \lambda$ ou $|(x-m)\sin\phi + (y-n)\cos\phi| \leq \lambda$. C'est-à-dire, $N = 2\lambda + 1$ dans (8) pour tous les ϕ.

Si le paramètre ϕ est choisi comme l'orientation de l'une des régions texturées avec un convenable λ, la réponse du filtrage sera maximale pour tous les pixels dans cette région parmi tous les filtres possibles. Ainsi, on peut définir un filtrage multi-directionnel (FMD) par un ensemble de $P_{\phi,\lambda}$:

$$\mathrm{FMD} = \{P_{\phi 1,\lambda 1}, P_{\phi 2,\lambda 2}, \ldots, P_{\phi K,\lambda K}\} \tag{10}$$

pour la segmentation de $f(x,y)$. Donc, le système de segmentation peut être donné comme le suivant :

Pour chaque pixel (x,y) dans l'image f,

$$\text{si } k = \arg\{\max_{1\leq i\leq K} P_{\phi i,\lambda i}[f(x,y)]\}, \text{ alors } (x,y) \in R_k . \tag{11}$$

Le résultat obtenu par le système (11) est souvent insuffisamment lissé pour une segmentation consistante. Pour l'améliorer, un postfiltrage est proposé d'appliquer à la réponse de chaque $P_{\phi,\lambda}$. Le filtre de type Gaussien est un bon postfiltre. Avec ce postfiltre, l'algorithme de segmentation devient :

Pour chaque pixel (x,y) dans l'image f,

$$\text{si } k = \arg\{\max_{1\leq i\leq K} G(x,y) * P_{\phi i,\lambda i}[f(x,y)]\}, \text{ alors } (x,y) \in R_k . \tag{12}$$

où G est le postfiltre du lissage Gaussien. L'organigramme du système défini dans (12) est illustré dans la figure 5.

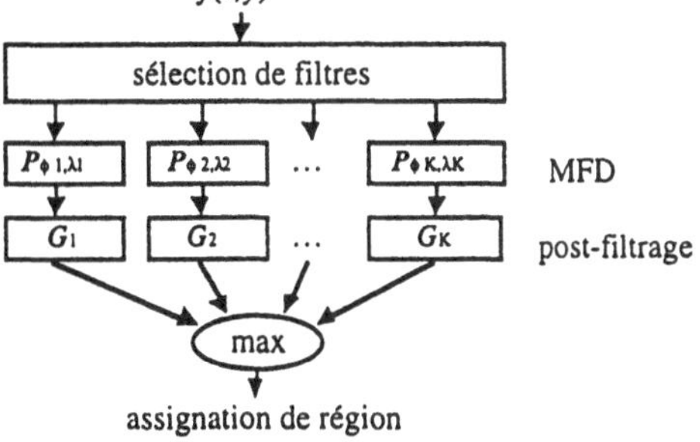

Fig. 5 – Organigramme du système de segmentation

La figure 6 (a) montre le résultat de segmentation de la figure 1(b) en utilisant le système (11) avec deux $P_{\phi,\lambda}$. On trouve qu'il y a des points mal classés dedans. Après avoir postfiltré les réponses de deux $P_{\phi,\lambda}$, on a obtenu une bonne segmentation (voir la figure 6(b)). La figure 7 est un autre exemple de la segmentation en 5 régions par le système (12). Le résultat est très satisfaisant.

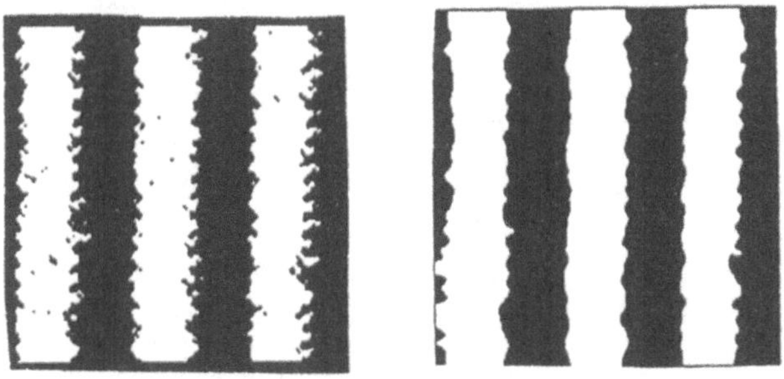

(a) (b)

Fig. 6 – Segmentation de Fig.1(b) en utilisant 2 filtres (ϕ_1=45°, λ_1=9; ϕ_2=135°, λ_2=9)
(a) sans postfiltrage; (b) avec postfiltrage.

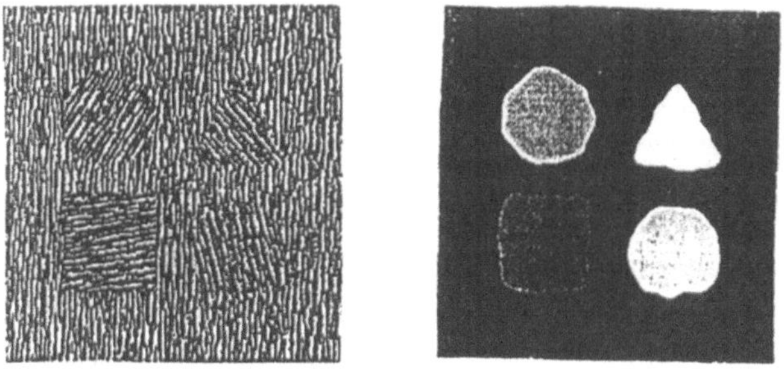

(a) (b)

Fig. 7 – Segmentation en 5 régions avec postfiltrage
(ϕ_1=10°, ϕ_2=60°, ϕ_3=90°, ϕ_4=105°, ϕ_5=130°; λ_1=λ_2=λ_3=λ_4=λ_5=7)
(a) image originale; (b) résultat de segmentation.

5. Conclusion et discussion

A l'aide de la transformation de Radon, une image peut être modélisée comme sa reconstruction à partir de ses projections. Dans cette communication, on a montré que la variance d'une projection donne une mesure de l'importance de cette projection pour la reconstruction. A partir de cette propriété, on peut caractériser une texture par la courbe de variance de projection (CVP) et segmenter une image texturée par le filtrage multidirectionnel (FMD). Les résultats d'essai sont efficaces tant sur les images synthétiques que les images réelles en utilisant le FMD suivi un postfiltrage.

Pour continuer cette étude, quelques axes sont envisagés. Ce sont de combiner cette approche avec les autres techniques pour améliorer les résultats, d'utiliser les rétroprojections au lieu des projections pour la segmentation et la classification et de détecter l'orientation pour chacun des pixels afin de les regrouper en fonction de l'orientation.

Références

[1] A. Gagalowicz - "Vers un modèle de textures", Thèse d'Etat, Univ. Pierre et Marie Curie, Paris VI, 1983.

[2] R. M. Haralick - *Statistical and stractual approaches to texture*, Proc. IEEE 67, pp 786-804, 1979.

[3] T. R. Reed and J. M. Hans du buf - *A review of recent texture segmentation and feature extraction techniques*, Image understanding, vol. 57, No. 3, May pp 359-372, 1993.

[4] A. C. Bovik *et al.* - *Multicannel texture analysis using localized spatial filters*, PAMI, vol. 12, pp 55-73, 1990.

[5] Th. Pavlidis - "Algorithms for graphics and image processing", Computer Science Press, 1982.

[6] A. K. Jain - "Fundamentals of digital image processing", Prentice-Hall International, 1989.

[7] F. Matus and J. Flusser - *Image representation via a finite Radon transform*, PAMI, vol. 15, pp 996-1006, 1993.

[8] J.-P. Cocquerez et S. Philipp - "Analyse d'images : filtrage et segmentation", Masson, 1995.

Adaptative image decompositions in a wavelets dictionary

François Bergeaud[1,2] and Stéphane Mallat[2,3]

[1] Ecole Centrale Paris, Groupe de Recherche M.A.S., Grande voie des vignes, 92290 Châtenay-Malabry, France
[2] Courant Institute of Mathematical Sciences, New York University, 251, Mercer Street, New York, NY 10012, U.S.A.
[3] Ecole Polytechnique, C.M.A.P., France

Summary. A crucial problem in image analysis is to construct efficient low-level representations of an image, providing precise characterization of features wich compose it, such as edges and texture components.

An image usually contains very different types of features, which have been successfully modelled by the very redundant family of 2D Gabor oriented wavelets, describing the local properties of the image: localization, scale, preferred orientation, amplitude and phase of the discontinuity.

However, this model generates representations of very large size. Instead of decomposing a given image over this whole set of Gabor functions, we use an adaptative algorithm (called Matching Pursuit) to select the Gabor elements wich approximate at best the image, corresponding to the main features of the image .

This produces compact representation in terms of few features that reveal the local image properties. Results proved that the elements are precisely localized on the edges of the images, and give a local decomposition as linear combinations of "textons" in the textured regions.

We introduce a fast algorithm to compute the Matching Pursuit decomposition for images. We address the problems of texture segmentation and image denoising based on these decompositions.

1. Introduction

The complexity of image structures including different types of textures and edges requires flexible image representations. Although an image is entirely characterized by its decomposition in a basis, any such basis is not rich enough to represent efficiently all potentially interesting low-level structures. Some image components are diffused across many bases elements and are then difficult to analyze from the basis representation. This is like trying to express oneself in a language including a small dictionary. Non available words must be replaced by long awkward sentences. To provide explicit information on important local properties, the image is represented as a sum of waveforms selected from an extremely redundant dictionary of oriented Gabor functions. As opposed to previous approaches, we do not decompose the image over the whole dictionary, but like in a sentence formation, we select the most appropriate Gabor waveforms to represent the image. Instead of increasing the representation by a large factor as in typical multiscale Gabor representations, the adaptive choice of dictionary vectors defines a compact representation that takes advantage of the flexibility offered by the dictionary redundancy.

There is an infinite number of ways to decompose an image over a redundant dictionary of waveforms. The selection of appropriate waveforms to construct the image representation is obtained by constructing efficient image approximations from few dictionary vectors. The optimization of the approximation is not intended for data compression but as a criteria for feature selection. If most of the image is recovered as a sum of few dictionary vectors, these vectors must closely match the local image properties. One can however prove that finding optimal approximations in redundant dictionaries is an NP complete problem. The redundancy opens a combinatorial explosion. This explosion is avoided by the matching pursuit algorithm that uses a non optimal greedy strategy to select each dictionary elements. For a dictionary of Gabor functions, the greedy optimization of the image approximation leads to an efficient image representation where each Gabor waveform reflects the orientation, scale and phase of local image variations. For textures, the selected Gabor elements can be interpreted as textons where as along edges, the multiscale properties of these Gabor elements reflect the edge properties. When the image is translated or rotated, the selected Gabor elements are translated and rotated. A fast implementation of this algorithm and numerical examples are presented.

2. The 2D Gabor Wavelets dictionary

Image decompositions in families of Gabor functions characterize the local scale, orientation and phase of the image variations. Gabor functions are constructed from a window $b(x,y)$, modulated by sinusoidal waves of fixed frequency ω_0 that propagate along different direction θ with two different phases $\phi = 0$ and $\phi = \frac{\pi}{2}$

$$b_{\theta,\phi}(x,y) = b(x,y)\cos(\omega_0(x\cos\theta + y\sin\theta) + \phi). \tag{2.1}$$

Each of these modulated windows can be interpreted as wavelets having different orientation selectivities. The window $b(x,y)$ is not chosen to be a Gaussian but is a compactly supported box spline that is adjusted so that the average of $b_{\theta,\phi}(x,y)$ is zero for all orientations and phases.

These oriented wavelets are then scaled by s and translated to define a whole family of Gabor wavelets $\{g_\gamma\}_{\gamma\in\Gamma}$ with:

$$g_\gamma(x,y) = \frac{1}{s}b_{\theta,\phi}(\frac{x-u}{s}, \frac{y-v}{s}) \tag{2.2}$$

where the multi-index parameter $\gamma = (\theta, \phi, s, u, v)$ carries the orientation, phase,scale and position of the corresponding Gabor function.

The Gabor transform of an image $f(x,y)$ is defined by the inner product

$$Gf(\gamma) =< f, g_\gamma >= \int\int_{R^2} f(x,y) g_\gamma(x,y)\, dx\, dy..$$

The orientation parameter θ allows to match the different orientation of image structures. The scaling factor s allows to zoom into singularities at fine scales but also to recover large scale image variations. The phase is a convenient parameter to modify the number of vanishing moments of the wavelet. When $\phi = 0$, the Gabor transform provides a second order partial derivative along θ of the image that is smoothed by a dilated window. When $\phi = \frac{\pi}{2}$, the Gabor transform can be interpreted as a first order partial derivative and thus responds particularly to discontinuities and edges.

The Fourier transform of a Gabor function is a waveform whose energy is well concentrated in the Fourier plane. In numerical computations, the scale is restricted to powers of two $\{2^j\}_{j\in Z}$ and the angles are discretized. The Gabor dictionary used in this paper includes 8 orientations. To define a complete representation, we guaranty that the whole Fourier plane is covered by dilations of the 8 elementary Gabor wavelets.

The decomposition of images in a Gabor dictionary defines a very redundant representation. For an image of 512 by 512 pixels, 8 orientations, 6 octaves and a two phases ($\phi = 0$ and $\phi = \frac{\pi}{2}$) representation would correspond to 96 images of 512 by 512 pixels. These images could be subsampled but we then lose the translation invariance of the representation. Instead of decomposing the image over the whole dictionary, we select specific Gabor waveforms that provide an efficient image approximation.

3. Matching Pursuit

We consider the general problem of decomposing a signal f over a dictionary of unit vectors $\{g_\gamma\}_{\gamma\in\Gamma}$ whose linear combinations are dense in the signal space $\mathcal{H}$. The smallest possible dictionary is a basis of $\mathcal{H}$; general dictionaries are redundant families of vectors. When the dictionary is redundant, unlike the case of a basis, we have some degree of freedom in choosing a signal's particular representation. This freedom allows us to choose few dictionary vectors, whose linear combinations approximate efficiently the signal. The chosen vectors highlights the predominant signal features. For any fixed approximation error ϵ, when the dictionary is redundant, we can show that finding the minimum number of dictionary elements that approximates the image with an error smaller than ϵ is an NP hard problem.

Because of the difficulty of finding optimal solutions, we use a greedy matching pursuit algorithm that has previously been tested on a one-dimensional signal.The matching pursuit [8] uses a greedy strategy that computes a good suboptimal approximation. It successively approximates a signal f with orthogonal projections onto dictionary elements.

Let $R^0 f = f$. Suppose that we have already computed the residue $R^k f$.

We choose $g_{\gamma_k} \in \mathcal{D}$ such that:

$$|< R^k f, g_{\gamma_k} >| = \sup_{\gamma \in \Gamma} |< R^k f, g_\gamma >|. \tag{3.1}$$

and project $R^k f$ on g_{γ_k}

$$R^{k+1} f = R^k f - < R^n k f, g_{\gamma_k} > g_{\gamma_k}. \tag{3.2}$$

which defines the residue at the order k+1. The orthogonality of $R^{k+1} f$ and g_{γ_k} implies

$$\|R^{k+1} f\|^2 = \|R^k f\|^2 - |< R^k f, g_{\gamma_k} >|^2. \tag{3.3}$$

By summing (3.2) for k between 0 and $n-1$, we obtain

$$f = \sum_{k=0}^{n-1} < R^k f, g_{\gamma_k} > g_{\gamma_k} + R^n f. \tag{3.4}$$

Similarly, summing (3.3) for k between 0 and $n-1$ yields

$$\|f\|^2 = \sum_{n=0}^{n-1} |< R^k f, g_{\gamma_k} >|^2 + \|R^n f\|^2. \tag{3.5}$$

The residue $R^n f$ is the approximation error of f after choosing n vectors in the dictionary.

In infinite dimensional spaces, the convergence of the error to zero is shown [8] to be a consequence of a theorem proved by Jones [5]:

$$f = \sum_{n=0}^{+\infty} < R^n f, g_{\gamma_n} > g_{\gamma_n}, \tag{3.6}$$

and we obtain an energy conservation

$$\|f\|^2 = \sum_{n=0}^{+\infty} |< R^n f, g_{\gamma_n} >|^2. \tag{3.7}$$

In finite dimensional signal spaces, the convergence is proved to be exponential [8].

4. Accelerated Matching Pursuit

Despite the apparent brute force strategy of a matching pursuit, this algorithm can be accelerated in an efficient way by using a set of Matching Pursuit decompositions in dictionaries dynamically computed from the residues.

The accelerated method is based on two major points: as the search for the maximal coefficient (scalar product of the residue with a dictionary vector) and the update with all the dictionary elements are very time-consuming, we first compute a much smaller dictionary $\mathcal{D}_I \subset \mathcal{D}$ "adapted" to the initial image I. It is constructed such that the vectors of $\mathcal{D}_I$ have a scalar product with I wich are greater than ϵ_0. A Matching Pursuit decomposition of I in $\mathcal{D}_0$ is run, where, at each iteration the vectors of $\mathcal{D}_I$ whose scalar product with the current residue is less than ϵ_0 are removed (Dynamic Matching Pursuit). After a finite number of iterations n_0, $\mathcal{D}_0$ is empty. A new subdictionary D_1 "adapted" to the current residue is then computed and the decomposition goes on. We proved the convergence of this algorithm. Since it consists in successive decompositions in small dictionaries, it is shown to be much faster.

The second point is the use of compact-supported functions which allows the update to be executed efficiently. Indeed, at each iteration n, once we have an estimation of the best vector g_{γ_n} such that:

$$|< R^n f, g_{\gamma_n} >| = \sup_{\gamma \in \Gamma} |< R^n f, g_\gamma >| \tag{4.1}$$

we compute the inner product of the new residue $R^{n+1}f$ with any $g_{\gamma'} \in \mathcal{D}'$, with a linear updating formula derived from equation (3.2)

$$< R^{n+1} f, g_{\gamma'} > = < R^n f, g_{\gamma'} > - < R^n f, g_{\gamma_n} > < g_{\gamma_n}, g_{\gamma'} > . \tag{4.2}$$

This updating equation can be interpreted as an inhibition of $< R^n f, g_\gamma >$ by the cross correlation of g_{γ_n} and g_γ. The only coefficients $< R^n f, g_{\gamma'} >$ of the Hash table wich are modified are such that $< g_{\gamma_n}, g_{\gamma'} >$ is non zero. As we use compact supported Gabor wavelets, this set is indeed very sparse, and the updating formula only apply to few coefficients ($\mathcal{O}(N \log^2 N)$ in the average), compared to the size of the table ($\mathcal{O}(N^2)$) .

5. Results

The matching pursuit algorithm applied to a Gabor dictionary selects iteratively the Gabor waveforms, also called atoms, whose scales, phases, orientations and positions best match the local image variations.

In order to display the edge information (localization, orientation, scale, amplitude), we adopt the following convention: each selected Gabor vector g_γ for $\gamma_n = (\theta, \phi, 2^j, u, v)$ is symbolized by an elongated Gaussian function of width proportional to the scale 2^j, centered at (u, v), of orientation θ. The mean gray level of each symbol is proportional to $|< R^n f, g_{\tilde{\gamma}_n} >|$.

The image reconstruction is simply obtained with the truncated sum of the infinite serie decomposition (3.6).

The figure 6.1 illustrates the texture discrimination properties of the representation: a rotated straw texture image is inserted into a bigger same texture image.The straw texture has horizontal and vertical structures. At fine scales, most structures are vertical because the horizontal variations are relatively smooth. At the intermediate scale 2^2 we see the horizontal and vertical image structures..

The symbolic representations at scale 2^1 and 2^2 (cf. figure 6.1) show the behaviour of the texture relative to the scale. In addition, we clearly distinguish the texture edge as the boundary between the two textures at scale 2^1.

As noticed by Turner in [?], the time-frequency analysis performed using Gabor functions corresponds to a tiling of the 4-dimentional space (2D for spatial position, 2D for the frequency position) called "information hyperspace". Each element of the dictionary occupies a particular "volume" or cell in this space, and combine information in space and frequency related to the structural description of the texture: the response to the spatially large Gabor filters and small spatial frequency extent give a precise information on the periodicity of the texture. On the other hand, the response to the spatially small filters and large spatially frequency extent distinguish texture elements and provide a mean to characterize the density of elements. The intermediate Gabor filters allow simultaneous measurements of textons and their distribution.

6. Texture segmentation

In the example of the rotated straw texture inserted into a bigger straw texture image, whe compute for each atom its local coefficient distribution relative to the orientation. The upper right image of figure 6.1 shows the image of the locally most probable orientation. The discrimination is thus achieved using the structural information embedded in the local texton orientations.

7. Image denoising

If we consider the image as a realization of a stochastic process, the denoising problem is defined as the estimation of the conditional expectation of the ideal image, given the noisy image. Recent non-linear estimation techniques consist in thresholding the wavelet transform : high coefficients, corresponding to a discontinuity of the signal (edge) are kept, whereas the small coefficients, uniformly spread in the smooth regions, are removed. The threshold

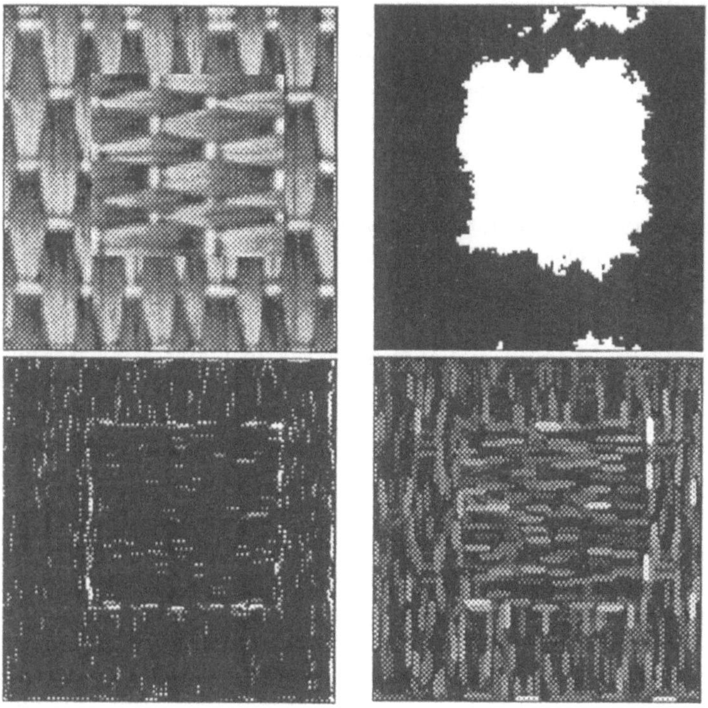

Fig. 6.1. Initial straw texture image (upper-left image) and best local orientation image (scales 2 + 3) (upper-right image). Symbolic representation of the Gabor vectors at scale 2^1 and 2^2.

T , introduced in the "wavelet shrinkage" method is the expectation of the maximum value of the amplitude of the coefficients of a white noise (variance σ^2) : $T = \sqrt{2\log(p)}\,\sigma$, where p is the dimension of the space.

Decompositions of noisy images with *Matching Pursuit* show that the computation of the "coherent structures" (atoms whose correlation with the dictionary is greater than a asymptotic value λ_∞) is equivalent to this non-linear estimation. This suggest the computation of the denoised image by computing the "coherent structures" of the noisy image.

The noisy lena image (SNR = 13.52 dB) of figure 7.1 was processed using this denoising scheme.

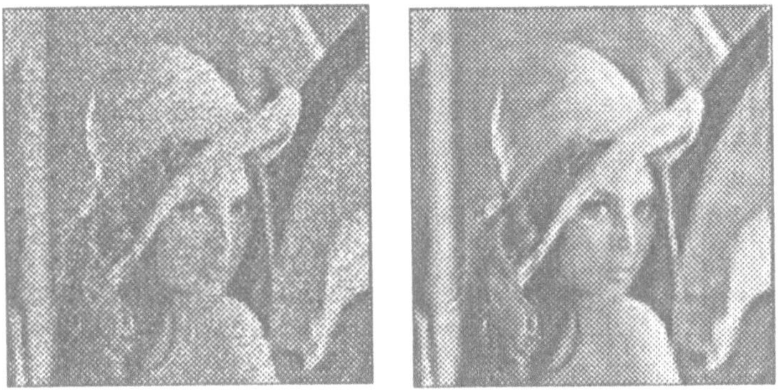

Fig. 7.1. Lena image with an addidive white Gaussian noise (SNR = 13.52 dB) and reconstruction from the thresholded atoms.

This result could be improved by computing the high-level geometrical links between the coherent structures given by the Matching Pursuit decomposition (by grouping atoms having the same orientations and similar amplitudes in an oriented neighbourhood). The geometrical relations between these structures are remaining in the presence of noise, whereas some structures are destroyed, and generally allow to recover the essential information embedded in the noise.

8. Vision applications

We introduced here a method to construct a decomposition of images into its main features. We showed that this transform provides a precise and complete characterization of the edges and texture components in terms of localization, orientation, scale and amplitude. By reconstructing high-visual quality images with very few atoms, we also showed that this representation is compact.

Another advantage of Matching Pursuit is the flexibility of the dictionary choice allowing to explicitely introduce *a priori* knowledge on the features of object classes into the dictionary to solve specific vision problems.

Experiments on image denoising suggest the use of a higher level information on edges obtained by atom linking to discriminate image structures from noise. The features carried by each atom such as the local orientation turn out to reveal the local configurations and allow structural discrimination between different textures.

A promising aspect of this method is its aptitude to model some early vision mechanisms in the visual cortex and especially those involved in texture discrimination. The Matching Pursuit with a Gabor dictionary acts as a linear filtering followed by a non-linearity (selection of the maximal coefficient) and some local inhibitions in the representation (update), which is very similar to the recent model of early vision mechanisms involved in texture discrimination and introduced by Malik and Perona [9].

References

1. J. Beck, "Textural segmentation", representation in perception, Beck J. ed., Organization and Lawrence Erlbaum Associates, Hillsdale, N.J., pp 285-317.
2. I. Daubechies, "Ten lectures on wavelets", CBMS-NSF Series in Appl. Math., SIAM, 1991.
3. J. Daugman, "Complete discrete 2D Gabor transform by neural networks for image analysis and compression", *IEEE Trans. on Acoustic, Speech, and Signal Processing*, ASSP-36, pp 1169-1179, 1988.
4. D. Gabor, "Theory of communication", *J. Inst. Elect. Ing.*, vol. 93, pp 429-457, 1946.
5. L. K. Jones, "On a conjecture of Huber concerning the convergence of projection pursuit regression", *The Annals of Statistics*, vol. 15, No. 2, pp. 880-882, 1987.
6. J. Jones, L. Palmer, "An evaluation of two-dimensional Gabor filters model of simple receptive fields in cat striate cortex", *J. Neurophisiol.*, vol. 58, pp. 538-539, 1987.
7. B. Julesz, J.R. Bergen, "Textons, the fundamental elements in preattentive vision and perception of textures", *The Bell System Tech. J.*, J. 62, pp. 1619-1645, 1983.
8. S. Mallat and Z. Zhang, "Matching Pursuit with time-frequency dictionaries", *IEEE Trans. on Signal Processing*, Dec. 1993.
9. J. Malik and P. Perona, "Preattentive texture discrimination with early vision mechanisms", *J. Opt. Soc. Am.*, vol. 7, No. 5, pp. 923-932, May 1990.
10. S. Marcelja, "Mathematical description of the response of simple cortical cells", *J. Opt. Soc. Am.*, vol. 70, No. 11, pp 1297-1300, November 1980.

Coiflets for Range Image Segmentation

† M. Djebali, ‡ K. Melkemi, † M. Melkemi and † D. Vandorpe

† LIGIM, Universite Claude Bernard Lyon1, Bat.710
43, Boulevad du 11 Novembre 1918
69622 Villeurbanne, Cedex
e-mail : mdjebali@ligim.univ-lyon1.fr

‡ LMC-IMAG, Universite Joseph Fourier
Domaine universitaire St. Martin d'Heres
38406 Grenoble Cedex.
e-mail : kmelkemi@imag.fr

Summary. In this paper, we propose a method for accelerating range image segmentation process which uses techniques of region-growing based on functions approximation and local neighborhood property. This method uses multiresolution wavelet transforms, where the used wavelets are *Coiflets*. The interesting property of this kind of wavelets is the interpolating characteristic of their associated scaling functions. This characteristic is due to the fact that both *Coiflets* and their scaling functions moments vanish. An overview on the proposed segmentation scheme gives the following description. First, input data is compressed to a fixed lower resolution (LR), then the partitioning process is applied on the compressed image. The result of this process is "projected" to the initial resolution (IR) and a global error criterion (GEC) is evaluated. The required result is obtained by increasing LR -if necessary- until the GEC is respected. Encouraging results are obtained for several range images.
Keywords : coiflets, curvatures, range image, segmentation, wavelets.

1. Introduction

Today there is a widespread interest in shape measurement and analysis of real objects because of the studies of pattern recognition, computer-aided design (CAD), medicine and apparel design. Industrial applications often require the definition of CAD models from objects. The design of various workpieces is still done manually using clay or plastic materials. CAD models often need to be constructed from such prototypes and the range images are the most appropriate tool for this aim. A range image RI is a digital surface, which represent a quantized discretely sampled version of objects surfaces. Each sample value represent the distance to a physical surface from a reference one. In several methods, the construction of CAD models pass through the segmentation of range images. The underlying structure of the sensed images is usually polynomial ; all physical objects of interest are polyhedrals or quadrics. The range images structure is a set of polynomial functions of degree not generally exceed in our case 2. The segmentation consists in partitioning each image into N regions. Each region R_l is often approximated in least squares sense by a polynomial function P_l

$$RI(x,y) = \sum_{l=1}^{N} P_l(x,y)\chi(x,y,R_l).$$

where

$$\chi(x,y) = \begin{cases} 1 & \text{if } (x,y) \in R_l \\ 0 & \text{otherwise} \end{cases}$$

The approximating functions are generally low-order bivariate polynomials. They can be written in the form of

$$P_l(x,y) = \sum_{i+j \leq m} a_{ij}x^i y^j$$

where $a_{ij} \in \mathbb{R}$, $(x,y) \in \mathbb{R}^+ \times \mathbb{R}^+$ and $m \in \mathbb{N}$, generally not exceed the value 4. The range image segmentation using techniques of region-growing based on function approximation and local neighborhood property, is computationally very expensive and many authors prefer a parallel formulation of such techniques to avoid huge times response. This solution presents certain drawbacks such as the necessity of particular equipments and the limit of the implementation transposability. The range images acquired for industrial applications are of huge size. They can overtake 1024×1024 pixels. The computational cost strongly depends on the amount of the input data. For example, using a sequential formulation, the necessary CPU time for segmenting images of size 128×128, vary between 240s and 540s.

It will be judicious, to make this computational cost beyond images sizes in input. We can reach this purpose by processing images at lower resolution. The use of multiple resolutions techniques can sometimes provides significant functional and computational advantages [5]. In this context, pyramids structures are the base of certain image processing works such as edge detection [4][6][7]. In our case, pyramidal approach is not a good scheme. Indeed, it presents weak points. We can cite particularly, the use of heuristics and the coarseness of the results, besides the loss of range images polynomial structures. In the past few years, wavelets have been developed both as a new analytic tool in mathematics and as a powerful source of practical tools for many applications, from differential equations to image processing. It's a kind of generalization of image pyramids concept. S. Mallat [3] proposes an algorithm to construct wavelet basis based on multiresolution analysis. This discrete wavelet transforms is apparently a good tool and to be useful, it must satisfy the condition that the segmentation results must be equivalent at both low and high resolution, in other words, bring images from higher to lower resolution may not alter the information contained in the image. Unfortunately, the most proposed wavelets basis does not present interpolating scaling functions. As a solution to this problem, I. Daubechies [8] has introduced the Coiflets : wavelets allowing a very good approximation of polynomial functions at different resolutions. The sampling is done in a way

that the polynomial structure at the initial resolution, is preserved with an insignificant error.
This paper is organized as follows. First we give a brief discussion on multiresolution analysis where we outline the interest of Coiflets for our application. Second we describe our segmentation scheme, which uses input images at lower scales by a discrete wavelet transforms based on Coiflets filter. Segmentation results using the proposed algorithm are very encouraging. Extensive experimentations are done to test the algorithm performances for several range images.

2. Multiresolution analysis

A wavelet orthonormal basis is an $L^2(\mathbb{R})$ basis constituted by a set of ψ_{jk} functions deducted from a single function ψ by translation and dilation and having regularity, decrease and oscillation (i.e. moments of ψ are nils) properties.
A multiresolution analysis [3] is a series of successive approximation closed subspaces $(V_j)_{j\in\mathbb{Z}}$ of $L^2(\mathbb{R})$ having the following properties

1. $\cup_{j\in\mathbb{Z}} V_j$ is dense in $L^2(\mathbb{R})$
2. $\cap_{j\in\mathbb{Z}} V_j = \{0\}$
3. $\forall f(x) \in L^2(\mathbb{R}), \forall j \in \mathbb{Z}, f(x) \in V_j \Leftrightarrow f(2x) \in V_{j+1}$
4. $\forall f(x) \in L^2(\mathbb{R}), \forall j, k \in \mathbb{Z}, f(x) \in V_0 \Leftrightarrow f(x-k) \in V_0$
5. A scaling function $\varphi(x) \in V_0$ exists such that the set $\{\varphi(x-l) \mid l \in \mathbb{Z}\}$ is an orthonormal basis of V_0.

We define also W_j as the orthogonal complementary space of V_j in V_{j+1}, such that $V_{j+1} = V_j \oplus W_j$, $f(x) \in W_j \Leftrightarrow f(2x) \in W_{j+1}$, and $f(x) \in W_0 \Leftrightarrow f(x+1) \in W_0$. Consequently $\bigoplus_{j=-\infty}^{+\infty} W_j = L^2(\mathbb{R})$.
We can show that exists a function $\psi \in V_0$ such as its whole translates from an orthonormal basis of W_0 verify the regularity, decrease and oscillation properties. Finally, the set $\{\psi_{jk}(x) = 2^{j/2}\psi(2^j x - l)\}_{l,j\in\mathbb{Z}}$ form an orthonormal basis of $L^2(\mathbb{R})$.
The interpretation of multiresolution analysis, can be stated as : given a function $f \in L^2(\mathbb{R})$, the orthogonal projection of f on spaces V_j represents more and more fine approximation while j increase. The orthogonal projection on W_j represents the details which appear at resolution $j+1$ and doesn't exist at the resolution j.
What about Coiflets ? : Coiflets are wavelets constructed by I. Daubechies [8] at the request of R. Coifman [9]. One can define Coiflets as follows. Let $L \in \mathbb{N}$, a Coiflet of L order is a wavelet ψ which verify :

1. $\int_{-\infty}^{+\infty} \phi(x)dx = 1.$
2. $\int_{-\infty}^{+\infty} x^l \phi(x)dx = 0$ for $l = 1, \ldots, L-1$.

3. $\int_{-\infty}^{+\infty} x^l \psi(x)dx = 0$ for $l = 0, \ldots, L-1$.

where ϕ is a scaling function associated to ψ. The Coiflets are wavelets coming from a multiresolution analysis such that the scaling function ϕ has a certain number of vanishing moments (except the first one because $\int_{-\infty}^{+\infty} \phi(x)dx = 1$ in every multiresolution analysis). The result of such construction is : for $K \in \mathbb{N}, L = 2K$, the support (supp) of ϕ is $[-L, 2L-1]$, and the ψ's one is $[-2L+2, L+1]$. The orthogonal projection of a function $f \in L^2(\mathbb{R})$ on the approximation space V_j is given by

$$f(x) = \sum_{l \in \mathbb{Z}} < f, \phi_{jl} > .\phi_{jl}(x) \tag{2.1}$$

with $\phi_{jl}(x) = 2^{j/2}\phi(2^j x - l)$, $j, l \in \mathbb{Z}$.

Property 1. *In the case of Coiflets, the polynomial functions of degree $d \leq L-1$ are in V_0, i.e. :*

$$\sum_{k \in \mathbb{Z}} k^l \phi(x-k) = x^l \tag{2.2}$$

where $l = 0, \ldots, L-1$.

Property 2. *Let $f \in L^2(\mathbb{R})$. If f is polynomial of degree less than L on the interval $[a, b]$ then*

1. $< f, \phi_{jk} >= 2^{-j/2} f(2^{-j}k)$ *for* $2^j a + L < k < 2^j b - 2L + 1$.
2. $< f, \psi_{jk} >= 0$ *if* $2^j a + 2L - 1 < k < 2^j b - L + 1$.

The proofs of the above properties are detailed in [8][9][10]. The Property 2 make Coiflets suited for range image segmentation processing. At each scale level, the sequence $\{< f, \phi_{jk} >\}_{k \in \mathbb{Z}}$ is the sequence of f samples $\{f(2^{-j}k)/2^{-j/2}\}_{k \in \mathbb{Z}}$. In other words, if $f(x)$ is polynomial at the initial scale level, then $f(x)$ is the same polynomial at lower resolution. The main idea of this paper is that the segmentation process must detect the same polynomial surfaces at reasonable scale levels.

3. Description of the segmentation algorithm

The key aspects of our segmentation algorithm represented in Figure 3.1, consist in bringing back range images in input at an initial resolution (IR) to a fixed lower resolution (LR). The obtained image is partitioned using a segmentation scheme based on an initial coarse surface estimation. This initial segmentation is guaranteed by a KH-mapping process. The resulting segments constitute seed regions for the fitting of second order bivariate polynomials. The result of the last step is projected to the initial resolution IR (initial scale). In order to validate the projection result, a global error criterion is evaluated. This scheme is repeated by increasing LR until the error

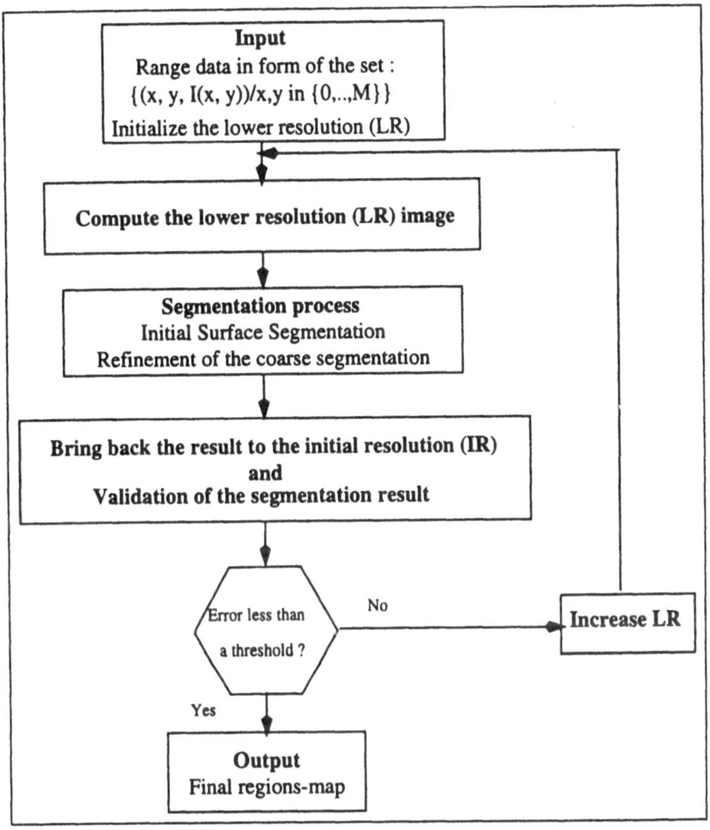

Fig. 3.1. Framework of the segmentation algorithm

criterion is respected.

Remark : The choice of LR has a capital importance. Indeed, it must be chosen "judiciously" to avoid the whole process break down by iterating many times. Intuitively, the lowest resolution can corresponds to the level where the details become non nils in interior of polynomial surfaces. We argue this by the fact that these details are nils while we handle polynomial surfaces.

3.1 Computing image of lower resolution LR

The implementation of discrete wavelet transforms (DWT) described in [3] is adapted in manner that we can use Coiflets filters. The choice of the Coiflet filter size is made according to the degree of the range image polynomials d. Indeed, $d \leq L-1$. We remind that L is the Coiflet order. Since the polynomial structure of images is approximated by a set of biquadrics i.e. $d \leq 2$ then we must have $L \geq 3$.

3.2 Initial surface segmentation

The aim of this step is to segment the range image surface into 8 basic fundamental forms : peak, pit, ridge, valley, saddle ridge, saddle valley, flat and minimal (for details one can refers to [2]). This segmentation is obtained by examining the mean (H) and Gaussian (K) curvatures at every points. Indeed, each point is labeled as a point in one of the 8 fundamental forms following the sign of K and H as shown in Table 3.1. To obtain the coarse segmentation, we use an eight-connectedness as connectivity relationship. A connected region is composed by adjacent pixels having same label.

We justify the use of surface curvature by the fact that the needed features for object localization and recognition purposes must be invariant under translation and space rotation. 3D range data considered as a sampled graph surface, is represented as the set $\{(x, y, I(x, y), 0 < x \leq N, 0 < y \leq N\}$. The mean and Gaussian curvatures are defined respectively as the mean and the product of the maximum and minimum normal curvatures at a point. Function of first and second order partial derivatives which scheme computation is widely described in [1], they may mathematically be determined by

$$K = \frac{I_{xx}I_{yy} - I_{xy}^2}{(1 + I_x^2 + I_y^2)^2} \tag{3.1}$$

$$H = \frac{I_{xx}(1 + I_y^2) + I_{yy}(1 + I_x^2) - 2I_xI_yI_{xy}}{2(1 + I_x^2 + I_y^2)^{\frac{3}{2}}} \tag{3.2}$$

Note also that surface curvature signs are also invariant with respect to rigid transformation. We use the following sign function to compute both K and H sign map :

$$sgn(x) = \begin{cases} +1 & \text{if } x > \epsilon \\ 0 & \text{if } |x| \leq \epsilon \\ -1 & \text{if } x < -\epsilon \end{cases}$$

where ϵ is a threshold value. A preliminary segmentation is then obtained by the association to every pixel of coordinates (i, j) in the image of a label $L(i, j)$ which represents one of the 8 surface types at that point. This labeling is performed by :

$$L(i, j) = 1 + 3(1 + sign(H(i, j))) + (1 - sgn(K(i, j))) \tag{3.3}$$

	$K > 0$	$K = 0$	$K < 0$
$H < 0$	peak	ridge,	saddle ridge
$H = 0$	non-determined	flat	minimal
$H > 0$	pit	valley	saddle valley

Table 3.1. Classification into 8 fundamental forms using curvatures sign

3.3 Refinement of the obtained segmentation

The refinement of the initial coarse segmentation is performed in two stages. First, all small regions inside regions of size greater than 30 pixels are merged. Second, the resulting regions are fitted with approximating polynomial surfaces. The degree of these surfaces is one for flat region type, two for others region types. For each region, we compute the error in a least squares sense. This error is used for region growing process. The obtained result is generally still raw, and must be refined to be rendered meaningful. This is accomplished by merging and re fitting actual adjacent "compatible" regions. The refinement procedure is formally detailed in next. Let first, introduce two definitions. Let $S = \{R_1, \ldots, R_N\}$ the initial segmentation, where each region R_i is fitted by a polynomial surface $P_i(i = 1, \ldots, N)$, with least squares error

$$E(R_i) = (\frac{1}{N_i}) \sum_{(x,y) \in R_i} (I(x, y) - P_i(x, y))^2$$

where N_i is the number of R_i points.

Definition 1. *The extrapolation error of the region R_i to the region R_j is defined by*

$$E_{ext}(R_i, R_j) = (\frac{1}{N_j}) \sum_{(x,y) \in R_j} (I(x, y) - P_i(x, y))^2 \tag{3.4}$$

where N_j is the number of R_j points.

Definition 2. *A region R is compatible connected to a region $Q \in S$, if exists a set $\{R_{i1}, R_{i2}, \ldots, R_{im}\} \subset S$ such that :*
$R_{i1} = R$, $R_{im} = Q$ and $\forall k \in \{2, \ldots, m-1\}$

- *R_{ik} is adjacent to R_{ik+1},*
- *$E_{ext}(R_{ik}, R_{ik+1}) < E(R_{ik+1})$,*
- *$E_{ext}(R_{ik}, R_{ik+1}) \leq E_{ext}(R_{ik+1}, R_{ik})$.*

The descriptive algorithm of the segmentation process is then as follows.

Input : A range image I.
Output : The segmentation result.
Step1 : Compute the initial segmentation ($S = \{R_1, \ldots, R_N\}$).
Step2 : For each region R_i

1. Construct the set $F_i = \{Q \in S / R_i$ is compatible connected to $Q\}$.
2. Merge to R_i all regions $Q \in F_i$.

3.4 Bring back the segmentation result to *IR* and its validation

Let $\{R_1^{(LR)}, R_2^{(LR)}, \ldots, R_N^{(LR)}\}$ the regions obtained after the segmentation process. Let also

$$P_i^{(LR)}(x, y) = \sum_{k+l \leq n_i} a_{kl}^{(LR)}(i) x^k y^l \tag{3.5}$$

where $n_i \leq L - 1$ and L is the vanishing moments, the polynomial approximation at the region $R_i^{(LR)}$. Without serious misunderstanding, we "project" the edges of regions $R_i^{(LR)} (i = 1, \ldots, N)$ at initial scale level IR. This leads to a partition $\{R_1^{(IR)}, R_2^{(IR)}, \ldots, R_N^{(IR)}\}$ of the image at IR. For each region $R_i^{(IR)}$, we compute the corresponding polynomial function in least squares sense

$$P_i^{(IR)}(x, y) = \sum_{k+l \leq n_i} a_{kl}^{(IR)}(i) x^k y^l \tag{3.6}$$

The validation of this segmentation result is related to the following criterion

$$\mathcal{E} = max_{i=1,\ldots,N} \parallel a^{(IR)}(i) - a^{(LR)}(i) \parallel \leq \varepsilon \tag{3.7}$$

where

$$a^{(IR)}(i) = (a_{00}^{(IR)}(i), \ldots, a_{n_i n_i}^{(IR)}(i)) \tag{3.8}$$

$$a^{(LR)}(i) = (a_{00}^{(LR)}(i), \ldots, a_{n_i n_i}^{(LR)}(i)) \tag{3.9}$$

ε is a threshold and $\parallel (x_1, \ldots, x_N) \parallel = max\{|x_1|, \ldots, |x_N|\}$

4. Experimental Results

The segmentation algorithm described above is implemented on HP9000/715/75 platform and tested for several range images. The obtained results indicate that multiresolution decomposition using Coiflets analysis offers an efficient method for compressing huge range images in manner they can be handled by segmentation process based on region-growing techniques. We assume that fast discrete Coiflets transforms (FDCT) must be a part of the segmentation process. The interest of this part, always claimed in this paper is the saving of time for the whole process.
As illustrating example, the "block2+harriscup" range image (Figure 4.1) is 512×512 from the MSU Pattern Recognition and Image Processing Lab's Technical Arts 100X scanner (aka 'White scanner'). It was converted to our own 'local' image format to be easy to handle. In results shown in Figure 4.2, the block sides are clearly delineated. The cup's profile is globally preserved and can be well approximated by a higher order surfaces. The detected regions are 8 in total. It's common that any real signal is corrupted by some noise and it's generally necessary to filter or de-noise data. The noise and it's nature are important factors on which depend the stability of the segmentation process. This problem is treated in detail in the extended version of this article [11].

5. Conclusion

We showed that the range image segmentation based on region-growing techniques can be accelerated to handle images of huge size. This has been made possible by a multiscale segmentation based on discrete Coiflet transforms. Both Coiflet and its associated scaling function moments vanish. This moment properties allow the preservation of polynomials i.e. the used Coiflet has N vanishing moments, polynomials of degree $N-1$ are not changed by successive approximations and the scaling coefficients of polynomials are also polynomial with the same degree. This Coiflets property permit the preservation of polynomial structure of range images, then, they can be processed at lower resolutions. This reduces the amount of input data and confer to the segmentation process real time response.
One important further improvement of this work concerns the adaptative choice of the initial values of the lower resolution LR, Indeed, as mentioned in the section describing the segmentation algorithm, this choice is fixed and consequentially slow down the segmentation process in certain cases of range images. Also, a thorough study of the segmentation stability must be done.

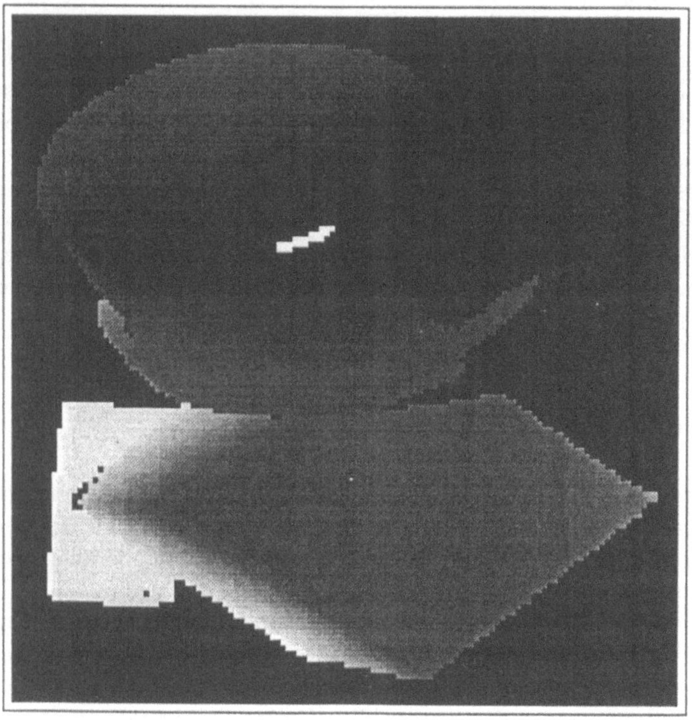

Fig. 4.1. The "block2+harriscup" range image

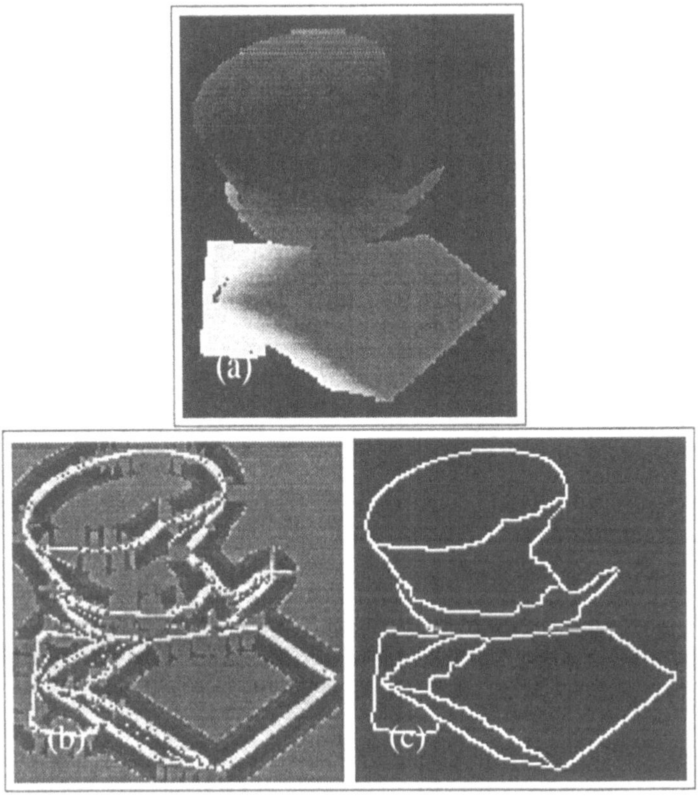

Fig. 4.2. (a) "block2+harriscup" Range Image at $LR = 2$, (b) its coarse segmentation and (c) its refined segmentation

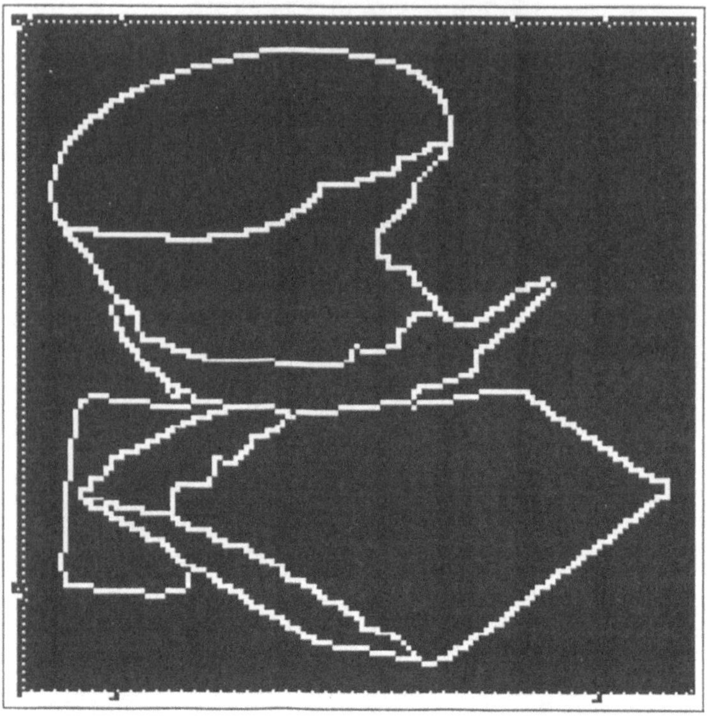

Fig. 4.3. Bringed back result to IR scale level

Acknowledgment

The authors would like to thank MSU PRIP Lab as the source of the images used to test our algorithms.

References

1. N. N. Abdelmalek (1990): *Surface Curvatures and 3-D range images segmentation*, Pattern Recognition, Vol. 23, No. 8, pp. 807-817.
2. P. J. Besl and R. C. Jain (1988): *Segmentation Through Variable-Order Surface Fitting*, IEEE Trans. Pattern Anal. and Mach. Intell. Vol. 10, No. 2, pp. 167-192.
3. S. Mallat (1989): *A theory for multiresolution signal decomposition : The wavelet representation*, IEEE Trans. Pattern Anal. and Mach. Intell, Vol. 11, No. 7, pp. 674-693.
4. S. Tanimoto and T. Pavlidis (1975): *A hierarchical data structure for picture processing*, CGIP,4 , 2, pp. 104-119.
5. H. P. Moravee (1977): *Towards automatic visual obstacle avoidance*, Proc. 5th IJCAL, pp. 584.
6. M. D. Levine (1978): *Acknowledge-based computer vision system* in CVS.
7. A. R. Hanson and E. M. Riseman (Eds) (1978): *Computer vision systems (CVS)*, New York : Academic Press.
8. I. Daubechies (1993): *Orthonormal bases of compactly supported wavelets. II. Variations on a theme*, SIAM J. MATH. ANAL., Vol. 24, No. 2, pp. 499-519.
9. G. Beylkin, R. Coifman and V. Rokhlin (1991):*Fast Wavelet Transforms and Numerical Algorithms I*, Communication on Pure and Applied Mathematics, Vol. XLIV, pp. 141-183.
10. K. Melkemi (1995): *Ondelette sur l'intervalle*, Memoire de DEA, Lab. LMC-IMAG, Universite Joseph Fourier Grenoble1.
11. M. Djebali M. Melkemi and D. Vandorpe (1996): *Segmentation d'Images de profondeur basée sur les Coiflets*, Technical Repport, Lab. LIGIM, Universite Claude Bernard Lyon1.

Image Restoration

A nonlinear primal-dual method for Total Variation-based image restoration *

Tony F. Chan**, Gene H. Golub*** and Pep Mulet†

Summary. We present a new method for solving total variation (TV) minimization problems in image restoration. The main idea is to remove some of the singularity caused by the non-differentiability of the quantity $|\nabla u|$ in the definition of the TV-norm *before* we apply a linearization technique such as Newton's method. This is accomplished by introducing an additional variable for the *flux* quantity appearing in the gradient of the objective function, which can be interpreted as the normal vector to the level sets of the image u. Experimental results show that the new method has much improved global convergence behavior than the primal Newton's method, with only a slight increase in cost per iteration.

1. Introduction

During some phases of the manipulation of an image some random noise and blurring is usually introduced. The presence of this noise and blurring makes difficult and inaccurate the latter phases of the image processing.

The algorithms for noise removal and deblurring have been mainly based on least squares. The output of these L^2-based algorithms will be a continuous function, which cannot obviously be a good approximation to our original image if it contains edges. To overcome this difficulty a technique based on the minimization of the Total Variation norm subject to some noise constraints is proposed in [15], where it is also proposed a time marching scheme to solve the associated Euler-Lagrange equations. Since this method can be slow due to stability constraints in the time step size, a number of alternative methods have been proposed, [18], [6], [12].

One of the difficulties in solving the Euler-Lagrange equations is the presence of a highly nonlinear and non-differentiable term, which causes convergence difficulties for Newton's method even when combined with a globalization technique such as a line search. The idea of our new algorithm is to

* This is a short version of the UCLA CAM report 95-43, where more details can be found.

** Department of Mathematics, University of California, Los Angeles. E-mail address: chan@math.ucla.edu. Supported by grants NSF ASC-92-01266 and ONR-N00014-96-1-0277.

*** Computer Science Department, Stanford University. E-mail: golub@sccm.stanford.edu. The work of this author was in part supported by the NSF :Grant # CCR-9505393.

† Departament de Matemàtica Aplicada, Universitat de València. E-mail address: mulet@math.ucla.edu. Supported by DGICYT grants EX94 28990695 and PB94-0987.

remove some of the singularity caused by the non-differentiability of the objective function *before* we apply a linearization technique such as Newton's method. This is accomplished by introducing an additional variable for the *flux* quantity appearing in the gradient of the objective function, which can be interpreted as the unit normal to the level sets of the image function. Our method can be viewed as a primal-dual method as proposed by Conn and Overton [7] and Andersen [2] for the minimization of a sum of Euclidean norms. Experimental results show that the new method has much improved global convergence behavior than the primal Newton method. The dramatic convergence improvement of this particular primal-dual implementation when compared to that experienced in linear programming problems seems to be due to the differential nature of the operators involved in the objective function, and not only to the fact that the algorithm uses the dual variables. It is hoped that the new approach can be applied to other geometry-based PDE methods in image restoration, such as anisotropic diffusion [14], affine invariant flows [16] and mean curvature flows [1], since the same singularity caused by $|\nabla u|$ occurs in these methods as well.

The organization of this paper is as follows: in section 2. we introduce the problem, the nonlinear equations associated to it and discuss how to solve them. In section 3. we present our new linearization technique for the (unconstrained) Tikhonov regularization form of the problem. Finally, in section 4. we present some numerical results for the denoising case.

2. Total Variation Regularization

An image can be interpreted as either a real function defined on a bounded and open domain of $\mathbb{R}^2$, Ω, (for simplicity we will assume Ω to be a rectangle henceforth) or as a suitable discretization of this continuous image. The notation $||u||$ ($u \in L^2(\Omega)$) stands for the 2-norm of the function u, $||u|| = (\int_\Omega u^2 \, dx \, dy)^{\frac{1}{2}}$, $|u|$ ($u = (u_1, \dots, u_d)$ a vector function) denotes the function $(\sum_1^d u_i^2)^{\frac{1}{2}}$ and $||y||$ ($y \in \mathbb{R}^m$) denotes the 2-norm of the vector y.

Our interest is to restore an image which is contaminated with noise and/or blur. The restoration process should recover the edges of the image. Let us denote by u_0 the observed image and u the real image. The model of degradation we assume is $Ku + n = u_0$, where n is a Gaussian white noise, of which we assume to know its level measured in the 2-norm, and K is a (known) linear *blur operator* (usually a convolution operator).

In general, the problem $Ku = z$, with K a compact operator, is ill-posed, so it is not worth solving this equation (or a discretization of it), for the data is assumed to be inexact, and the solution would be highly oscillatory. But if we impose a certain *regularity* condition on the solution u, then the method becomes stable. We can consider two related techniques of regularization: Tikhonov regularization and noise level constrained regularization.

Tikhonov regularization consists in solving the unconstrained optimization problem:

$$\min_u \alpha R(u) + \frac{1}{2}||Ku - u_0||^2, \tag{2.1}$$

for certain functional R which measures the irregularity of u in a certain sense and a suitably chosen coefficient α which will measure the tradeoff between a good fit to the data and a *regular* solution.

Another approach consists in solving the following constrained optimization problem:

$$\begin{aligned} &\min_u R(u) \\ &\text{subject to } ||Ku - u_0||^2 = \sigma^2, \end{aligned} \tag{2.2}$$

Here we seek a solution with minimum irregularity from all candidates which match the known noise level.

Examples of regularization functionals that can be found in the literature are, $R(u) = ||u||, ||\Delta u||, ||\nabla u \cdot \nabla u||$, where ∇ is the gradient and Δ is the Laplacian. The drawback of using these functionals is that they do not allow discontinuities in the solution, and since we are interested in recovering features of the image, they are not suitable for our purposes.

In [15], it is proposed to use as regularization functional the so-called *Total Variation norm* or *TV-norm*:

$$TV(u) = \int_\Omega |\nabla u|\, dx\, dy = \int_\Omega \sqrt{u_x^2 + u_y^2}\, dx\, dy. \tag{2.3}$$

The TV norm does not penalize discontinuities in u, and thus allows us to recover the edges of the original image. For simplicity we use in this section the Tikhonov formulation of the problem. Hence the restoration problem can be written as

$$\min_u \int_\Omega \left(\alpha\sqrt{u_x^2 + u_y^2} + \frac{1}{2}(Ku - u_0)^2\right) dx\, dy, \tag{2.4}$$

that is

$$\min_u \alpha||\,|\nabla u|\,||_1 + \frac{1}{2}||Ku - u_0||_2^2, \quad |\nabla u| = \sqrt{u_x^2 + u_y^2}. \tag{2.5}$$

The Euler-Lagrange equation for this problem, assuming homogeneous Neumann boundary conditions, is:

$$0 = -\alpha\nabla \cdot \left(\frac{\nabla u}{|\nabla u|}\right) + K^*(Ku - u_0). \tag{2.6}$$

where K^* is the adjoint operator of K with respect to the L^2 inner product. This equation is degenerate due to the presence of the term $1/|\nabla u|$. A commonly used technique to overcome this difficulty is to slightly perturb the Total Variation norm functional to become:

$$\sqrt{|\nabla u|^2 + \beta}, \tag{2.7}$$

where β is a small positive parameter. So now the problem is:

$$\min_u \alpha \int_\Omega \sqrt{|\nabla u|^2 + \beta}\, dx\, dy + \frac{1}{2}||Ku - u_0||^2, \tag{2.8}$$

and the corresponding Euler-Lagrange equation is:

$$0 = -\alpha\nabla \cdot \left(\frac{\nabla u}{\sqrt{|\nabla u|^2 + \beta}}\right) + K^*(Ku - u_0) = g(u). \tag{2.9}$$

The main difficulty that this equation poses is the linearization of the highly nonlinear term $-\nabla \cdot \left(\frac{\nabla u}{\sqrt{|\nabla u|^2+\beta}}\right)$.

A number of methods have been proposed to solve (2.9). L. Rudin, S. Osher and E. Fatemi [15] used a time marching scheme to reach the steady state of the parabolic equation $u_t = -g(u)$ with initial condition $u = u_0$:

$$u_t = \alpha\nabla \cdot \left(\frac{\nabla u}{\sqrt{|\nabla u|^2 + \beta}}\right) - K^*(Ku - u_0), \quad u(x,0) = u_0(x).$$

This method can be slowly convergent due to stability constraints. C. Vogel and M. Oman [18] proposed the following fixed point iteration to solve the Euler-Lagrange equation:

$$-\alpha\nabla \cdot \left(\frac{\nabla u^{k+1}}{\sqrt{|\nabla u^k|^2 + \beta}}\right) + K^*(Ku^{k+1} - u_0) = 0.$$

At each step, a linear differential-convolution equation has to be solved. This method can be viewed as a continuous analog of Weiszfeld's algorithm [19], which solves a sequence of minimization problems with quadratic objective functions. Using the framework developed in [19], it can be seen that this method is globally but only linearly convergent.

Due to the presence of the highly nonlinear term $\nabla \cdot \left(\frac{\nabla u}{\sqrt{|\nabla u|^2+\beta}}\right)$, Newton's method does not work satisfactorily, in the sense that its domain of convergence is very small. This is especially true if the regularizing parameter β is small. On the other hand, if β is relatively large then this term is well behaved. So it is natural to use a continuation procedure starting with a large value of β and gradually reducing it to the desired value. T. Chan, R. Chan and H. Zhou proposed in [6] such an approach. Although this method is locally quadratically convergent, the continuation step can be difficult to control.

3. A new linearization based on a dual variable

We propose here a better technique to linearize the term $\nabla \cdot \left(\frac{\nabla u}{|\nabla u|}\right)$. This technique bears some similarity to techniques from primal-dual optimization methods and gives a better global convergence behavior than that of the usual Newton's continuation method.

The method is based on the following simple observation. While the singularity and non-differentiability of the term $w = \nabla u/|\nabla u|$ is the source of the numerical problems, w itself is *smooth* because it is in fact the unit normal vector to the level sets of u. The numerical difficulties arise only because we linearize it in the wrong way.

The idea of the new method is to introduce

$$w = \frac{\nabla u}{|\nabla u|} \tag{3.1}$$

as a new variable and replace (2.9) by the following equivalent system of nonlinear partial differential equations:

$$\begin{aligned} |\nabla u|w - \nabla u &= 0 \\ -\alpha\nabla \cdot w + K^*(Ku - u_0) &= 0. \end{aligned} \tag{3.2}$$

We can then linearize this (u, w) system, for example by Newton's method. This approach is similar to the technique of introducing a *flux variable* in the *mixed finite element method* [3].

For completeness we compare the linearization of the u system:

$$\left[-\alpha\nabla \cdot \left(\frac{1}{|\nabla u|}(I - \frac{\nabla u \nabla u^T}{|\nabla u|^2})\nabla\right) + K^*K\right] \delta u = -g(u), \tag{3.3}$$

to the linearization of the (w, u)-system:

$$\begin{bmatrix} |\nabla u| & -(I - \frac{w\nabla u^T}{|\nabla u|})\nabla \\ -\alpha\nabla\cdot & K^*K \end{bmatrix} \begin{bmatrix} \delta w \\ \delta u \end{bmatrix} = -\begin{bmatrix} f(w,u) \\ g(w,u) \end{bmatrix}. \tag{3.4}$$

Equation (3.4) can be solved by first eliminating δw and solving the resulting equation for δu:

$$\left[-\alpha\nabla \cdot \left(\frac{1}{|\nabla u|}(I - \frac{w\nabla u^T}{|\nabla u|})\nabla\right) + K^*K\right] \delta u = -g(u). \tag{3.5}$$

After δu is obtained we can compute δw by:

$$\delta w = \frac{1}{|\nabla u|}(I - \frac{w\nabla u^T}{|\nabla u|})\nabla\delta u - w + \frac{\nabla u}{|\nabla u|}. \tag{3.6}$$

We note that the cost per iteration of our new linearization technique is only slightly higher than for the standard Newton's method (3.3), because the main cost is the solution of the differential-convolution equations (3.3) and (3.5) for δu. The cost per iteration is also comparable to that of the fixed point method, in that an elliptic-integral equation is solved at each step. The equation in the fixed point method has slightly simpler coefficients and is symmetric. On the other hand, the matrix in (3.5) can be symmetrized, yielding and iteration that still preserves local quadratic convergence.

Under reasonable assumptions on K, it can be shown that equation (3.5) is non-singular provided $|w| \leq 1$. This can be ensured by restricting the step length for w: if $|w_k(x)| < 1\ \forall x$, then take $w_{k+1} = w_k + \gamma\, \delta w_k$, where

$$\gamma = \rho \min\{s\colon |w_k(x) + s\, \delta w_k| < 1, \forall x\} \quad 0 < \rho < 1. \tag{3.7}$$

Although this step length γ could be used for the u variable as well, we have used unit step length for it.

The motivation is that the (w,u) system is somehow better behaved than the u system. Although at this point we do not have a complete theory to support this, we will now give a scalar example that can explain the better convergence behavior of the new approach. We compare Newton's method applied to the equivalent equations $f(x) = a - \frac{x}{\sqrt{x^2+\beta}} = 0$ (which resembles (3.1)) and $g(x) = a\sqrt{x^2+\beta} - x = 0$ (which resembles $w\sqrt{|\nabla u|^2+\beta} - \nabla u = 0$), where $a \approx 1$ and $\beta \approx 0$. In Fig 3.1 we can see that g looks more "linear" that f over much of the x-axis. In particular, if we start Newton's method with whatever initial guess not very close to the actual solution, it will diverge for f but it will converge for g. This is confirmed by the numerical results shown in Table 3.1. We believe that the reason why the primal-dual algorithm presented here shows such a dramatic convergence improvement over the standard Newton's method is precisely this better linearization.

	Newton's iteration for $g(x)=0$					Newton's iteration for $f(x)=0$				
$x_0 \backslash \beta$	10^{-1}	10^{-2}	10^{-3}	10^{-4}	10^{-5}	10^{-1}	10^{-2}	10^{-3}	10^{-4}	10^{-5}
1	7	6	4	3	5	*	9	6	6	*
2	6	5	2	4	6	*	8	4	*	*
3	6	4	3	5	6	10	7	5	*	*
4	5	4	3	5	7	9	6	*	*	*
5	5	3	4	6	7	8	5	*	*	*

Table 3.1. Comparison of the number of iterations required by Newton's method to solve $f(x) = 0$ and $g(x) = 0$, for $a = 0.9999$, for different β (horizontally) and different initial guesses x_0 (vertically). A $*$ means that the corresponding iteration failed to converge.

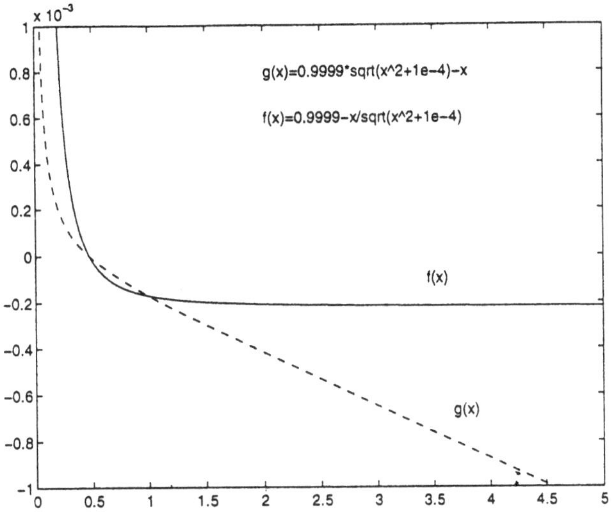

Figure 3.1. Plot of $f(x)$ and $g(x)$. The y-axis has been scaled by a factor of 10^{-3}.

3.1 Dual formulation

The definition (2.3) is not valid when u is non-smooth. The general definition of the Total Variation of a general (not necessarily smooth) u is as follows (see[9]):

$$TV(u) = \max\{\int_\Omega u \nabla \cdot w \, dx\, dy \colon w = (w_1, w_2),\ w_i \in C_0^\infty(\Omega),\ |w|_\infty \leq 1\} \tag{3.8}$$

Note that we can recover definition (2.3) when u is smooth by simply using the Cauchy-Schwartz inequality and $w = \frac{\nabla u}{|\nabla u|}$. Therefore, another formulation for the Total Variation restoration problem is

$$\min_u \max_{|w|\leq 1} \alpha \int_\Omega u \nabla \cdot w \, dx\, dy + ||Ku - u_0||^2. \tag{3.9}$$

It can be formally shown that the first order conditions for this problem lead to equation (3.2).

4. Numerical results

In these experiments we will use denoising and the unconstrained formulation.

Our first experiment consists in the comparison of the primal Newton and the primal-dual Newton methods under the following circumstances:

1. Continuation on β and no line search.
2. Continuation on β and line search on y.

3. No continuation on β and use line search on y.
4. No continuation and use line search on y and x for the primal-dual method.

The original image, which is 256×256 pixels and has dynamic range $[0, 255]$, appears in Fig 4.1. A Gaussian white noise with variance $\sigma^2 \approx 1200$ ($\sigma_h^2 = \sigma^2 \times 256^2$) is added to it, resulting in the image displayed in Fig 4.2, with $SNR = \frac{||u-\overline{u}||}{\sigma} \approx 1$. Fig 4.3 depicts the solution obtained by the primal-dual Newton method.

We have set the parameter α in the Tikhonov formulation to the inverse of the Lagrange multiplier yielded by a previous run of the constrained problem solver, in this case $\alpha = 1.18$. The parameter β_h has been set to 0.01. For the primal-dual method we have used a quasi-Newton approach which consists in replacing the matrix in (3.5) by its symmetrization; since the matrix in (3.5) converges to the (symmetric) matrix of (3.3), this replacement preserves the quadratic convergence of Newton's method. Furthermore, we have used *truncated* versions of Newton's algorithm, based on the conjugate gradient method with incomplete Cholesky as preconditioner. The stopping criterion for the (outer) Newton's iteration is a relative decrease of the nonlinear residual by a factor of 10^{-4}. The stopping criterion for the n-th inner linear iteration is a relative decrease of the linear residual by a factor of η_n, where we follow the suggestion of [11, Eq. 6.18] and set

$$\eta_n = \begin{cases} 0.1 & \text{if } n = 0, \\ \min(0.1, 0.9||g_n||^2/||g_{n-1}||^2) & \text{if } n > 0 \end{cases} \tag{4.1}$$

where $g_n = g(u_n)$ denotes the gradient of the objective function as appears in (2.9), at the n-th iteration. In Table 4.1 we compare the primal-dual and the primal versions of Newton's method for the experiments described above.

The conclusions that can be drawn from this experiment are:

- The most crucial factor for the primal-dual method is controlling the dual variables via the step length algorithm appearing in (3.7). In fact, our experience is that this algorithm with the dual step length is globally convergent for the parameters α and β in a reasonable range. A line search for the primal variables almost always yields unit step lengths.
- The primal-dual method with the dual step length algorithm does not need continuation to converge, although using it might be slightly beneficial in terms of work.
- The primal-dual method with the dual step length has a much better convergence behavior than the primal method.

In our second experiment, we compare the primal-dual Newton, fixed point and time marching methods. For the primal-dual Newton method, we use the step length algorithm for the dual variables, no continuation and the same parameters as in the previous experiment. The same parameters are used for the fixed point method, except that we have used a fixed linear relative residual decrease $\eta_n = 0.1$ (it is in this case *optimal* according to our experience). We have used a line search based on sufficient decrease for the time marching method. The stopping criterion for the time marching method is based on the iteration count since we have not been able to achieve the prescribed accuracy in a reasonable amount of time. In Figs. 4.4, 4.5 and 4.6 we plot the convergence history of this experiment.

The conclusions we draw from this experiment are:

- The primal-dual algorithm is quadratically convergent, whereas the others are at best linearly convergent.
- The primal-dual algorithm behaves similarly to the fixed point method in the early stages, but in a few iterations can attain high accuracy.
- The cost per iteration of the primal-dual method is between 30 and 50 per cent more than for the fixed point iteration. The memory requirements roughly satisfy this as well. The cost of an iteration of the time marching method is roughly the same as that of an inner CG iteration for the primal-dual method, but both the primal-dual method and the fixed point method require far less (CG) iterations than the time marching method.

Figure 4.1. Original image, 256×256 pixels.

Acknowledgements

We would like to thank Andy Conn and Michael Overton for making their preprint [7] available to us, and Jun Zou for many helpful conversations on mixed finite elements with the first author. The third author wants to heartily thank Paco Arándiga, Vicente Candela, Rosa Donat and Antonio Marquina for their encouragement and continuous support.

Figure 4.2. Noisy image, SNR$\approx$ 1, $||g|| = 2.07$.

Figure 4.3. Denoised image, 13 Newton's iterations, 58 CG iterations, $||g|| = 5.57 \times 10^{-5}$.

References

1. L. Alvarez, P. Lions, and J. Morel. Image selective smoothing and edge detection by nonlinear diffusion II. *SIAM J. Numer. Anal.*, 29:845–866, 1992.
2. K. D. Andersen. *Minimizing a sum of norms (large scale solution of symmetric positive definite linear systems)*. PhD thesis, Odense University, 1995.
3. F. Brezzi. A survey of mixed finite element methods. In *Finite elements*, ICASE/NASA LaRC, pages 34–49. Springer, 1988.
4. P. H. Calamai and A. R. Conn. A stable algorithm for solving the multifacility localtion problem involving euclidean distances. *SIAM Journal on Scientific and Statistical Computing*, 1:512–526, 1980.
5. P. H. Calamai and A. R. Conn. A second-order method for solving the continuous multifacility location problem. In G. A. Watson, editor, *Numerical analysis: Proceedings of the Ninth Biennial Conference, Dundee, Scotland*, volume 912 of *Lecture Notes in Mathematics*, pages 1–25. Springer-Verlag, 1982.

	primal-dual Newton		primal Newton	
	NWT	CG	NWT	CG
continuation, no line search	25	186	70	265
continuation, line search on y	25	187	53	210
continuation, line search on $y\&x$	13	51	53	210
no continuation, line search on $y\&x$	12	58	Not converged	

Table 4.1. Comparison of primal and primal-dual Newton methods.

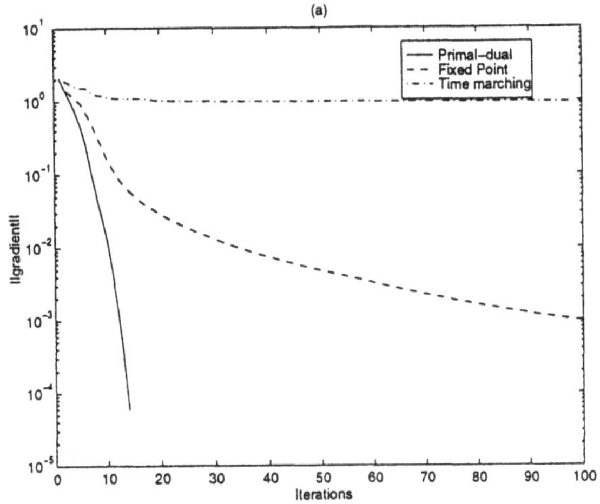

Figure 4.4. Plot of the L_2-norm of the gradient $g(u)$ of the objective function versus iterations for the different methods.

6. R. H. Chan, T. F. Chan, and H. M. Zhou. Continuation method for total variation denoising problems. Technical Report 95-18, University of California, Los Angeles, 1995.
7. A. R. Conn and M. L. Overton. A primal-dual interior point method for minimizing a sum of euclidean norms. preprint.
8. D. Dobson and F. Santosa. Recovery of blocky images from noisy and blured data. Technical Report 94-7, Center for the Mathematics of Waves, University of Delaware, 1994.
9. E. Giusti. *Minimal Surfaces and Functions of Bounded Variations*. Birkhäuser, 1984.
10. G. Golub and C. van Loan. *Matrix computations, 2nd ed.* The Johns Hopkins University Press, 1989.
11. C. T. Kelley. *Iterative Methods for Linear and Nonlinear Equations*, volume 16 of *Frontiers in Applied Mathematics*. SIAM, 1995.
12. Y. Li and F. Santosa. An affine scaling algorithm for minimizing total variation in image enhancement. Technical Report 12/94, Center for theory and simulation in Science and Engineering, Cornell University, 1994.
13. M. E. Oman. Fast multigrid techniques in total variation-based image reconstruction. to appear in the Preliminary Proceedings of the 1995 Copper Mountain Conference on Multigrid Methods.

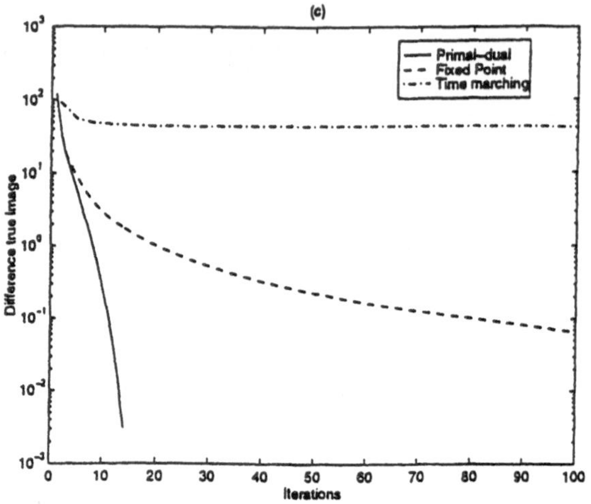

Figure 4.5. Plot of the L_2-norm of the difference between the current iterate and the solution for the problem computed by Newton's method with high accuracy versus iterations for the different methods.

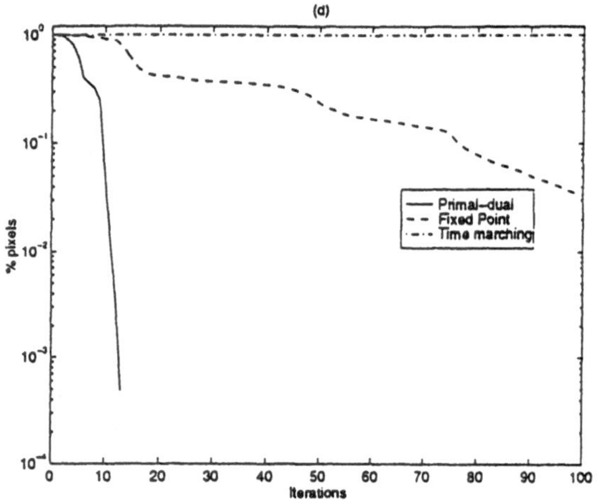

Figure 4.6. Plot of # pixels which differ more than .001 (relatively) from the solution for the problem computed by Newton's method with high accuracy versus iterations for the different methods.

14. P. Perona and J. Malik. Scale space and edge detection using anisotropic diffusion. *IEEE Trans. Pattern Anal. Mach. Intelligence*, 12:629–639, 1990.
15. L. Rudin, S. Osher, and E. Fatemi. Nonlinear total variation based noise removal algorithms. *Physica D*, 60:259–268, 1992.
16. G. Sapiro and A. Tannenbaum. Area and length preserving geometric invariant scale-space. In *Proc. 3rd European Conf. on Computer Vision, Stockholm, Sweden, May 1994, vol. LNCS*, volume 801, pages 449–458, 1994.
17. C. R. Vogel. A multigrid method for total variation-based image denoising. in Computation and Control IV, conference proceedings to be published by Birkhauser.
18. C. R. Vogel and M. E. Oman. Iterative methods for total variation denoising. *SIAM J. Sci. Statist. Comput.*, to appear.
19. H. Voß and U. Eckhardt. Linear convergence of generalized Weiszfeld's method. *Computing*, 25:243–251, 1980.

A new variational technique for Shape from Shading

Gabriele Ulich

Fachbereich Mathematik, TU Berlin, Germany

Summary. The Shape from Shading Problem is mathematically equivalent to a non-linear first order PDE. Horn[10] first solved this equation with a variational approach. We design an extended version by adding a smoothing term and present selected test results with synthetic and real images. The results are compared with the results of the original Horn approach.

1. Introduction

The basic problem in Shape from Shading (SFS) is to recover the shape z(x,y) of a surface from its variation in brightness. If we denote the reflectance map by R(p,q), where (p,q) is the surface gradient with $p = z_x$ and $q = z_y$ at a point (x,y) and the image brightness at this point by E(x,y), then, under certain assumptions (see. [9]) we have the image irradiance equation

$$R(p,q) = E(x,y). \tag{1.1}$$

The reflectance map depends on the properties of the surface material of the object, and the distribution of light sources. For a Lambertian surface illuminated by a single point source far away from the object we can use following nonlinear function:

$$R(p,q) = max\{0, \rho(\frac{1 + p_0 p + q_0 q}{\sqrt{1 + p_0^2 + q_0^2}\sqrt{1 + p^2 + q^2}})\}, \tag{1.2}$$

where ρ is the surface albedo and $(p_0, q_0, -1)$ is the light source direction. Equation (1.2) is a nonlinear, first order, PDE; it has been studied with a variety of different techniques (see e.g. [11] [12] [14] [15] [9]). The idea behind the variational technique is to find the surface which best satisfies (1.2) rather then to solve the image irradiance equation directly. This leads to following problem:
Find functions p(x,y), q(x,y) which minimizes the functional

$$F(p,q) = \int_{\Omega} (E - R(p,q))^2 d\Omega \tag{1.3}$$

over the domain Ω. There are two advantages of this "Ansatz". First regularization terms can be added, second one can choose to solve the corresponding Euler equations. The disadvantage is that when solving the Euler equations uniqueness of the solution does not follow [6] [11] [4]. Some first attempts at solving the SFS problem numerical with the variational approach have been made by Horn&Brooks [9] Leclerc&Bobick [13] and Ascher&Carter [1]. Ascher&Carter presented a multigrid algorithm for minimizing following functional directly:

$$\int_\Omega \lambda(p_x^2+p_y^2+q_x^2+q_y^2)+\mu((z_x-p)^2+(z_y-q)^2)+((E(x,y)-R(p,q))^2 d\Omega \quad (1.4)$$

where λ and μ are Lagrange multipliers, $(p_x^2+p_y^2+q_x^2+q_y^2)$ is a smoothness term and $(z_x-p)^2+(z_y-p)^2$ is an integrability term for the surface height. Ascher&Carter tested their algorithm with synthetic images and obtained good results when they used multiple images of the same object, otherwise poor convergence has been reported. In this paper we add a regularization term in the variational approach and then solve the corresponding Euler equations numerically with an overrelaxation Gauss-Seidel multigrid method with Natural or Dirichlet boundary conditions. This approach will be tested with synthetic and real images. For our approach, we add an integrability term for the surface slope $(p_y-q_x)^2$ to (1.4); this leads to following problem: Find p(x,y), q(x,y) and z(x,y), which minimizes the functional

$$\begin{aligned}\int_\Omega &\lambda(p_x^2+p_y^2+q_x^2+q_y^2)+\mu((z_x-p)^2+(z_y-p)^2)\\ &+((E(x,y)-R(p,q))^2+\nu(p_y-q_x)^2 d\Omega\end{aligned} \quad (1.5)$$

where ν is another Lagrange multiplier. A necessary condition for the existence of a solution is that the following Euler equations (see [5]) are satisfied:
$\lambda\Delta p+\mu(z_x-p)+(E-R)R_p+\nu(p_{yy}-q_{xy})=0$
$\lambda\Delta q+\mu(z_y-q)+(E-R)R_q+\nu(q_{xx}-p_{yx})=0$
$\Delta z-(p_x+q_y)=0$ with following natural boundary conditions: $\frac{\partial p}{\partial n}=\frac{\partial q}{\partial n}=0$ and $\frac{\partial z}{\partial n}=\mathbf{n}(p,q)$. where $\mathbf{n}$ denotes the outward normal at the boundary curve. Natural boundary data will be used when no boundary data is available in any form, otherwise Dirichlet boundary data will be used by fixing the unknowns at the boundary.

2. Numerical methods

We solve these Euler equations using a Gauss-Seidel iteration with overrelaxation. In comparison with Horn who used a Jacobi iterative method, the Gauss-Seidel method provides less storage requirements. Ascher&Carter [1] presented a multigrid technique to speed up the convergence rate. The multigrid method provides a fast and efficient method for solving elliptic partial differential equations (see [2] [7] [3]). Instead of solving the equation $L\mathbf{u}=0$ directly, where L is a non-linear, elliptic differential operator, we solve the residual equation $L\mathbf{e}=\mathbf{r}$ at a hierachy of coarser grids, where $\mathbf{r}$ is the residual

$r = -Lv$, v is the approximated solution of u and e is the error $e = u - v$. For our approch we will start at the finest grid level while dividing the domain Ω into square cells with vertices with length h, e.g. $x_i = x_{i-1} + h, y_i = y_{i-1} + h$ and the usual approxiamtion $z_{ij} = z(x_i, y_j)$. The Euler equations are discretized using central differential approximation:
$u_x = \frac{u_{i+1,j} - u_{i-1,j}}{2h}, u_y = \frac{u_{i,j+1} - u_{i,j-1}}{2h}$ the Laplacian is approximated by: $\Delta p = \frac{1}{4h^2}((p_{i+1,j+1} - 2p_{i+1,j} + p_{i+1,j-1}) + 2(p_{i,j+1} - 2pi, j + p_{i,j-1}) + (p_{i-1,j+1} - 2p_{i-1,j} + p_{i-1,j-1}))$.
For the multigrid scheme we use a V-cycle on 2 levels, and solve on the coarsest grid with a Gauss-Seidel iteration with overrelaxation. The overrelaxation factor and the Lagrange multipliers ν and μ are fixed during the calculations; the Lagrange Multiplier λ is reduced at each finer grid. This means that $\lambda = \frac{\lambda_f}{h_f^2}$ is constant at every grid with step size f (see[1]).

3. Test results

In this section we present several test results with synthetic and real images. For selected images we compare the results with the results of Horn&Brooks and Ascher&Carter.

3.1 Synthetic images

We have tested our algorithm on synthetic images illuminated from different light directions. If the whole object can be seen in the image, the background is held at a constant colour, and the iteration takes place at the silhouette of the object. The first input image is a sphere illuminated from (0,0,-1) and from (0.,0.5,-1) (see Figure 1a) and 1b)). The results with natural boundary conditions are shown in Figure 2a) and 2b).

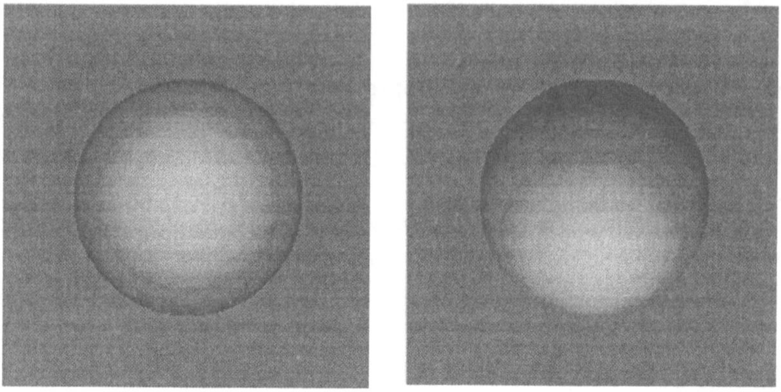

$(p_0, q_0, -1) = (0, 0, -1)$
Fig.a)

$(p_0, q_0, -1) = (0, 0.5, -1)$
Fig.b)

Fig. 3.1. input images

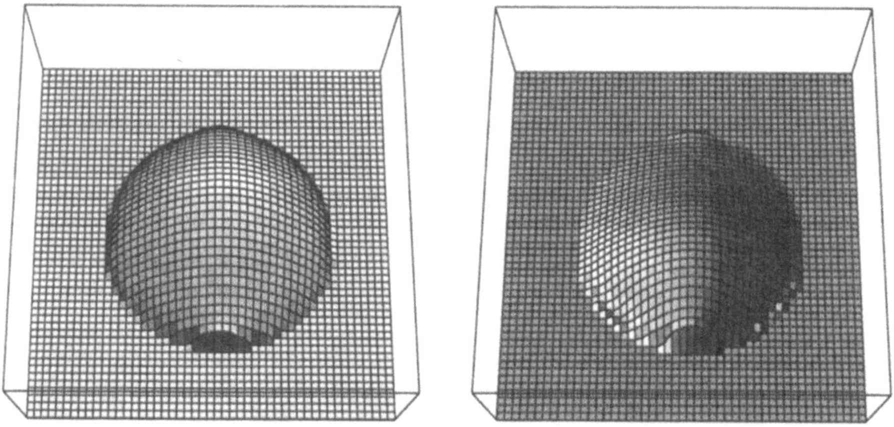

height for 3.1*a*)
Fig.a)

height for 3.1*b*)
Fig.b)

Fig. 3.2. test results for sphere

Horn&Brooks and Ascher&Carter presented in [1] and [8] the following approach to the Shape from Shading problem: Minimize

$$\int_{\Omega} \lambda(p_x^2+p_y^2+q_x^2+q_y^2)+\mu((z_x-p)^2+(z_y-q)^2)+((E(x,y)-R(p,q))^2 d\Omega \quad (3.1)$$

with the corresponding Euler-equations:

$\lambda \Delta p + \mu(z_x - p) + (E - R)R_p = 0$

$\lambda \Delta q + \mu(z_y - q) + (E - R)R_q = 0$

$\Delta z - (p_x + q_y) = 0$

We now want to compare this approach with our new approach given by (1.5). We use the same discretization as we used for solving (1.5). As an example we took a "mexican hat surface". The input image is presented in Figure 3. In contrast with our new results (see Figure 5) the results with the Horn&Brooks approach shows oscilations after the same number of iterations (see Figure 4).

Fig. 3.3. input image

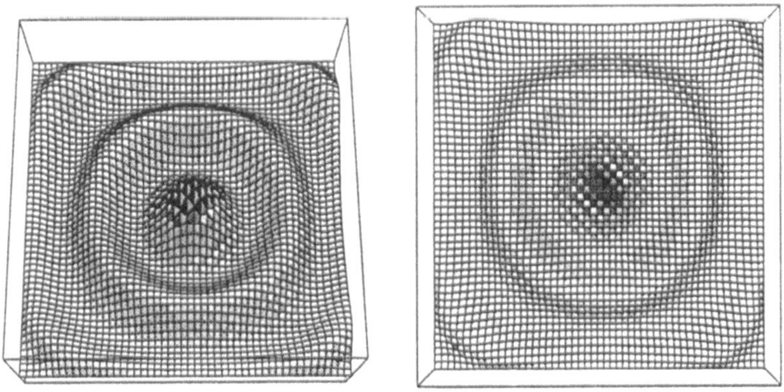

Fig.a) *Fig.b)*

Fig. 3.4. results for mexican hat surface

The scheme from equation (1.5) applied to the surface height produces the following result:

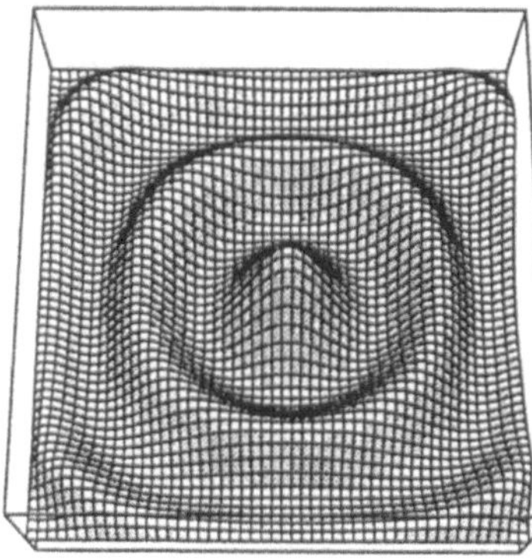

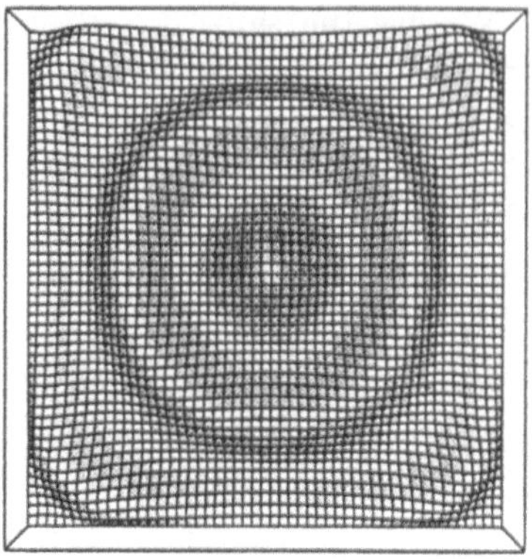

Fig.a) *Fig.b)*

Fig. 3.5. results for mexican hat surface with the new technique

The integrability term adds smoothing to the solution, so that oscillations are smoothed out.

3.2 Real images

Consider now an image of a vase illuminated by(-0.939,1.867,-1)(see Figure 6a)). The results with our approach are shown in Figure 6b).

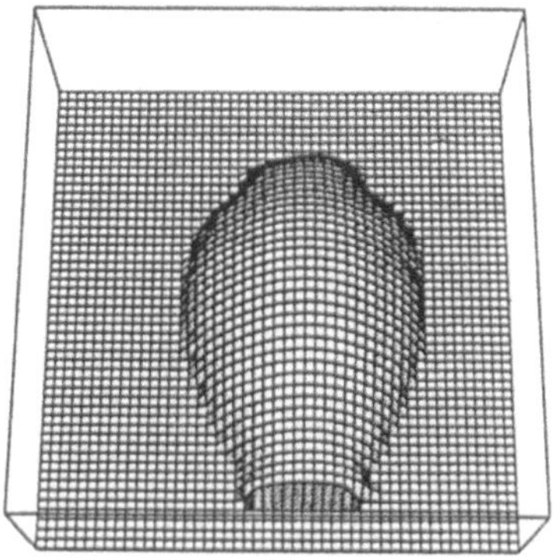

input image
Fig.a)

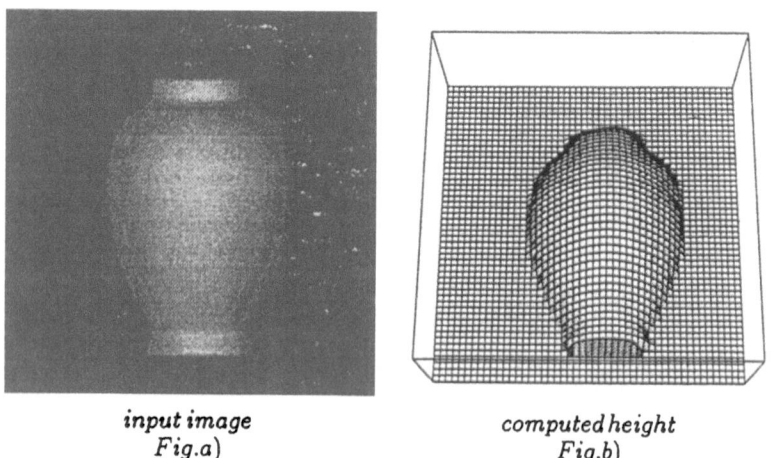

computed height
Fig.b)

Fig. 3.6. vase illuminated by (-0.939,1.867,-1)

References

1. **Ascher, U., Carter, M.**, *A Multigrid Method for Shape from Shading*, SIAM J. Numerical Analysis, Februar 1993.
2. **Brandt, A., Dinar, N.**, *Multigrid Solutions to elliptic flow problems*, Numerical Methods for Partial Differential Equations, S.Parter, ed., Proceedings, Oktober 1978, Madison, WI, Academic Press, NY, 1979.
3. **Brandt, A.**, *Guide to Multigrid Developement* , Multigrid Methods, ed. W.Hackbusch and U.Trottenberg, Proceedings of Conference on Multigrid Methods, Köln-Porz, 1981.
4. **Brooks, Chojnacki, N.**, *MultigridSolutions to Elliptic Flow Problems*, Multigrid Methods for Partial Differential Equations, ed. S.Parter, Academic Press, 1979.
5. **Courant, R., Hilbert, D.**, *Methods of Mathematical Physics*, Interscience, 1953.
6. **Chabrowski,J.,Zhang,K.**, *On variational approach to photometric stereo, Annales H Poincare 1994*
7. **Hackbusch, W.**, *Multigrid Methods and Applications* , Springer Verlag 1985.
8. **Horn, B., Brooks, M.**, *A Variational Approach to Shape from Shading*, Computer Vision, Graphics and Image Processing, 1986.
9. **Horn, B., Brooks, M.**, *Shape from Shading*, MIT Press, Boston, MA, 1989.
10. **Horn, B.**, *Robot Vision* MIT Press, Boston, MA, 1986.
11. **Kozera, R.**, *A Note on Existence and Uniqueness in Shape from Shading*, IEEE Conferenc on Computer Vision and Pattern Recognition, 1993.
12. **Kozera, R.**, *On Shape Recovery from two shading patterns*, Intern. Jour. of Pattern Rec. and Artifficial Intelligence, 6 (4), 1992.
13. **Leclerc, Y.G., Bobick, A.F.**, *The direct computation of Height from shading*, CVPR 1991
14. **Lions, P., Rouy, E., Tourin, A.**, *Shape From Shading , Viscosity Solutions and Edges*, Numerische Mathematik, 64 (3), 1993.
15. **Rouy, E., Tourin, A.**, *A Viscosity Solutions Approach to Shape from Shading*, SIAM J. Numer. Anal. ,Juni 1992.

Generalized Convergence Theorem for Wavelet Approximate Solutions of Elliptic Partial Differential Equations

E. B. Lin and X. Zhou

Department of Mathematics University of Toledo, Toledo, OH 43606 U.S.A.

Abstract

We present a higher order interpolation result for square integrable functions by using locally finite series of coiflets. Applications to wavelet - Galerkin numerical solutions of elliptic partial differential equations can also be obtained.

1 Introduction

Wavelets have been used for solutions of differential equations in many different aspects, namely, directed as bases with Galerkin's method [7], for algebraic manipulations to simplify the results using existing discretization methods [10] and after adaptation to specific differential operators when Galerkin's method is used [14]. Numerical results on the accuracy and complexity of algorithms are also discussed in several papers [1,2,8,12,13]. A crucial role in the design of these methods is played by the good localization properties that wavelets display both in space and frequency, that allow to preview the behaviour of the solution from the localization properties of the solution at the previous timestep in a simple way. Compactly supported wavelets which are differentiable were introduced by Daubechies in her celebrated paper [4], which has had applications in a number of areas. Depending on the applications they had in mind, several scientists have requested possible variations on the construction of Daubechies' wavelet [6]. There are several most recurrent wish items. Motivated by the need of more vanishing moments, coiflets of finite order were introduced [3]. Furthermore, we introduced generalized coiflets which give rise to better results in approximation of solutions of Elliptic partial differential equations. We refer the detailed version of this paper to [8].

2 Daubechies Wavelets and Coiflets

In this section, we will briefly review the basic ideas and descriptions of wavelets, multiresolution analysis, Daubechies wavelets, Mallat algorithm and coiflets.

Consturction of wavelet functions can start from the building of scaling function, $\phi(x)$, and a set of related coefficients, $\{a_k\}_{k\in Z}$, which satisfy the *two-scale relation* or *refinement equation*,

$$\phi(x) = \sum_k a_k \; \phi(2x-k) \tag{2.1}$$

and some additional conditions. The scaling function $\phi(x)$ has compact support if and only if finitely many coefficients a_k are non-zero.

Translations of the scaling function, $\{\phi(x-k)\}$, form a Riesz or unconditional basis of a subspace $V_0 \subset L^2(R)$. Furthermore, through translation and dilation of ϕ, a Riesz basis $\{\phi_{n,k}(x)\}_{k\in Z}$ is obtained for the subspace $V_n \subset L^2(R)$, where

$$\phi_{n,k}(x) = 2^{\frac{n}{2}}\phi(2^n x - k) \tag{2.2}$$

If translation of the scaling function, $\{\phi(x)\}$, is orthonormal at the same level n, i.e.,

$$\int_{-\infty}^{\infty} \phi_{n,k}\,(x)\; \phi_{n,l}\,(x)dx = \delta_{k,l}, \quad n,k,l \in Z \tag{2.3}$$

Then the best approximation of an $L^2(R)$ function, $f(x)$, by a function $P_n f(x)$ in the subspace V_n of $L^2(R)$ is given by the orthogonal projection of f on V_n, as follows:

$$P_n f(x) = \sum_k \alpha_{n,k}\; \phi_{n,k}(x) \tag{2.4}$$

where $\alpha_{n,k}$ is given by the inner product of $f(x)$ and $\phi_{n,k}(x)$,

$$\alpha_{n,k} = < f, \phi_{n,k} > = \int_{-\infty}^{\infty} f(x)\phi_{n,k}(x)dx \tag{2.5}$$

The operator $P_n : L^2(R) \rightarrow V_n$ is a linear orthogonal projection resulting in the best approximation of $f(x)$ in V_n.

Approximation of a function, $f(x)$, can be conducted at different resolution levels, and the approximations in the subspaces, $\cdots, V_{n-1}, V_n, V_{n+1}, \cdots$, have the following properties:

(i) $V_n \subset V_{n+1}$

(ii) $f(x) \in V_n \Leftrightarrow f(2x) \in V_{n+1}$

(iii) $f(x) \in V_n \Leftrightarrow f(x + 2^{-n}k) \in V_n, \forall k \in Z$

(iv) $\lim_{n\rightarrow\infty} \; V_n = \cup_n \; V_n$ is dense in $L^2(R)$

(v) $\cap_{n\in Z} V_n = \{0\}$

(vi) $\{\phi(x-n)\}_{n\in Z}$ is an orthonormal basis of V_0. Here, we call $\phi(x) \in L^2(R)$ a scaling function that generates a *multiresolution analysis* (MRA) with the above properties.

For every $j \in Z$, define W_j to be the orthogonal complement of V_j in V_{j+1}. We have

$$V_{j+1} = V_j \oplus W_j \tag{2.6}$$

and

$$W_j \perp W_{j'} \quad \text{if} \quad j \neq j' \tag{2.7}$$

(If $j > j'$, e.g. then $W_{j'} \subset V_j \perp W_j$.)
It follows that, for $j > J$

$$V_j = V_J \oplus (\oplus_{k=0}^{J-j+1} W_{J-k}) \tag{2.8}$$

where all these subspaves are orthogonal.
By property (iv) and (v) above, we have

$$L^2(R) = \oplus_{j\in Z} W_j, \tag{2.9}$$

a decomposition of $L^2(R)$ into mutually orthogonal subspaces. It turns out [4,7] that a basis for W_n can be obtained by dilating and translating a single function, $\psi(x)$, called *basic (mother, analyzing)* wavelet, which is defined by

$$\psi(x) = \sum_k b_k \phi(2x-k) \tag{2.10}$$

where b_k are a set of coefficients for the two-scale relationship of wavelet basis. In fact, $\{\psi_{j,k}(x) = 2^{\frac{j}{2}}\ \psi(2^j x - k);\ j,k \in Z\}$ is an orthonormal basis for $L^2(R)$.

We can define an operator $Q_n : L^2(R) \to W_n$ similar to P_n. Then

$$Q_n f(x) = \sum_k \beta_{n,k}\ \psi_{n,k}(x) \tag{2.11}$$

where $\{\beta_{n,k}\}$ are coefficients :

$$\beta_{n,k} =< f, \psi_{n,k} >= \int_{-\infty}^{\infty} f(x)\psi_{n,k}(x)dx \tag{2.12}$$

Thus, for a function, $f(x)$, we have

$$P_n f(x) = \sum_k \alpha_{n,k}\ \phi_{n,k}(x)$$

$$= \sum_k \alpha_{n-1,k}\phi_{n-1,k}(x) + \sum_k \beta_{n-1,k}\psi_{n-1,k}(x) \tag{2.13}$$

where the first sum is $P_{n-1}f(x) \in V_{n-1}$, the second sum is $Q_{n-1}f(x) \in W_{n-1}$.

The relationships between the approximation coefficients at different levels, e.g. $\{\alpha_{n,k}\}$ at level n, and $\{\alpha_{n-1,k}\}$, $\{\beta_{n-1,k}\}$ at level $n-1$, are determined through some sequences of constants which depend only on the wavelet basis being used. This is the *Mallat algorithm* [9], and the formulas will be given below.

$$\alpha_{n+1,\ell} = \sum_k a_{\ell-2k}\alpha_{n,k} + b_{\ell-2k}\beta_{n,k} \tag{2.14}$$

where $\{a_k\}$ and $\{b_k\}$ are called *reconstruction sequences.* So, if we know $\{\alpha_{n-1,k}\}$ and $\{\beta_{j,k}\}_{j=n-\ell}^{n-1}$, we can construct the approximation at level n from the lower level $n-l$.

The construction of *Daubechies wavelets* [4] starts from finding a sequence with finite non-zero terms a_k, $k = 0, \cdots, 2N-1$, for the two-scale relation of the scaling function (2.1), and the coefficients $\{a_0, a_1, \cdots, a_{2N-1}\}$ satisfy

$$\sum_{k=0}^{2N-1} a_k = 2 \tag{2.15}$$

$$\sum_{k=0}^{2N-1} a_k a_{k+2\ell} = 2\delta_{0,\ell}, \quad \ell \in Z \tag{2.16}$$

The vector $(a_0, \cdots, a_{2N-1})$ is called the *scaling vector* and we define the associated wavelet vector $(b_0, \cdots, b_{2N-1})$ by defining

$$b_k = (-1)^{k+1}\, a_{2N-1-k}. \tag{2.17}$$

Then wavelet function $\psi(x)$ is

$$\psi(x) = \sum_{k=0}^{2N-1} b_k \phi(2x-k). \tag{2.18}$$

With this convention, both the scaling function $\phi(x)$ and the corresponding wavelet function $\psi(x)$ have the same compact support $[0, 2N-1]$. Moreover, we have the following properties of moments of these functions.

$$\int \phi(x)dx = 1 \tag{2.19}$$

$$\int x\phi(x)dx = \frac{1}{2}\sum_{k=0}^{2N-1} ka_k \tag{2.20}$$

$$\int x^l \psi(x)dx = 0, \quad l = 0, \cdots, N-1 \tag{2.21}$$

let us denote by

$$c := \frac{1}{2}\sum_{k=0}^{2N-1} ka_k, \tag{2.22}$$

the *first moment* of the function $\phi(x)$.

Smooth scaling functions arise as a consequence of the degree of approximation of the translates. The conditions

$$\sum_k (-1)^k k^m a_k = 0 \quad m = 0, 1, \cdots, N-1 \tag{2.23}$$

implies that $\{1, x, \cdots, x^{N-1}\}$ can be expressed as linear combinations of the translates of $\phi(x-k)$[11].

Depending on the application one had in mind, Daubechies constructed coiflets of order N [6]. Namely, to construct ϕ, ψ, *coiflet* of order N, which satisfy

$$\int x^m \psi(x) dx = 0, \quad m = 0, 1, \cdots, N-1. \tag{2.24}$$

$$\int x^m \phi(x) dx = 0, \quad m = 0, 1, \cdots, N-1. \tag{2.25}$$

The need for orthonormal bases with this property first came up in the application of wavelet bases to numerical analysis in the work of [3]. Imposing such vanishing moments on ϕ also increases its symmetry, but there is a price to pay, namely, a coiflet with $2k$ vanishing moments typically has support width $6k-1$, as compared to $4k-1$ for Daubechies wavelet. Because these orthonormal wavelet bases with vanishing moments for both ϕ and ψ were requested by Coifman, these wavelets were name coiflets by Daubechies. [5,6]

3 Wavelet Interpolation

We state the following interpolation theorem in R^2 which is also true in R^n. Let ϕ, ψ be the orthonormal multiresolution analysis with compact support and ψ satisfies(2.21). Let $\{M_l\}$ denote the moments of the scaling function ϕ, i.e.

$$M_l := \int x^l \phi(x) dx, \quad l = 1, 2, \cdots \tag{3.1}$$

In particular,

$$M_1 = c = \frac{1}{2} \sum_k k a_k \tag{3.2}$$

<u>*Theorem 3.1*</u> Assume the function $f \in C^N(\overline{\Omega})$, where Ω is a bounded open set in R^2, $N \geq 2$.
Let, for $j \in Z$

$$f^j(x, y) := \frac{1}{2^j} \sum_{p,q \in \wedge} f\left(\frac{p+c}{2^j}, \frac{q+c}{2^j}\right) \phi_p^j(x) \phi_q^j(y), \quad (x, y) \in \Omega, \tag{3.3}$$

where the index set $\wedge = \{(p, q) | (\mathrm{supp}(\phi_p^j) \otimes \mathrm{supp}(\phi_q^j)) \cap \Omega \neq \phi\}$
In addition the moments M_l satisfy

$$M_l = (c)^l \quad l = 1, 2, \cdots, N-1 \tag{3.4}$$

Then

$$\|f - f^j\|_{L^2(\Omega)} \leq C\|f^{(N)}\|_\infty (\frac{1}{2^j})^N \tag{3.5}$$

$$\|f - f^j\|_{H^1(\Omega)} \leq C\|f^{(N)}\|_\infty (\frac{1}{2^j})^{N-1} \tag{3.6}$$

where C is a constant depending only on N, diameter of Ω and

$$\|f^{(N)}\|_\infty := \max_{(x,y)\in\Omega,\ m=0,1,\cdots,N} \left|\frac{\partial^N f}{\partial x^m \partial y^{N-m}}(x,y)\right| \tag{3.7}$$

Remark 3.2 The coiflets of order N satisfies (3.4) since $M_1 = M_2 = \cdots = M_{N-1} = 0$, and hence the theorem holds for coiflets.
Remark 3.3 Under the assumption $\phi, \psi \in C^N$, one can derive the following estimates.

$$\|f - f^j\|_{H^m(\Omega)} \leq C\|f^{(N)}\|_\infty (\frac{1}{2^j})^{N-m}, \quad m = 0, 1, \cdots, N-1 \tag{3.8}$$

4 Estimates for Elliptic Partial Differential Equations

In this section we prove an estimate for a Galerkin-Wavelet solution of a specific boundary value problem as an example of the application of Theorem 3.1. Suppose $\{V^j\}_{j\in Z}$ is the multiresolution analysis generated by the scaling function ϕ which satisfies the assumption in Theorem 3.1.

Let Ω be a bounded open set in R^2 with a Lipschitz boundary. Let

$$W^j := \{f \in V^j : \text{supp } f \cap \bar{\Omega} \neq \phi\}. \tag{4.1}$$

Since Ω is bounded, it's clear that W^j is a finite-dimensional subspace of V^j, and $W^j \subset H^1(\Omega)$. This will be our Galerkin-Wavelet approximation space.

Consider the elliptic equation

$$-\triangle u + u = f, \quad \text{in } \Omega \tag{4.2}$$

with the Neumann boundary condition

$$\frac{\partial u}{\partial n} = g \quad \text{on } \partial\Omega \tag{4.3}$$

where n is the unit outward normal vector of $\partial\Omega$. If u is a solution of (4.2) and (4.3) and if h is a test function in $H^1(\Omega)$, then multiplying (4.2) by h and integrating by parts over Ω, one has from (4.3) that

$$\int_\Omega \nabla u \nabla h dxdy + \int_\Omega uhdxdy = \int_\Omega fhdxdy + \int_{\partial\Omega} ghds \tag{4.4}$$

Solving (4.2) and (4.3) is equivalent to finding $u \in H^1(\Omega)$; so that (4.4) is satisfied for all $h \in H^1(\Omega)$. Let u^j be a solution of (4.4) where $u^j \in W^j$ and (4.4) is satisfied for all $h \in W^j$ (the Galerkin-Wavelet approximation).
Theorem 4.1 If u is a solution to (4.2) and (4.3), and if $u \in C^N(\Omega)$, then

$$||u - u^j||_{H^1(\Omega)} \leq C||u^{(N)}||_\infty(\frac{1}{2^j})^{N-1} \tag{4.5}$$

where C depends on N and diameter of Ω.

We now generalize the above result by replacing the data f and g by perturbations of f and g before we make the Galerkin approximation. In fact, this is what happens in real world, as it is necessary to use numerical approximations for f and g in a given problem. Suppose that we are given f and g as in (4.2) and (4.3), where we suppose both f and g are C^N functions in R^2 which restrict to Ω and $\partial\Omega$ respectively, and let f^j and g^j be wavelet interpolations for f and g of order j as in Theorem 3.1. Let now u^j be the wavelet-Galerkin solution to

$$\int_\Omega \nabla u \nabla h dx dy + \int_\Omega u h dx dy = \int_\Omega f^j h dx dy + \int_{\partial\Omega} g^j h ds \tag{4.6}$$

and then we have the following result.

Corollary If u is a solution to (4.2) and (4.3), and u^j is the wavelet-Galerkin solution to (4.6), then

$$||u - u^j||_{H^1(\Omega)} \leq C(\frac{1}{2^j})^{N-1},$$

where C depends on the diameter of Ω, on N and the maximum modulus of Nth derivatives of u, f, g.

5 Numerical Examples

In this section, we will present examples in two dimensional cases.

We verify the result by using coiflets of order 4 (2.24 , 2.25) to approximate some smooth functions in the form of (3.3). Suppose $\Omega = [0,1] \times [0,1]$, $f \in C^4(\Omega)$, then the estimate (3.5) becomes

$$||f - f^j||_{L^2(\Omega)} \leq C||f^{(4)}||_\infty(\frac{1}{2^j})^4 \tag{5.1}$$

This means that (3.3) is an exact sampling function if $f(x,y)$ is a polynomial of order less than 4.

Consider the test functions

$$f_1(x,y) = (x+y)^4, f_2(x,y) = \cos(\pi xy), f_3(x,y) = \cosh(x^2y+1).$$

Let

$$f_i^j(x,y) = \frac{1}{2^j} \sum_{p,q\in\Lambda} f_i(\frac{p}{2^j}, \frac{q}{2^j})\phi_p^j(x)\phi_q^j(y) \tag{5.2}$$

be the sampling function of $f_i(x,y)$ and

$$e_i^j(x,y) = f_i(x,y) - f_i^j(x,y). \tag{5.3}$$

We calculate $||e_i^j||_{L^2(\Omega)} = ||f_i - f_i^j||_{L^2(\Omega)}$ for some j. The results are in the following table:

	$\|\|e_1^j\|\|_{L^2}$	$\|\|e_2^j\|\|_{L^2}$	$\|\|e_3^j\|\|_{L^2}$
j=1	0.0165214	0.012311	0.542946
j=2	7.18854E-4	2.06741E-4	3.3421E-4
j=4	4.77539E-6	1.63849E-6	1.15966E-6
j=6	1.86539E-8	7.27896E-9	4.19861E-9

This shows that $||e_i^j||_{L^2}$ decreases at rate of $(\frac{1}{2^j})^4$ as j increases.

References

[1] E. Bacry, S. Mallat and G. Papanicolaou, A wavelet based space-time adaptive numerical method for partial differential equations, Mathematical Modelling and Numerical Analysis, 26 (1992), pp 793-834.

[2] G. Beylkin, On wavelet-based algorithms for solving differential equations, Wavelets: Mathematics and Applications, J. J. Benedetto and M. Frazier, eds., CRC Press, Boca Raton, FL (1993)

[3] G. Beylkin, R. Coifman, and V. Rokhlin, Fast wavelet transforms and numerical algorithms. I, Comm. Pure Appl. Math., 44 (1991), pp 141-183.

[4] I. Daubechies, Orthonormal basis of compactly supported wavelets, Comm. Pure Appl. Math., 41 (1988), pp 909-996.

[5] ____, Ten lectures on wavelets, CBMS-NSF Regional Conf. Ser. in Appl. Math., Society for Infustrial and Applied Mathematics, Philadelphia, PA 1992.

[6] ____, Orthonormal bases of compactly supported wavelets II., Variations on a theme, SIAM J. Math. Anal. vol 24, no 2, (1993), pp 499-519.

[7] R. Glowinski, W. Lawton, M. Ravachol, and E. Tenenbaum, in Proceedings, 96h International Conf.on Numerical Methods in Applied Sciences and Engineering, SIAM, Philadelphia, PA 1990, pp 55-119.

[8] E. B. Lin and X. Zhou, Wavelet Interpolation and Approximation solution of Partial Differential Equations, preprint.

[9] S. Mallat, Multiresolution approximation and wavelet orthonormal bases of $L^2(R)$, Trans. Amer. Math. Soc., 315, (1989), pp 69-87.

[10] S. Qian and J. Weiss, Wavelets and the numerical solution of partial differential equations, Journal of Computational Physics, 106, (1993), pp 155-175.

[11] G. Strang, Wavelets and dialation equations: a brief introduction, SIAM Review, vol 31, no 4 (1989), pp 614-627.

[12] R. O. Wells, Jr. and X. Zhou, wavelet interpolation and approximate solutions of elliptic partial differential equations, preprint.
[13] R. O. Wells, Jr. and X. Zhou, Wavelet solutions for the Dirichlet problem, preprint, Rice University.
[14] Xu, J. C. and W. C. Shann, Galerkin-wavelet methods for two-point boundary value problems, Numer. Math., 63(1), 123-142 (1992).

Regularization Methods for the Image Restoration of Electromagnetic and Optical Complex Systems

C.Bonifazzi[1], G.Maino[2*] and A.Tartari[3]

[1] *Istituto di Fisiologia Umana, Universitá di Ferrara, via Fossato di Mortara 17/19, 44100 Ferrara, Italy.*
[2] *ENEA, Dipartimento Innovazione, Divisione Fisica Applicata, via Don Fiammelli 2, 40129 Bologna, and INFN, Sezione di Firenze, Italy.*
[3] *Dipartimento di Fisica, Universitá di Ferrara, via Paradiso 12, 44100 Ferrara, Italy.*

Summary. We present a regularization technique for solving inverse ill–posed problems which arise in many fields of applied optics and electrodynamics when one has to deal with restoration of images. Numerical results for electron scattering and free–electron–laser data are shown and the problem of reconstructing a planar image from its line integrals in computerized Compton and X–ray diffraction tomographies is addressed.

1. Introduction

In physics, engineering and natural sciences it is usual to deal with experimental situations where the information about the observed phenomena, in general related to complex systems, has to be extracted from measured data affected by uncertainties arising from the experimental set–up, the detector efficiency, its spatial and time resolutions, etc. Moreover, quantities of effective interest may not be directly linked to the experimental results but have to be deduced from them. It is the case of the imaging technique in medical and industrial applications where the *real* image is reconstructed from suitable projections of density profiles, electron distributions, etc.

In these terms we consider a class of image restoration problems represented by the derivation of *original* images from the *observed* ones assuming that the experimental detection always implies a modification or a perturbation of the *real* image to be recovered or reconstructed from the measured data. It is the same philosophy underlying the predictional approach to the image restoration but proposed in a more extended context[1].

It is our intent to show that the regularization method and the relevant algorithm presented in sect.2 when applied to processing experimental data may

* e–mail address: maino@risc990.bologna.enea.it

significantly improve the accuracy with which the parameters of the involved physical objects are determined. The resolving power of an experimental apparatus can thus be substantially increased simply by using computer–data processing and avoiding expensive or complex modifications of the set–up itself. Moreover, basic information can be recovered from the measured data without resorting to expensive computer simulations as in the cases of the nuclear mass densities from electron scattering cross sections and the energy distribution of electron beams in Free Electron Laser (FEL) devices and linear colliders, discussed in sect.3. Sect. 4 is then devoted to preliminary results of image reconstructions in Compton and X–ray computerized tomographies. Finally, conclusions and perspectives for future work are drawn in sect.5.

2. Regularization methods

In both applied sciences and technology the problem frequently arises of solving integrals equations of first kind which relate a unknown function, $u(y)$, to another function, $f(x)$, experimentally determined,

$$f(x) = \int_a^b k(x,y)\, u(y)\, dy, \tag{2.1}$$

where the integration limits can be both fixed or one fixed, the other variable and (x,y) are in general multidimensional variables but — for simplicity's sake — will be assumed monodimensional in the following discussion. The generalization to the N–dimensional case is straightforward. The kernel of eq.(2.1), $k(x,y)$, is a piecewise continuous function and represents the response function of the experimental apparatus.

The problem represented by eq.(2.1) is ill–posed according to Hadamard's well–known definition[2] since arbitrarily small perturbations of $f(x)$ due, for instance, to the experimental uncertainty, the noise, etc., may cause arbitrarily large perturbations in the solution, $u(y)$, as Riemann-Lebesgue theorem states[3] for integrable kernel functions. Therefore, eq.(2.1) cannot be directly inverted and a stable approximation to the solution must be obtained by means of suitable techniques or *a priori* assumptions about the form of the solution itself.

We adopt a particular regularization procedure based on the general approach developed by Tikhonov and coworkers[3,4], starting from the following Volterra integral equation of first kind,

$$f(x) = K(x,u) \equiv \int_x^R k(x,y)\, u(y)\, dy, \qquad (0 \le x \le R), \tag{2.2}$$

where the piecewise continuous kernel, $k(x,y)$, and the square sommable function, $f(x)$, are known. When the lower limit of integration is fixed, generally equal to zero, eq.(2.2) reduces to a Fredholm integral equation. Even in this case the classical method of solution based on the reduction to a corresponding differential equation cannot be always exploited since the observed $f(x)$ function is not exactly known from experiments and given over all the domain.

Assuming that the unknown funtion, $u(y)$, is differentiable up to fourth order, eq.(2.1) can be converted into a well-posed variational problem by imposing suitable constraints on the boundary conditions of the solution[3,4] which is then stable. One has to minimize the following functional in the class of square sommable functions:

$$U_\alpha(f,u) \equiv \int_0^R [K(x,u) - f(x)]^2 \, dx + \alpha^2 \int_0^R |u''(y)|^2 \, dy. \tag{2.3}$$

The second term in the r.h.s. of eq.(2.3) is a regularizing functional of the second order[3,5] — it contains the second derivative of $u(y)$ — and α, $(\alpha > 0)$, is a numerical parameter. It can be shown[3] that for any positive α value and any square sommable function, $f(x)$, there is a unique solution, $u_\alpha(y)$, yielding a minimum of eq.(2.3). This solution is stable against perturbations on $f(x)$ and converges uniformly to the searched solution, $u(y)$, of eq.(2.1) when $\alpha \to 0$ and simultaneously the data error vanishes.

Therefore, the problem of solving eq.(2.1) is reduced to the equivalent but well-posed problem of finding a minimum for the functional (2.3), a task that can be easily accomplished by solving the relevant Euler equation[6],

$$\int_0^y \int_x^R k(x,y)\, k(x,z)\, u(z)\, dz\, dx + \alpha^2\, u^{IV}(y) = \int_0^y k(x,y)\, \tilde{f}(x)\, dx, \tag{2.4}$$

with suitable boundary conditions once a discretization procedure has been applied. Here, $u^{IV}(y)$ is the fourth derivative of $u(y)$ and $\tilde{f}(x)$ is an approximation of $f(x)$ in the sense that

$$\int_0^R |f(x) - \tilde{f}(x)|^2 \, dx \le \int_0^R \delta^2(x)\, dx = \Delta^2. \tag{2.5}$$

because of the experimental uncertainties. This inequality provides also a upper bound for parameter α,

$$\alpha \le \frac{\Delta}{\left[\int_0^R |\tilde{u}''(y)|^2 \, dy\right]^{1/2}}. \tag{2.6}$$

It is worth remarking that the *best* value of the regularization parameter, α, depends on both the regularizing functional (2.3) and the uncertainties on the measured data, $\tilde{f}(x)$. In practice, α is determined by a numerical iteration procedure taking into account the relation (2.6).

3. Applications to electromagnetic systems

3.1 Electron scattering by nuclei A classical probe for the investigation of nuclear structure properties is provided by high-energy electron beams[7]. In particular, nuclear charge densities can be determined in great detail by electron–nucleus scattering experiments because of the high spatial resolution in present measurements arising from the broad range of momentum transfer values available in actual factories. However, nuclear mass–density distributions are not directly observed in electron scattering experiments and must be extracted from the deduced charge densities as an important source of information to be compared with microscopic theoretical models of nuclear structure.

In many cases, including heavy–mass nuclei, charge and centre–of–mass proton densities may differ significantly. These quantities are related together by an integral equation of convolution type which, for nuclei with even numbers of neutrons and protons, reads as:

$$\tilde{\rho}(\vec{r}) = \int_V \rho_p(\vec{r} - \vec{s})\, \rho(\vec{s})\, d\vec{s}, \qquad (3.1.1)$$

where $\tilde{\rho}(\vec{r})$ and $\rho(\vec{r})$ are the nuclear charge and proton centre–of–mass distributions, respectively. The single–proton charge density, $\rho_p(\vec{r})$, is provided by suitable assumptions on the basis of given nucleon models[1] and a widely adopted expression is

$$\rho_p(\vec{r}) = \left(\frac{\mu_p^3}{4\pi}\right) f(\mu_p r)\, \frac{e^{-\mu_p r}}{\mu_p r} + c_p \delta(\vec{r}), \qquad (3.1.2)$$

with $f(\mu_p r) = a_p - 2b_p + b_p \mu_p r$, and $c_p = 1 - a_p$ (see Table I for the numerical values corresponding to some common nucleon models), and $\delta(\vec{r})$ Dirac distribution.

The three–dimensional problem represented by eq.(3.1.1) can be reduced to a monodimensional one, depending on the radial coordinate only, in the case of spherical–shape nuclei[8,9] :

$$\frac{4\pi r}{\mu_p^2} \tilde{\rho}(r) = \frac{4\pi}{\mu_p} \int_0^{r_0} k(\mu_p r, \mu_p r')\, \rho(r')\, r'\, dr', \qquad (3.1.3)$$

TABLE I Parameters of different proton model

Proton model	a_p	b_p	$\mu_p(\mathrm{fm}^{-1})$
pointlike	0.0	0.0	∞
Drell	1.0	0.0	2.341
Hofstadter	1.0	0.5	4.016
Clementel Villi	1.2	0.0	3.115

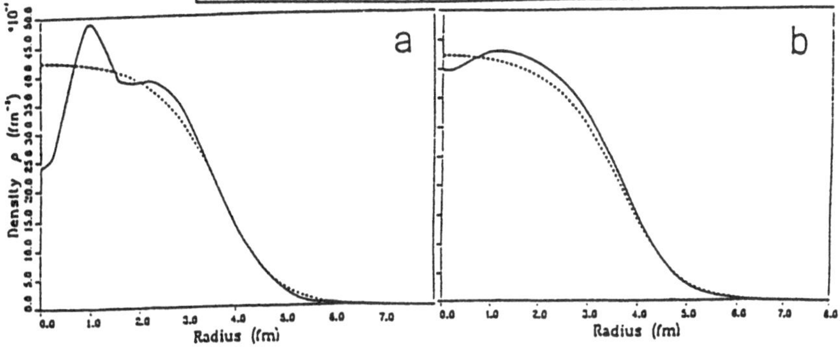

Fig.1 — Nuclear mass density of ^{40}Ca (solid line) derived from the observed charge density[10] (dashed line) by solving eq.(3.1.3) with a direct inversion method (a) and the Tikhonov's regularization procedure (b), assuming the Hofstadter's proton model.

where r_0 is the upper limit of nuclear and charge densities such as $\rho(r > r_0) = 0$, $k(x,y) = \tilde{k}(x,y)$ if $y \le x$ and $k(x,y) = \tilde{k}(y,x)$ otherwise, and

$$\tilde{k}(x,y) \;=\; \mu_p\, e^{-x}\, [(1 - b_p + b_p x)\sinh y \;-\; b_p y \cosh y]. \tag{3.1.4}$$

Moreover, if $a_p = 1$, as for Drell's and Hofstadter's proton models of Table I, the integral equations (3.1.1) and (3.1.3) are ill–posed and a regularization procedure has to be adopted in order to recover a stable solution. In fig.1 we show the evaluated nuclear mass density of ^{40}Ca from the empirical charge density[10] obtained by (a) direct inversion of eq.(3.1.3) without regularization and (b) using the regularization technique of sect.2 on the assumption of the Hofstadter's proton model. In the first case, (a), unphysical oscillations of the nuclear mass distributions appear because of the ill–posedness of the problem.

Starting from different proton models, results very close to that of fig.1(b) are obtained, minor discrepancies arising in the innermost nuclear region where experimental charge densities, $\tilde{\rho}(\bar{r})$, are determined with less precision than in outer regions since high–momentum transfer values are concerned, thus confirming the reliability of the present approach in general cases.

3.2 Free Electron Lasers and electron beams

Free Electron Lasers (FEL)[11] are unconventional sources of coherent radiation, where the active medium is represented by a beam of free electrons in interaction with a magnetic undulator consisting of alternating north–south poles, giving the necessary transverse component in the electron motion and allowing for the coupling to an external wave. The electron beam is then modulated in energy and density at the same wavelength as the input radiation, resulting into the coherent emission from each electron packet and the amplification of the initial wave.

The electron beam emits radiation with inhomogeneously broadened spectrum traversing a wiggler or an undulator (i.e. a magnetic field) when it reaches ultrarelativistic energies and does not have perfect qualities, namely non–negligible energy spread and emittance. This distortion of the spectrum, referred in the literature as *brightness degradation*, depends on the beam–energy spread and emittances above all, but also on the specific spatial, angular and energetic beam distributions.

We confine the present analysis to the inhomogeneous broadening of the energy–distribution shape, neglecting the spatial and angular distributions which can be treated by means of analogous techniques. The spectral shape distortion induced by an electron beam characterized by an energy distribution $f(\epsilon)$, is given by[12]

$$P(\nu') = \int_0^{E_{max}} \left[sinc\left(\frac{\nu' + \delta\nu_\epsilon}{2} \right) \right]^2 f(\epsilon)\, d\epsilon, \qquad (3.2.1)$$

where $sinc(x) = \sin(x)/x$, as usual, and $\delta\nu_\epsilon = 4\pi N\epsilon$, with N number of periods of the undulator.

If the brightness, $P(\nu')$, is experimentally known, eq.(3.2.1) represents an ill–posed integral equation in the unknown function, $f(\epsilon)$. The brightness in then used as a diagnostic tool for determining the electron–beam qualities.

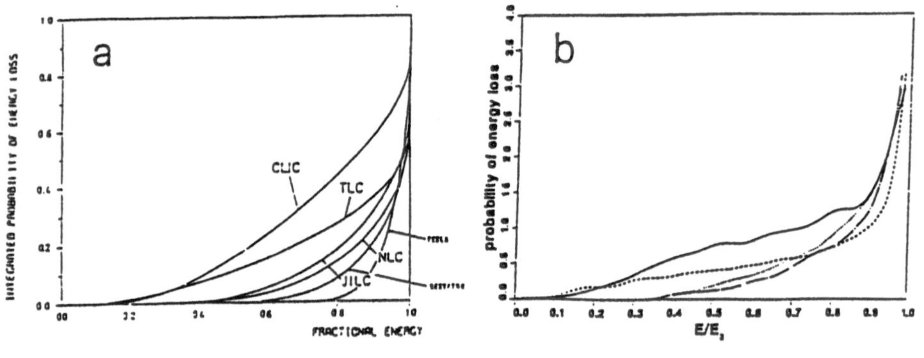

Fig.2 — (a) Integrated energy spectrum, P, of the disrupted electron beam in the interval $(0, E/E_0)$. Different curves refer to various collider studies[13] and represent the probability that the fractional energy is smaller than E/E_0. (b) Probability of energy loss, $f(\epsilon)$, obtained as a regularized solution of eq.(3.2.1).

We have applied the regularization technique of sect.2 to the analysis of the observed energy spectra of electron beams integrated between 0 and E/E_0, taken from ref.[13] where the possibility of using the wiggler brightness in order to produce a high-intensity positron source has been studied. The proposed set-up [13] consists of a 250 GeV electron beam passing through a special optics section. The beam is subsequently injected into a wiggler where it produces 30 MeV energy photons which are converted into electron-positron pairs in a thin target. Since the high-energy electron beam is obtained from the collision of a primary beam, it is disrupted by the interaction and some measured energy-integrated spectra are shown in fig.2(a). The deduced relative energy distributions, evaluated[14] by solving eq.(3.2.1), are given in the corresponding fig.2(b). As indicated by the rather long tail, the electron spectrum is largely broadened and reduced even in the "best" cases. This interesting result follows from the use of a powerful yet simple regularizing technique in the analysis of the experimental data.

4. Applications to optical systems

4.1 Compton scattering tomography

In the diagnostic γ- or X-ray energy ranges from about 30 keV to 150 keV and for materials with low to intermediate atomic numbers, Compton scattering is by far the dominant interaction mechanism in the matter. This fact has led many Authors[15,16] to explore the possibility of imaging directly the energy-dispersive properties of Compton-scattered radiation.

A two-dimensional (2-D) imaging system is designed to generate cross-sectional images in a plane through the object onto a one-dimensional (1-D) detector[15]. The representation of the object density by its line integral in transmission tomography is given by the following Beer's law:

$$I = I_0 \int_L \mu(x)\, dx, \tag{4.1.1}$$

where I_0 and I are, respectively, the incident and emerging X-ray intensities and $\mu(x)$ is the linear absorption coefficient. In eq.(4.1.1), subscript L indicates that the integral is to be evaluated along all the linear paths in the considered plane.

In Compton tomography this task is attained by means of line integrals measured over circular paths as shown in fig. 3. For a γ ray of energy E_0 emitted from a source point, S, Compton scattered once and then detected at point $D(x_d)$, it can be proved that, assuming the scattering process confined to a plane, the locus of scattering points having the same recorded energy, $E_\alpha = E_0[1 + (E_0/mc)(1 + \cos\alpha)]$, where α, $(0 < \alpha < \pi)$, is the scattering angle, is a circle whose center and radius are uniquely determined by S, D and E_α.

Therefore, for an emitting source, S, in (x_0, y_0) the number, $I(x_d, E_\alpha)$, of Compton-scattered photons recorded at x_d with energy E_α is a line integral of the electron density, $f(x, y)$, weighted over this circular path, and the weighting function, $w(x, y; x_d, E_\alpha)$, is related to the Compton-scattering

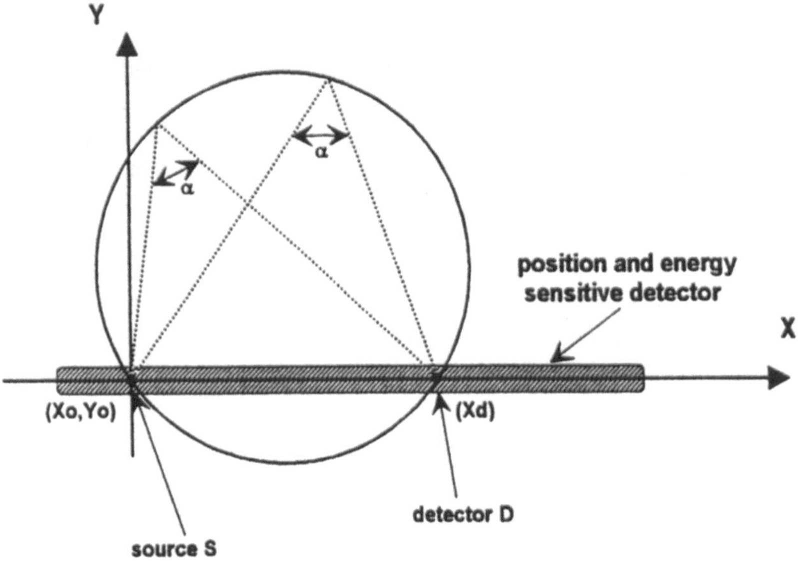

Fig.3 — The circle defined as the locus of all the scattering points in the plane, emitted by the same source, detected at the same point and scattered by the same angle, α.

cross section assumed to be dependent on the material mass density and composition:

$$I(x_d, E_\alpha) = \int_{C(x_d,E_\alpha)} f(x,y)\; w(x,y;x_d,E_\alpha)\; d\ell. \qquad (4.1.2)$$

In this equation, $C(x_d, E_\alpha)$ identifies the scattering circle, $d\ell$ is an element of the circular path, and the detector is supposed to have both a good position and energy sensitivity.

In the weighting function, $w(x,y;x_d,E_\alpha)$, the following effects have to be taken into account:

1) The angular dependence of the emitting source;

2) The differential cross section for various angles, α;

3) The detector's finite size;

4) The attenuation of the γ rays along the path from the source to the scattering point and back to the detector.

The image reconstruction from projections is a typical inverse ill–posed problem[17] requiring for its solution the numerical derivation of $f(x,y)$ from a family of circular paths.

Many techniques have been developed in the past[17,19] and few of them applied to the Compton scattering tomography. Recently[18], an analytic formula relating $f(x,y)$ to the photon intensity, $I(x_d, E_\alpha)$, has been derived by means of a back–projection filtered approach like that used in the conventional transmission computerized tomography (CT). This solution has been obtained assuming an ideal detector in both energy and spatial resolution powers and considering the multiple scattering as a background noise or arbitrarily reducible during the experiment. The result[18] is then an analytic function if no attenuation is considered or an approximated series development if a uniform attenuation is assumed.

In a more realistic situation, the detector has finite spatial and energy resolutions, and the attenuation is not uniform. Moreover, the multiple–to–single scattering ratio can not be neglected but — as recently shown[20] for energies in the diagnostic range — it is strongly dependent on the sample dimension and the scattering angle, α. This departure from the ideality introduces errors and approximations in the measurement of $I(x_d, E_\alpha)$ and eq.(4.1.2) has to be regularized in order to avoid numerical instabilities in the inversion procedure.

Considering that a Compton scattering imaging is very attractive because of the free choice of measurement geometry, back–scattering included, the possibility of improving the image resolution by using the regularization techniques discussed in sect.2 is under study by the Authors and preliminary attempts to evaluate the multiple scattering contamination on a Compton electron density measurements are in progress.

4.2 X–ray diffraction tomography

X–ray scattering is considered a severe problem in radiological imaging techniques causing image degradation. Two physical effects account for X–ray scattering, namely inelastic Compton and elastic Rayleigh scattering, both of them being treated as a noise source in transmission CT, and a large amount of effort is still performed in order to overcome this difficulty. As described in sect.4.1, several proposals have been made to directly image the density dependence of the Compton scattering and, recently, it has been shown[21] that combining diffractometry with CT a new imaging method can be developed.

In applications of medical interest, the X–ray scattering intensity, I, for given solid angle, $d\sigma$, and scattering angle, θ, depends on the total differential cross section:

$$\frac{d\sigma_{tot}}{d\Omega(\theta)} = \frac{d\sigma_{in}}{d\Omega(\theta)} + \frac{d\sigma_{co}}{d\Omega(\theta)} = \frac{d\sigma_{el}}{d\Omega(\theta)} \left[F^2(x) + S(x)\right], \qquad (4.2.1)$$

where the subscripts, *in* and *co*, refer to incoherent and coherent cross sections, respectively. In the right–hand–side of eq.(4.2.1), $d\sigma_{el}/d\Omega(\theta)$ is the classical free–electron cross section, $F^2(x)$ is the atomic form factor and $S(x)$ the coherent scattering function accounting, respectively, for the diffracted and diffused components. The x variable is the momentum transfer given by $x = (E/hc)\sin(\theta/2)$, as usual, with E incident photon energy.

It has been shown by Monte Carlo simulations that for X–ray energies in the diagnostic range (less than 100 keV) and scattering angles $\theta \leq 12°$, the single and multiple Compton scattering are strongly inhibited by the electron binding effects, at least for objects of small dimensions[22,23]. As a consequence, the elastic component dominates the total differential cross section (4.2.1) and — since the coherent scattering exhibits interference effects which depend on the atomic arrangement in the scattering volume — the diffracted X–ray quanta may in principle be used as a sensitive technique for imaging spatial variations in molecular structures[24].

Interference effects are commonly used in low–energy diffractometry for material characterization. In these experiments the specimen dimension, the set–up geometry and the X–ray energy are chosen in such a way that the diffracted intensity, I, is directly correlated to the atomic form factor and scales with the momentum transfer,

$$I(x) \propto F^2(x). \tag{4.2.2}$$

It is worth observing that in the diffractometric technique the approximation (4.2.2) is correct for a wide range of scattering angles, depending on E, so that the diffraction profile can be exactly resolved. In the *in vivo* applications of X–ray diffraction CT, there are several effects which limit the resolution of $I(x)$. The range of θ is strongly reduced due to the high energy needed to penetrate large objects. Monochromatic high–energy X–ray sources are difficult to obtain. The large dimensions of the specimen produce X–ray attenuation and, because of the requirement of reduced size of the apparatus, the resolution of the scattering angle, θ, is poor.

All these effects introduce degradation in the pattern of diffraction and the effectively measured scattering intensity is the weighted average of $I(x)$ over a blurring function, $\omega(x,y)$:

$$\tilde{I}(x) = \int_0^{y_{max}} I(y)\, \omega(x,y)\, dy, \tag{4.2.3}$$

where $(0, y_{max})$ is the range of allowed values of the momentum transfer. Among the effects limiting the momentum resolution, the uncertainty in θ appears to be the hardest to be removed. Many efforts have been made to reduce y_{max}, either in experiments where a polychromatic X–ray beam, diffracted at fixed angle $\bar{\theta}$, is analyzed with an energy resolving detector[24], or by using a monochromatic γ source and measuring $\tilde{I}(x)$ at different angles[25,26,27]. In both cases, in order to obtain a moment resolution near about 5–10%, a long measurement time is necessary to achieve a sufficient counting statistics, i.e. a large dose is delivered to the specimen.

An increase in the solid angle subtended by the object at the detector's position will obviously improve the counting statistics but the resolution of momentum tranfer would be drammatically reduced and the measured diffraction profile confused.

In fig.4, a design is shown of an experimental set–up planned in such a way to characterize biological tissues of similar molecular composition by means of X–ray diffraction CT. This experiment is designed to maintain the delivered dose to acceptable levels and, therefore, the angular acceptance $(\theta \pm \Delta\theta)$ is

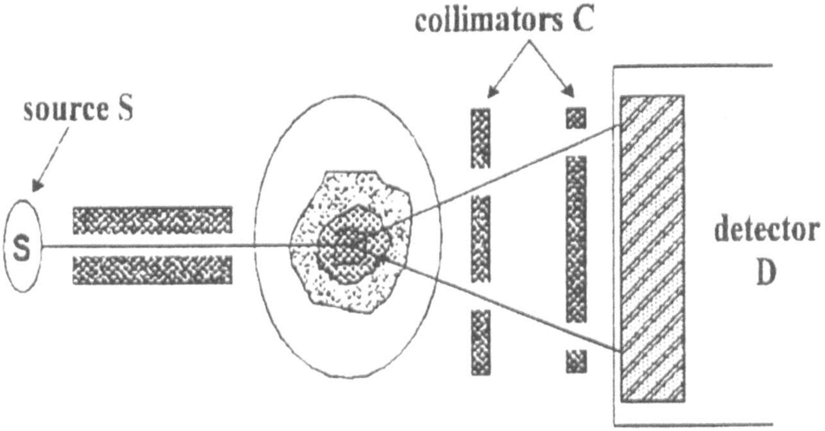

Fig.4 — Layout of the experimental CT–diffraction apparatus. The collimator assembly is symmetric with respect to the incident beam direction.

large. The poor resolution in the momentum transfer, x, makes impossible to distinguish the differences in the molecular structure from the $\tilde{I}(x)$ profile.

A possible improvement can be obtained by solving eq.(4.2.3) for $I(y)$ but — as previously observed — the experimental uncertainties introduce discontinuities in $\tilde{I}(x)$ and this equation represents an ill–posed problem that does not admit a stable solution. A Monte Carlo simulation of the relevant experimental conditions has been undertaken in order to verify the effectiveness of the regularization procedure of sect.2 to solve eq.(4.2.3). The preliminary results look promising. The numerical regularization of measured data increases the available resolution and the material composition — at least in the most important cases — can be derived, then allowing the X–ray diffraction CT to be used as a valuable tool to image differences between healthy and patologic tissues.

5. Conclusions

A wide set of problems concerning image restoration from experimental data affected by relevant uncertainties or image reconstruction from projections on linear or circular paths can be expressed in terms of integral equations of Volterra or Fredholm type. These equations generally belong to the class of ill–posed problems whose solution is unstable with respect to arbitrarily small perturbations.

This difficulty implies that many integral equations describing imaging of electromagnetic and optical complex systems of main interest for basic sciences and technological applications cannot be solved by direct inversion but suitable regularization procedures or *ad hoc* assumptions about the properties of the solution itself have to be adopted.

As stated by R.E.Burge, "it appears to be possible to design an imaging system for *every radiation* as has been done for effectively *the whole of the electromagnetic spectrum,* divided up into regions for experimental reasons, to start with dielectric (medical) images and microwaves (radar) at the long wavelength end of the spectrum and finish with γ–cameras for ultrashort waves. Similarly we have electron, neutron, acoustic images, etc....."[28]. We have then applied the Tikhonov's general regularization technique discussed in sect.2 to some imaging systems of this kind, starting from electron beams to γ and X rays, with promising results. It seems that a convenient unified framework to deal with restoration of radiation images over a broad range of wavelengths can thus be defined.

References

1. R.C.Gonzalez and P.Wintz, *Digital Image Processing,* Addison–Wesley, New York (1977).
2. J.Hadamard, Bull. Univ. Princeton **13** (1902); *Le problème de Cauchy et les équations aux dérivées partielles linéaires hyperboliques,* Hermann, Paris (1932).
3. A.N.Tikhonov and V.Arsenine, *Méthodes de resolution de problémes mal posées,* MIR Publ., Moscow (1976).
4. A.N.Tikhonov and A.V.Goncharsky, eds., *Ill–Posed Problems in the Natural Sciences,* MIR Publ., Moscow (1987).
5. D.L.Phillips, ACM J. **9** (1962) 84.
6. B.Hofmann, *Regularization for Applied Inverse and Ill–Posed Problems. A Numerical Approach,* Teubner — Texte zur Mathematik **85**, Leipzig (1986).
7. R.Hofstadter, *Electron Scattering and Nuclear and Nucleon Structure,* Academic Press, New York (1963).
8. T.A.Minelli, A.Pascolini and C.Villi, Nuovo Cim. **A86** (1985) 42.
9. S.Lorenzutta and G.Maino, submitted for publication.
10. B.B.P.Sinha, G.A.Peterson, R.R.Whitney, I.Sick and J.S.McCarthy, Phys. Rev. **C7** (1973) 1930.
11. G.Dattoli, A.Renieri and A.Torre, *Lectures on the Free Electron Laser Theory and Related Topics,* World Scientific, Singapore (1993).
12. F.Ciocci and G.Dattoli, Nucl. Instr. Meth. (1992) .
13. K.Flöttmann and J.Roßbach, *A High Intensity Positron Source for Linear Colliders,* DESY report M–91–11, Hamburg (1991).
14. G.Maino, *Mathematical aspects of the image restoration in applied optics: a general and historical view,* invited talk to the Int. Workshop on *Common Methodologies for Image Synthesis and Analysis,* Rome, Dec. 5–7, 1994.
15. H.H.Barret and W.Swindell, *Radiological Imaging. The Theory of Image Formation, Detection and Processing,* Academic Press, New York (1981).
16. S.R.Gautam, F.F.Hopkins, R.Kliksiek and I.L.Morgan, IEEE Trans. Nucl. Sci. **NS–30** (1983) 1680; G.Harding and R.Tischler, Phys. Med. Biol. **31** (1986) 477.

17. G.T.Herman, *Image Reconstruction from Projections,* Academic Press, New York (1980); A.C.Kak and M.Slaney, *Principles of Computerized Tomographic Imaging,* IEEE Press, New York (1988).
18. S.J.Norton, J. Appl. Phys. **76** (1994) 2007.
19. A.M.Cormack, J. Appl. Phys. **34** (1963) 2722.
20. A.Tartari, C.Bonifazzi, J.Felsteiner and E.Casnati, Nucl. Instr. Meth. Phys. Res. (submitted).
21. G.Harding, J.Kosanetky and U.Neitzel, Med. Phys. **14** (1987) 515.
22. U.Neitzel, J.Kosanetky and G.Harding, Phys. Med. Biol. **30** (1985) 1289.
23. A.Tartari, E.Casnati, C.Bonifazzi, J.Felsteiner and J.E.Fernandez, in Proc. Int. Conf. on *Röntgen Centennial,* Würsburg, Oct. 23–27 (1995) p.C30.
24. G.Harding, M.Newton and J.Kosanetky, Phys. Med. Biol. **35** (1990) 33.
25. D.A.Bradley, D.R.Dance, S.H.Evans and C.H.Jones, Med. Phys. **16** (1989) 851.
26. S.H.Evans, D.A.Bradley, D.R.Dance, J.E.Bateman and C.H.Jones, Phys. Med. Biol. **35** (1991) 33..
27. J.A.Grant, M.J.Morgan, D.R.Davis and P.Wels, Meas. Sci. Technol. **4** (1993) 83.
28. R.E.Burge, J. Phys. **D19** (1986) i.

Coding

Biorthogonal Brushlet Bases for Directional Image Compression

François G. Meyer and Ronald R. Coifman

Department of Mathematics, Yale University, New Haven CT, 06520, USA.

Abstract–We construct new biorthogonal bases that provide precise frequency localization and good spatial localization. We develop a compression algorithm that exploits the bases to obtain a representation of the image in terms of textured patterns with different orientations, frequencies, sizes, and positions. The technique directly works in the Fourier domain and has potential applications for highly textured images.

1. Introduction

Edges and textures in an image can exist at all possible locations, orientations, and scales. The ability to efficiently analyze and describe textured patterns is thus of fundamental importance for image analysis and image compression. Edges can be characterized using a wavelet transform [1]. However, wavelets provide only an octave based decomposition of the Fourier plane with a poor angular resolution. Wavelet packets make it possible to adaptively construct an optimal tilling of the Fourier plane, and they have been used for image compression [2]. However a wavelet is always associated with two peaks in frequency that does not allow to selectively localize a unique frequency. We propose to segment the Fourier plane to obtain a precise representation of the image in terms of oriented textures with all possible directions, frequencies, and locations. To achieve this we have constructed new biorthogonal bases that we call *brushlets.* Each brushlet is accurately localized in the Fourier plane and is well localized in the position plane as well. The method consists in expanding the Fourier transform of a function into bases of smooth localized exponentials.

The paper is organized as follows. In the next section we review the construction of smooth localized orthonormal exponential bases. In section 3, we construct biorthogonal localized exponential bases. The new biorthogonal brushlet bases are described in section 4. The image compression algorithm that exploits a brushlet expansion, with results of experiments are given in section 5.

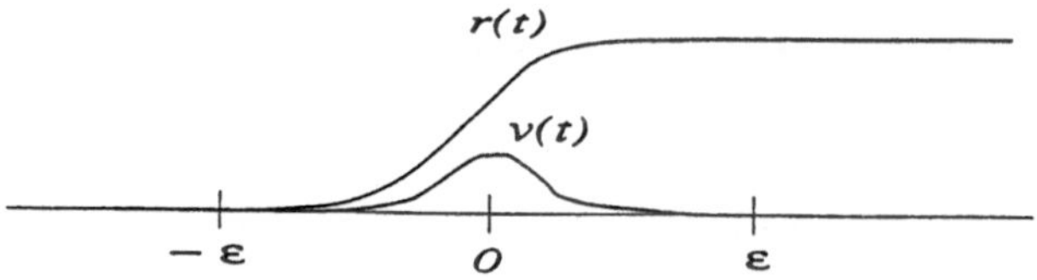

Fig. 2.1. Ramp function r, and bump function v.

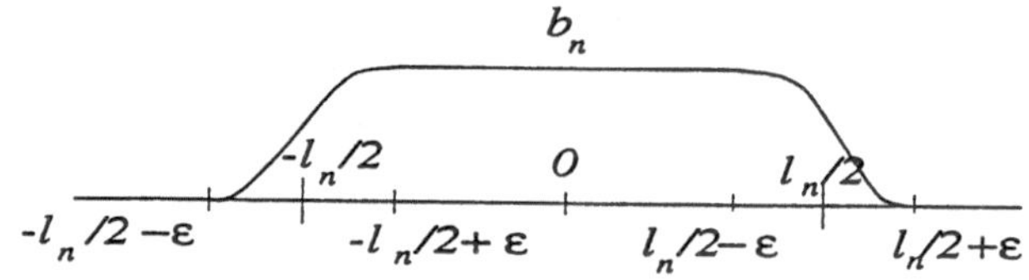

Fig. 2.2. Windowing function b_n.

2. Local trigonometric bases

First we review the construction of smooth localized orthonormal exponential basis [3, 4]. These functions are exponentials with good localization in both position and Fourier space. We consider a cover $\mathbb{R} = \bigcup_{n=-\infty}^{n=+\infty} [a_n, a_{n+1}[$. We write $l_n = a_{n+1} - a_n$, and $c_n = (a_n + a_{n+1})/2$. Around each a_n we define a neighborhood of radius ε. Let r be a ramp function such that

$$r(t) = \begin{cases} 0 & \text{if} \quad t \leq -\varepsilon \\ 1 & \text{if} \quad t \geq \varepsilon \end{cases} \tag{2.1}$$

and

$$r^2(t) + r^2(-t) = 1, \qquad \forall t \in \mathbb{R} \tag{2.2}$$

Let v be the bump function supported on $[-\varepsilon, \varepsilon]$ (see Fig. 2.1)

$$v(t) = r(t)r(-t) \tag{2.3}$$

Let b_n be the windowing function supported on $[-l_n/2 - \varepsilon, l_n/2 + \varepsilon]$ (see Fig. 2.2)

$$\begin{aligned} b_n(t) &= r^2(t + l_n/2) \quad \text{if} \quad t \in [-l_n/2 - \varepsilon, -l_n/2 + \varepsilon] \\ &= 1 \quad \text{if} \quad t \in [-l_n/2 + \varepsilon, l_n/2 - \varepsilon] \\ &= r^2(l_n/2 - t) \quad \text{if} \quad t \in [l_n/2 - \varepsilon, l_n/2 + \varepsilon] \end{aligned} \tag{2.4}$$

We consider the collection of exponential functions $e_{j,n} = e^{-2i\pi j(\frac{x-a_n}{l_n})}$, and we construct a basis of smooth localized orthonormal exponential functions $u_{j,n}$. Each $u_{j,n}$ is supported on $[a_n - \varepsilon, a_{n+1} + \varepsilon]$ and is given by [4]

$$u_{j,n}(x) = b_n(x - c_n)e_{j,n}(x) + v(x - a_n)e_{j,n}(2a_n - x) - v(x - a_{n+1})e_{j,n}(2a_{n+1} - x) \tag{2.5}$$

We have the following result [3, 4]

Theorem 2.1. [3, 4] *The collection $\{u_{j,n}\ j,n \in \mathbb{Z}\}$ is an orthonormal basis for $L^2(\mathbb{R})$*

We note that this basis uses exponentials, other smooth local bases that use sines, or cosines only can be also constructed [3, 5].

If we take the inverse Fourier transform $\check{u}_{j,n}$ of $u_{j,n}$ we obtain:

$$\check{u}_{j,n}(x) = e^{2i\pi c_n x}\left\{(-1)^j\widehat{b}_n(x - \frac{j}{l_n}) - 2i\,sin(\pi l_n x)\hat{v}(x + \frac{j}{l_n})\right\} \tag{2.6}$$

where $\widehat{b}_n$ is the Fourier transform of b_n, and $\hat{v}$ is the Fourier transform of v. The function $\check{u}_{j,n}$ is composed of two terms, localized around j/l, and around $-j/l$, that are oscillating with the frequency $c_n = (a_n + a_{n+1})/2$. In order to obtain a representation in terms of windowed exponentials we would like $\widehat{b}_n$ to be a positive windowing function. However, this constraint and the constraint (2.2) are incompatible. Therefore, we will relax condition (2.2), and construct two biorthogonal bases.

3. Biorthogonal Local trigonometric bases

We construct here biorthogonal bases of smooth localized exponentials. One basis can be used for the decomposition or analysis, and the other one for the reconstruction, or synthesis. We will proceed in a way similar to [6].

We consider here a cover $\mathbb{R} = \bigcup_{n=-\infty}^{n=+\infty}[a_n, a_{n+1}[$. All intervals have the same size $l = a_{n+1} - a_n$. Let $c_n = (a_n + a_{n+1})/2$. Let r be a ramp function such that

$$r^2(t) + r^2(-t) > 0, \qquad \forall t \in \mathbb{R} \tag{3.1}$$

Let v be the bump function supported on $[-l/2, l/2]$

$$v(t) = r(t)r(-t) \tag{3.2}$$

and let b be the windowing function supported on $[-l, l]$ (see Fig. 3.1).

$$\begin{aligned} b(t) &= r^2(t + l/2) \quad \text{if} \quad t \in [-l, 0] \\ &= r^2(l/2 - t) \quad \text{if} \quad t \in [0, l] \end{aligned} \tag{3.3}$$

We choose r such that $\hat{b}$, the Fourier transform of b, is positive. An example of such b is the cubic spline:

$$C(x) = \chi_{[-1/2,1/2]} * \chi_{[-1/2,1/2]} * \chi_{[-1/2,1/2]} * \chi_{[-1/2,1/2]} \tag{3.4}$$

where $\chi_{[-1/2,1/2]}$ is the characteristic function of $[-\frac{1}{2}, \frac{1}{2}]$. C is compactly supported on $[-2, 2]$, as shown in Fig. 3.1. Since the Fourier transform of $\chi_{[-1/2,1/2]}$ is $\frac{\sin(\pi\xi)}{\pi\xi}$, the Fourier transform of C is

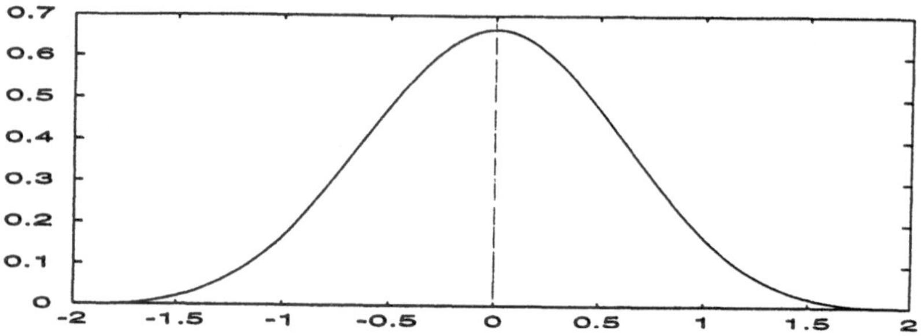

Fig. 3.1. Windowing function b, with $l = 2$

$$\hat{C}(\xi) = \left(\frac{\sin(\pi\xi)}{\pi\xi}\right)^4$$

The tail of $\hat{C}$ is rapidly decreasing to zero.

Let s be the l periodic function

$$s(x) = \frac{1}{\sum_{k\in\mathbb{Z}} r^2(x - a_k) + r^2(a_k - x)} \tag{3.5}$$

As above, we consider the collection of exponential functions

$$e_{j,n} = e^{-2i\pi j(\frac{x-a_n}{l})}$$

First we construct a basis of smooth localized orthonormal exponential functions $u_{j,n}$. Each $u_{j,n}$ is supported on $[a_n - l/2, a_{n+1} + l/2]$

$$u_{j,n}(x) = b(x-c_n)e_{j,n}(x)+v(x-a_n)e_{j,n}(2a_n-x)-v(x-a_{n+1})e_{j,n}(2a_{n+1}-x) \tag{3.6}$$

Then the dual basis is defined as

$$\tilde{u}_{j,n}(x) = s^2(x)u_{j,n}(x) \tag{3.7}$$

We have the following result

Lemma 3.1.

$$\int u_{j,n}(x)\tilde{u}_{k,m}(x)dx = \delta_{j,k}\delta_{n,m} \tag{3.8}$$

$\forall f \in L^2(\mathbb{R})$,

$$f(x) = \sum_{j,n} f_{j,n}u_{j,n}(x) \qquad \textit{with} \quad f_{j,n} = \int f(x)\tilde{u}_{j,n}(x)dx \tag{3.9}$$

$$f(x) = \sum_{j,n} \tilde{f}_{j,n}\tilde{u}_{j,n}(x) \qquad \textit{with} \quad \tilde{f}_{j,n} = \int f(x)u_{j,n}(x)dx \tag{3.10}$$

$u_{j,n}$ and $\tilde{u}_{j,n}$ are Riesz biorthogonal bases.

Proof. We first normalize the ramp function $r(t)$. Let

$$r^0(x) = \frac{r(x)}{\sqrt{r^2(x) + r^2(-x)}} \tag{3.11}$$

be the normalized ramp function that satisfies (2.2). We can then apply theorem 2.1 and construct a local exponential basis u^0

$$\begin{aligned} u^0_{j,n}(x) = b^0(x - c_n)e_{j,n}(x) \quad &+ \quad v^0(x - a_n)e_{j,n}(2a_n - x) \\ &- \quad v^0(x - a_{n+1})e_{j,n}(2a_{n+1} - x) \end{aligned} \tag{3.12}$$

where $b^0(x)$ is the window function associated with r^0 defined by (2.4). Similarly, v^0 is the bump function associated with r^0 defined by (2.3).

If $x \in [a_n - l/2, a_n + l/2]$ then

$$\begin{aligned} u^0_{j,n}(x) &= \frac{r^2(x - a_n)}{r^2(x - a_n) + r^2(a_n - x)} e_{j,n}(x) \\ &+ \frac{r(x - a_n)r(a_n - x)}{r^2(x - a_n) + r^2(a_n - x)} e_{j,n}(2a_n - x) \end{aligned} \tag{3.13}$$

thus

$$u^0_{j,n}(x) = s(x)\left\{r^2(x - a_n)e_{j,n}(x) + r(x - a_n)r(a_n - x)e_{j,n}(2a_n - x)\right\} \tag{3.14}$$

and finally

$$u^0_{j,n}(x) = s(x)u_{j,n}(x) \tag{3.15}$$

Similarly, if $x \in [a_{n+1} - l/2, a_{n+1} + l/2]$ we get

$$u^0_{j,n}(x) = s(x)u_{j,n}(x) \tag{3.16}$$

We now use (3.15,3.16) and the fact that $u^0_{j,n}$ is an orthonormal basis to prove (3.8). We have

$$\begin{aligned} \int u_{j,n}(x)\tilde{u}_{j,n}(x)dx &= \int s^2(x)u_{j,n}(x)u_{j,n}(x)dx \\ &= \int u^0_{j,n}(x)u^0_{k,m}(x)dx = \delta_{j,k}\delta_{n,m} \end{aligned}$$

Let $f \in L^2(\mathbb{R})$, in order to prove (3.9) we expand $f(x)s(x)$ into the basis $u^0_{j,n}$. We have

$$f(x)s(x) = \sum_{j,n}\{fs\}_{j,n}\ u^0_{j,n}(x) = \sum_{j,n}\{fs\}_{j,n}u_{j,n}(x)s(x) \tag{3.17}$$

with

$$\{fs\}_{j,n} = \int f(x)s(x)u^0_{j,n}(x)dx = \int f(x)s^2(x)u_{j,n}(x)dx = \int f(x)\tilde{u}_{j,n}(x)dx \tag{3.18}$$

thus from (3.17) and (3.18) we obtain

$$f(x) = \sum \int f(x)\tilde{u}_{j,n}(x)dx \;\; u_{j,n}(x). \tag{3.19}$$

In a similar way, if we expand f/s into the basis $u^0_{j,n}$ we obtain (3.10).

4. Biorthogonal brushlet bases

4.1 One-dimensional case

We now construct the biorthogonal brushlet bases. We first construct two biorthogonal trigonometric bases $\{u_{j,n}, \tilde{u}_{k,m}\}$, with a ramp function $r(x)$ such that the Fourier transform of b is positive. Let $w_{j,n}$ be the Fourier transform of $u_{j,n}$ and let $\tilde{w}_{k,m}$ be the Fourier transform of $\tilde{u}_{k,m}$. Since the Fourier transform is a unitary operator, we have

Lemma 4.1. *$\{w_{j,n}, \tilde{w}_{k,m} j, k, m, n \in \mathbb{Z}\}$ are biorthogonal bases for $L^2(\mathbb{R})$*

We call $\{w_{j,n}, \tilde{w}_{j,n}\}$ the biorthogonal brushlet bases. We have

$$w_{j,n}(x) = e^{2i\pi c_n x} \left\{ (-1)^j \widehat{b}(x - \frac{j}{l}) - 2i\, sin(\pi l\, x)\hat{v}(x + \frac{j}{l}) \right\} \tag{4.1}$$

As explained above $w_{j,n}$ is composed of two terms, localized around j/l, and around $-j/l$, that are oscillating with the frequency $c_n = (a_n + a_{n+1})/2$. The first term is an exponential multiplied by the window $\widehat{b}_n$. The window of the second term is not positive. However the magnitude of the second term is smaller than the first one. Therefore we will consider that $w_{j,n}$ is mainly localized around $-j/l$. Figure 4.1 shows the graph of the real part of $w_{j,n}$ when we use the cubic spline for b.

4.2 Two-dimensional case

In the two-dimensional case we define two partitions of $\mathbb{R}$, $\bigcup_{n=-\infty}^{n=+\infty}[a_n, a_{n+1}[$, and $\bigcup_{m=-\infty}^{m=+\infty}[b_m, b_{m+1}[$. We write $l = a_{n+1} - a_n$, and $h = b_{m+1} - b_m$. We then consider the tilling obtained by the tensor products $[a_n, a_{n+1}[\otimes[b_m, b_{m+1}[$. We consider the separable tensor products of bases $w_{j,n}$, and $w_{k,m}$. We have

Lemma 4.2. *The sequences $\{w_{j,n} \otimes w_{k,m}, \tilde{w}_{j',n'} \otimes \tilde{w}_{k',m'}\}$ are Riesz biorthogonal bases for $L^2(\mathbb{R}^2)$*

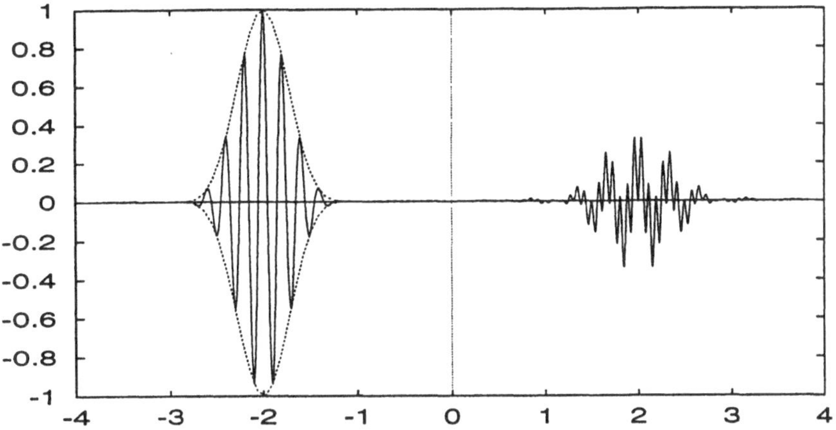

Fig. 4.1. Graph of the real part of $w_{j,n}$. On the left is the principal part of the brushlet: a windowed exponential. On the right is the part necessary to obtain perfect localization in Fourier space. In the two-dimensional case, with tensor products of brushlets, this part can be neglected.

The tensor product $w_{k,m}(x) \otimes w_{j,n}(y)$ is an oriented pattern oscillating with the frequency $((a_m + a_{m+1})/2, (b_n + b_{n+1})/2)$ and localized at $(k/h, j/l)$, as shown in Fig. 4.2. The size of the pattern is inversely proportional to the size of the analyzing window: $h \times l$ in the Fourier space. We note that the decomposition achieved by wavelet packets does not permit us to localize a unique frequency, for instance in the positive part of the Fourier space. Indeed two symmetric windows are always associated with a wavelet. As a result a wavelet packet expansion will require many more coefficients to describe a pattern with an arbitrary orientation; whereas the same pattern can be coded with a single brushlet coefficient.

5. Image compression

We have developed a compression algorithm that exploits the biorthogonal brushlet bases. We use $\tilde{w}_{j,n}$ for the decomposition and $w_{j,n}$ for the reconstruction. We use the cubic spline for b. We describe here a monoresolution of the algorithm, where all windows $[a_n, a_{n+1}[\otimes[b_m, b_{m+1}[$ have the same size. We are currently working on an adaptive multiresolution algorithm where the size of the window is selected according to the frequency content of the image.

5.1 Brushlet decomposition

The Fourier transform $\hat{f}$ of the image f is computed using a FFT. $\hat{f}$ is hermitian-symmetric, therefore we only retain the upper half of the Fourier plane $\{(\nu, \xi), \xi \geq 0\}$ for coding. We divide the upper half into two quadrants.

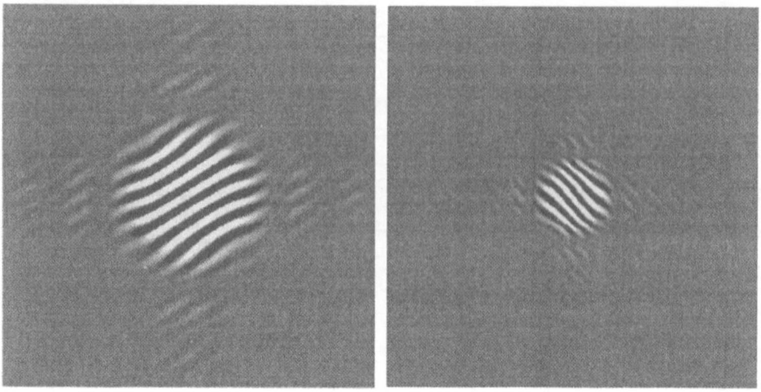

Fig. 4.2. Two dimensional brushlet basis functions $\{w_{j,n} \otimes w_{k,m}\}$ A good frequency resolution corresponds to a b with a small support, and is thus associated with a poor spatial resolution as shown on the left. A good spatial resolution corresponds to a $\hat{b}$ with a small support, and is thus associated with a poor frequency resolution as shown on the right.

For each quadrant we calculate the brushlet coefficients with square windows $[a_n, a_{n+1}[\otimes[b_m, b_{m+1}[$ of same size. Instead of calculating the inner product of $\hat{f}$ with $u_{j,n} \otimes u_{k,m}$ we fold the image around the horizontal and vertical lines associated with the tilling [4]. We then calculate the 2-D FFT of each folded block, and obtain the brushlet coefficients. Each block associated with this segmentation corresponds to a set of brushlet coefficients. These coefficients describe the intensity of one single "brush stroke" at different locations in the image, as illustrated in Fig. 4.2. This "brush stroke" has a particular frequency, orientation, and size, that are given by the position of the block, and its size.

5.2 Quantization and frequency scanning of the coefficients

The brushlet coefficients are quantized with uniform quantizers. In order to exploit the correlation between brushlet coefficients in different subbands, we order all the coefficients associated with the same spatial location by increasing frequency order. Since the magnitude of the terms in the sequence decreases with an exponential decay, we encode a terminating symbol after the last non-zero coefficient to indicate that the remaining coefficients are zeros. This represents a zero-tree like extension of the algorithm proposed in [7].

5.3 Entropy coding

After frequency ordering, the coefficients are entropy coded using variable length coding and an adaptive arithmetic coder. The first term of a frequency scan corresponds to a DC coefficient, and is therefore differentially encoded.

	Barbara
Compression	PSNR (dB)
7:1	31.03
14:1	26.98
31:1	24.15
56:1	22.65
63:1	22.31
74:1	21.95
86:1	21.54

	Mandrill
Compression	PSNR (dB)
5:1	28.68
10:1	24.64
28:1	21.84
45:1	21.1
63:1	20.71
76:1	20.52
95:1	20.35

Table 5.1. Coding results for 8bpp. 512x512 Barbara (left) , and 8bpp. 512x512 Mandrill (right)

5.4 Experiments

We have implemented the coder and decoder, and an actual bit stream was created for each experiment. We present the results of the algorithm using two test images that are difficult to compress: 512x512 "Barbara", and 512x512 "Mandrill". The performance of the algorithm are summarized in Table 1, and results for compression ratios of 74:1 and 76:1 are shown in Fig. 3. We note that even at a compression ratio of 76:1 the mandrill still keeps its high frequency features such as the whiskers.

References

1. S. Mallat and S. Zhong. Characterization of signals from multiscale edges. *IEEE Trans. on Pattern Analysis and Machine Intelligence*, Vol 14, No. 7:pp 710–732, July 1992.
2. K. Ramchandran and M. Vetterli. Best wavelet packet bases in a rate-distortion sense. *IEEE Trans. on Image Processing*, pages pp 160–175, April 1993.
3. R.R. Coifman and Y. Meyer. Remarques sur l'analyse de fourier à fenêtre. *C.R. Acad. Sci. Paris I*, pages pp. 259–261, 1991.
4. M.V. Wickerhauser. *Adapted Wavelet Analysis from Theory to Software*. A.K. Peters, 1995.
5. I. Daubechies, S. Jaffard, and J.L. Journé. A simple wilson orthonormal basis with exponential decay. *SIAM J. Math. Anal.*, 22:554–572, 1991.
6. G. Matviyenko. Optimized local trigonometric bases. *Yale University, Department of Computer Science, Research report YALEDU/DCS/RR-1041*, 1994.
7. J.M. Shapiro. Embedded image coding using zerotrees of wavelet coefficients. *IEEE Trans. on Signal Processing*, pages 3445–3462, Dec. 1993.

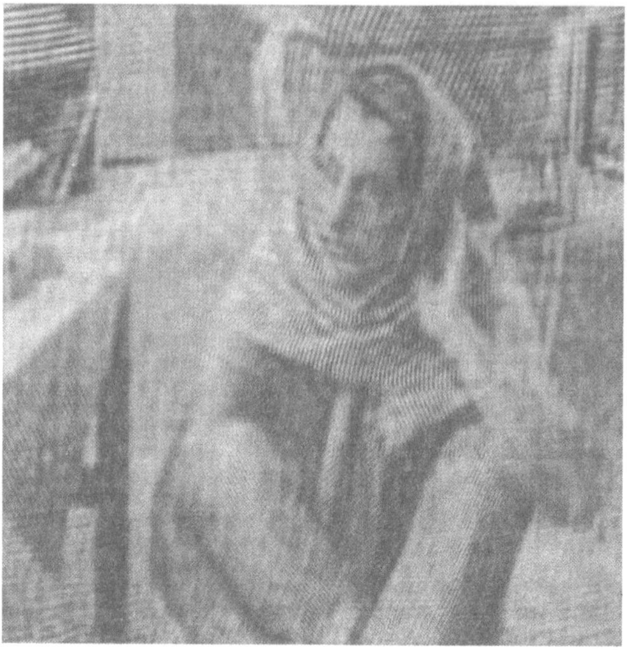

Fig. 5.1. Barbara, compression 74:1

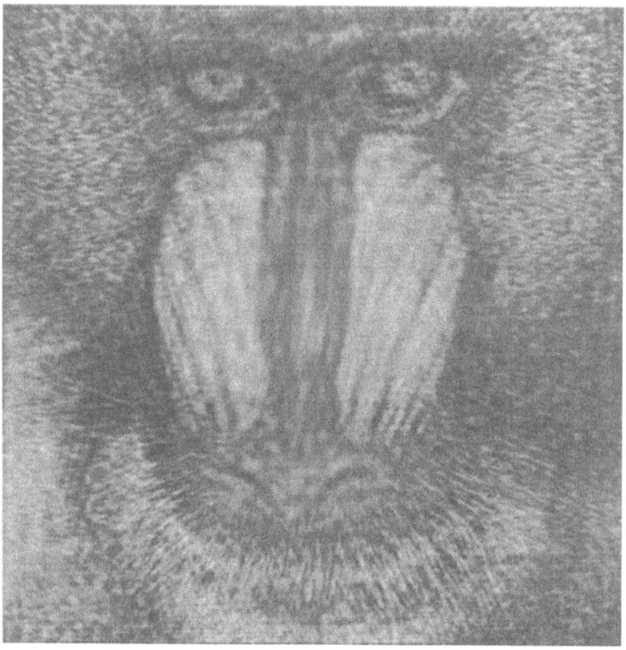

Fig. 5.2. Mandrill, compression 76:1.

Morphological Interpolation for Image Coding

Josep R. Casas*

Universitat Politècnica de Catalunya
Campus Nord D5, Gran Capità s/n, 08034 Barcelona, Spain
E-Mail: josep@gps.tsc.upc.es

Summary. The aim of this paper is to present a new interpolation technique intended for spatial interpolation from sparse data sets. The proposed implementation, which is based on non-linear morphological operators, overperforms linear interpolation by means of diffusion processes performing iterative space-variant filtering on the initial image. Morphological interpolation is applied to sketch-based image coding. We put forward a perceptually motivated two-component image model that strongly relies on morphological operators. The watershed is used to detect strong edge features in the first component of the model. The smooth areas of the image are recovered from the extracted edge information by morphological interpolation. The residual component, containing fine textures, is separately coded by a subband coding scheme.

1. Introduction

Interpolative coding techniques are based on the coding and transmission of a subset of pixels of the original image so that, on the receiver side, the remaining pixels have to be interpolated from the transmitted information alone. The reconstructed image is usually approximated by continuous functions with some permissible error at the interpolated positions. The subset of transmitted pixels, called the *initial set* in the following, may be either a regular subsampling grid or any arbitrary set of points. In the latter case, both the amplitude values and positions of the pixels of the initial set should be coded and transmitted.

1.1 Interpolation and sketch-based image coding

A number of different approaches using interpolation techniques have been reported in the literature for 'perceptually motivated' coding applications [1, 4, 9]. The underlying image model is based on the concept of the "raw primal sketch" [6]. The image is assumed to be made mainly of areas of constant or smoothly changing intensity separated by discontinuities represented by strong edges. The coded information, also known as *sketch data*, consists of the geometric structure of the discontinuities and the amplitudes at the edge pixels. In very low bit-rate applications, the decoder has to reconstruct the smooth areas in between using just this information. This can be posed as a scattered data interpolation problem from arbitrary initial sets (the

* This work was supported by the European RACE/Morpheco project 2053

sketch data) under certain smoothness constraints. For higher bit-rates, the residual texture information is separately coded by means of a waveform coding technique; for instance, pyramidal or transform coding.

The performance of such perceptual model has been thoroughly investigated [9], proving its utility for most coding applications and showing subjective improvements over DCT-based methods, as JPEG, at low bit-rates. However, one of the important drawbacks is the large computation time spent in the interpolation process. The techniques proposed to solve the problem of interpolation from sparse data are based on the solution of an energy minimization (variational) problem, governed by the heat or diffusion equation. The practical implementations make use of iterative space-variant filtering operations that converge rather slowly to the final interpolated image.

The aim of this paper is to present a morphological technique intended to perform spatial interpolation from any set of initial pixels. The algorithm, described in section 2., is based on morphological operators: namely geodesic dilation and the morphological Laplacian, resulting in a highly efficient process compared to those that perform linear filtering on the initial image. Comparative figures of computation time will be given to assert the efficiency of the process. Its application to sketch-based coding is illustrated in section 3., where a two-component model is proposed for perceptual image coding. Finally, some coding results and the conclusions are presented in the last section of the paper.

2. Morphological interpolation technique

The target of the morphological interpolation algorithm is to approximate the amplitudes of unknown pixels of the image by fitting a surface on a subset of pixels of known values (the initial set). Such surface is constrained to be maximally smooth between the known pixels in the sense that pixel to pixel variations in the interpolated area should be minimized.

A suitable strategy for spatial interpolation is to compute at each point the average of the amplitudes of the initial pixels weighted by the inverse of the distances to each of them [11]. The nearest pixels have stronger influence than the distant ones, and the interpolated amplitudes change slowly in the areas in between.

2.1 Geodesic distance weighting

The distance measure is taken as the *geodesic distance* within the set of unknown pixels, that is, the length of the shortest path joining two points which is completely included within the set of unknown pixels. The use of the geodesic distance allows the preservation of the transitions imposed by the initial set. This is illustrated in Fig. 2.1. The set of initial pixels is indicated by thick solid and dashed lines. Let us suppose that the dashed line represents the upper edge of a spatial transition and the solid line represents the lower edge. The influence of the amplitude values of the upper edge (dashed line) at pixel x is given by the inverse of the geodesic distance d_{3g}, which is larger than the Euclidean distance d_{3e}. Therefore, the interpolated values at pixel x will be mainly influenced by the initial pixels of the solid line because the weights of the pixels located on the other side of the transition, at a larger geodesic distance, will be much smaller. As a result, the use of the geodesic distance allows the preservation of the transitions.

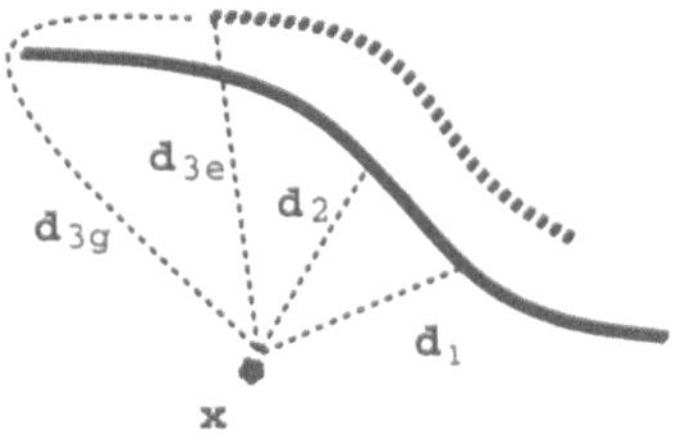

Fig. 2.1. Geodesic distance measure for the interpolation of pixel x

2.2 Two-step iterative algorithm

Starting from the set of initial pixels the *morphological interpolation* technique is implemented by an efficient two-step algorithm. The two steps, namely geodesic propagation step and smoothing step, are successively iterated until convergence.

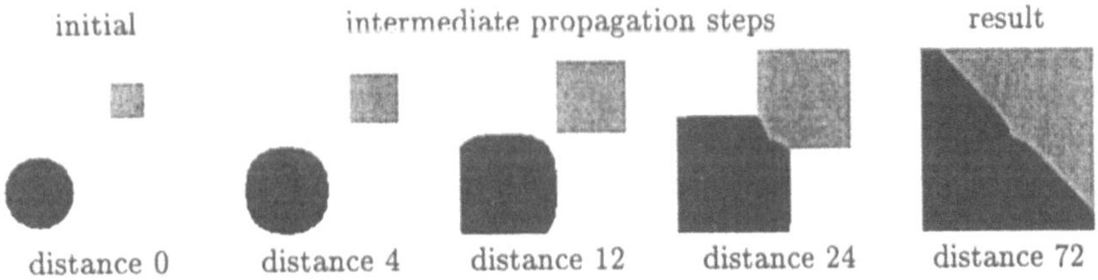

Fig. 2.2. Geodesic propagation: initial pixels and some intermediate images

- *Geodesic propagation step*
  In the ***geodesic propagation step***, instead of computing geodesic distances from all the unknown pixels to every point of the initial set, the amplitude values of the known pixels are propagated by geodesic dilation to fill the empty areas of the image. This is performed by a fast algorithm, using FIFO queues, so that each pixel is treated only once in a complete propagation process. Figure 2.2 shows some intermediate images corresponding

to the propagation process from a given initial set (a synthetic initial image consisting of two small geometric figures).

- *Progressive smoothing step*
  At the positions where two or more propagation fronts originated from initial pixels of different amplitudes meet, the process stops and a false transition is created. The false transitions appearing outside the set of initial pixels are smoothed in the second step. The morphological Laplacian[1] is used as a transition detector in order to obtain these false transitions. Pixels on both sides of the false transitions compose the set of *secondary pixels*. A grey level value equal to the average of the intensity values on both sides of the transition is assigned to each secondary pixel. This is the *smoothing step*. Secondary pixels will be used in the next iteration of the algorithm in order to smooth out these transitions.
- *Iteration*
  Then, a second iteration is performed: the propagation step propagates the grey level values from the sets of initial as well as secondary pixels. The propagation creates new false transitions which define a new set of secondary pixels where grey level values are smoothed. Note that this new set of secondary pixels generally does not include the first set of secondary pixels. This process of 1) propagation of values from the initial and secondary pixels, and 2) smoothing of the grey levels at the false transitions, is iterated until idempotence. Figure 2.3 illustrates several iterations of the algorithm. Please observe the progressive smoothing of the false transitions. After a few number of iterations, the algorithm quickly converges to the final interpolated image.

2.3 Algorithm efficiency

The efficiency of the morphological interpolation algorithm in terms of computational load is illustrated in table 2.1. Comparative figures of execution time[2] are given for the previous example of morphological interpolation and solved by applying linear diffusion by means of iterated space variant filters. The result is very similar in both cases, as can be observed in Fig. 2.4. Please notice the drastic reduction in the number of iterations needed for the

[1] The morphological Laplacian, $L(f)$, is defined as the residue of the gradient by dilation, $g^+()$, and the gradient by erosion, $g^-()$:

$$g^+(f) = \delta(f) - f \qquad g^-(f) = f - \varepsilon(f) \tag{2.1}$$

$$L(f) = g^+(f) - g^-(f) \tag{2.2}$$

The morphological Laplacian is greater than zero at the lower edge of the transitions and smaller than zero at the upper edge. In flat surfaces or slanted planes without convexity changes, it cancels out. Indeed, it can be shown that the morphological Laplacian is an approximation of the signal second derivative.

[2] Note: CPU times were computed on a Sun SPARC10 workstation

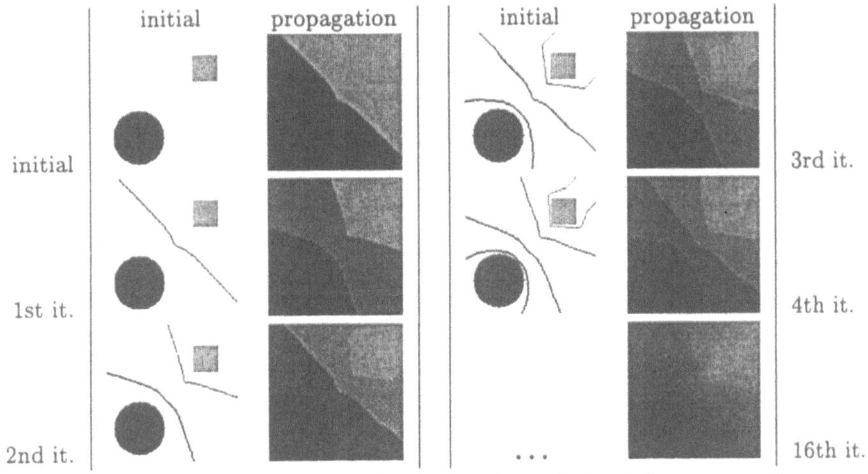

Fig. 2.3. Smoothing: left, initial and secondary pixels; right, propagation

morphological technique. Each pixel of the image to interpolate is treated hundreds of times less. Furthermore, each iteration of the morphological interpolation does not require any multiplication, decreasing the time of each individual iteration compared to the linear filtering technique. This explains the reduced execution time of the described nonlinear interpolation process. Clearly, there is no need of multigrid techniques for speeding up convergence when the morphological interpolation algorithm is used.

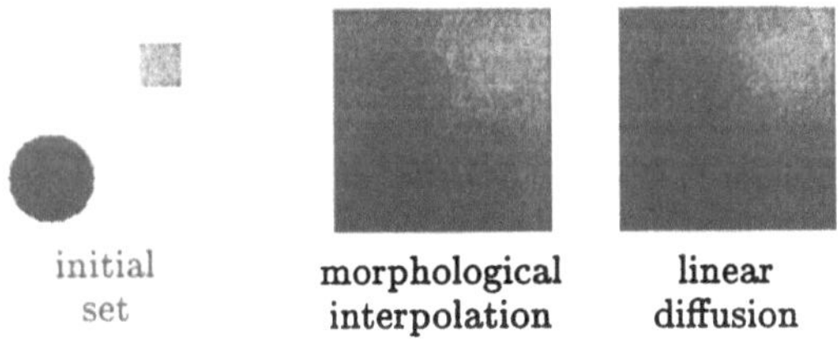

Fig. 2.4. Comparison of morphological interpolation and linear diffusion results

Table 2.1. Execution times of interpolation techniques for the previous example

Interpolation technique	Execution time [s]	No. iterations
Linear diffusion	312,8	4980
Multigrid diffusion	53,3	equivalent to 795
Morphological	2,8	16

3. Application to "sketch-based" image coding

A smooth approximation of an image may be obtained by interpolation from the set of pixels with large curvature values. The following experiment has been carried out in order to investigate its possible application to image compression. In the left image of Fig. 3.1, a set of pixels having large absolute values of the second derivative (actually, the morphological Laplacian) is shown. If we attempt to interpolate the remaining pixels of the smooth areas in between, the result will be the one presented in the right image. About one tenth of the pixels of the image have been used as initial points for the interpolation algorithm. The peak to noise ratio of the interpolated image of Fig. 3.1 is only 23 dB but its subjective quality is not bad, because our attention is primarily drawn to the strong transitions which have been correctly placed and reproduced.

Fig. 3.1. Morphological interpolation from pixels with large Laplacian values: left, initial image (about 10% pixels); right, interpolation result

This experiment proves that it is possible to obtain a fair approximation of the original image from the amplitudes and positions of some pixels having large curvature values. The technique described in the previous section has been used for the interpolation process. Furthermore, the morphological Laplacian performs as an effective enhancement operator for the detection of such set of initial pixels. Obviously, the application of this idea to image coding relies on the selection of a proper set of initial pixels for the interpolation process. The *initial set* should lead to a compact representation and, at the same time, allow a good approximation of the original image by interpolation.

3.1 Image coding by maximum and minimum curvature lines

The extrema of the second derivative locate the points with largest curvature values. The lines of largest curvatures are placed at the upper and lower side of each transition, bringing information about the transition width and the intensity change. These lines are called upper and lower *edge brims* by some authors [8] and may be obtained as the 'crest' and 'valley' lines of a second derivative operator. Edge brims do look promising for the characterization of visual information from a perceptual point of view. Robinson [10] claims that brim lines are less noisy than Laplacian zero-crossings, which follow the edge midpoints, and have been often used for contour extraction. Edge brims do not show so many random fluctuations because they do not represent a very rapid change in value with respect to position as edge midpoints do.

In the left image of Fig. 3.2, the white and black lines correspond, respectively, to the crest and valley lines of the Laplacian or, likewise, to the positions of the lower and upper edge brims of the initial image. Edge brims may be detected by computing the watershed [7] of the Laplacian and of its dual with an appropriate set of markers. In order to obtain the lower brims (crest lines of the Laplacian), the set of markers is formed by the union of two sets: the flat areas of the original image larger than a given size and the connected components of negative Laplacian values indicating the presence of valleys. For the upper brims (valley lines of the Laplacian), the second set is formed by connected components of positive Laplacian values indicating peaks and ridges. Please notice that some pieces of contour have been removed from the watershed result, either because the Laplacian was not significant enough at these positions or because the lines were too short. The necessary thresholds have been chosen on an empirical basis.

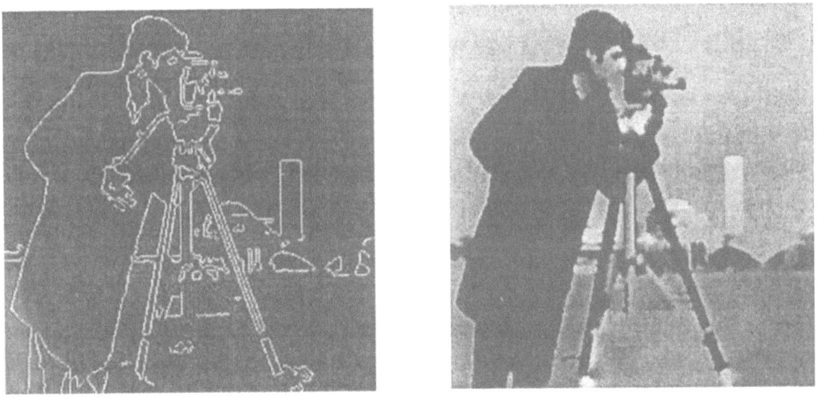

Fig. 3.2. Interpolation from lower and upper edge brims: left, brims' positions; right, interpolation result at 0.18 bpp

The geometric structure of the brim lines may be coded at low cost by means of a contour-following technique. The amplitudes of the initial pixels in these lines should also may be coded with a few number of bits. Given that intensity values along the edge brims should keep rather constant, some approximation may be employed to code the values within each brim line.

If the initial set is composed of the pixels at the positions indicated by the watershed lines shown in Fig. 3.2 with the approximated intensity values, the interpolation results in the right image of the same figure.

3.2 Two-component model

We put forward a two-component model for perceptual image coding that strongly relies on morphological operators. The interpolation result of Fig. 3.2 corresponds to the ***primary*** component of the perceptual model. The residual component, or ***texture*** component, contains the fine textures, which will be separately coded by a subband coding scheme.

- *Primary component*
  The first component of the model consists of the strong edges and smooth areas of the image. The smooth areas are generated by interpolation from the positions and amplitudes of the pixels of the initial set, i.e. the lower and upper brims of strong edges. A derivative chain code technique [5] is used to code the pixels' positions, whereas the amplitude values have been coded by polynomial approximation. More precisely, the network of brim lines is broken at each triple point (points with more than two branches). Then, the amplitudes of the pixels located under the resulting curves are approximated by a first order polynomial. The two coefficients defining each polynomial are quantized, entropy coded and transmitted. In the example of Fig. 3.2, the overall bit-rate needed for the primary component is 0.18 bits per pixel. About 16% of this rate is spent in the coding of amplitudes, 70% for the chain-code information and the remaining 9% for the initial positions of each brim line.
- *Texture component*
  The coding residue of the first component –computed as the difference between the original image and the interpolation result– mainly consists of fine textures. This second component of the model is shown in the left image of Fig. 3.3. It lacks of significant transitions and may be approximated by a waveform coding technique. A coded reconstruction at low bit-rate (0.15 bpp) is shown in the right image. It has been obtained by the application of the linear subband coding scheme presented in [2]. Information about the edge structure –available from the first component– is used for the texture coding of the second component, so that the masking effect of strong transitions may be considered. Amplitude errors in the neighborhood of these areas are less noticeable for the human eye than in other parts of the image [3]. Therefore, the quantization process is allowed to introduce large errors near the transitions by employing adaptive quantizers and bit allocations over arbitrarily shaped sub-edge regions in order to reduce the total number of bits.

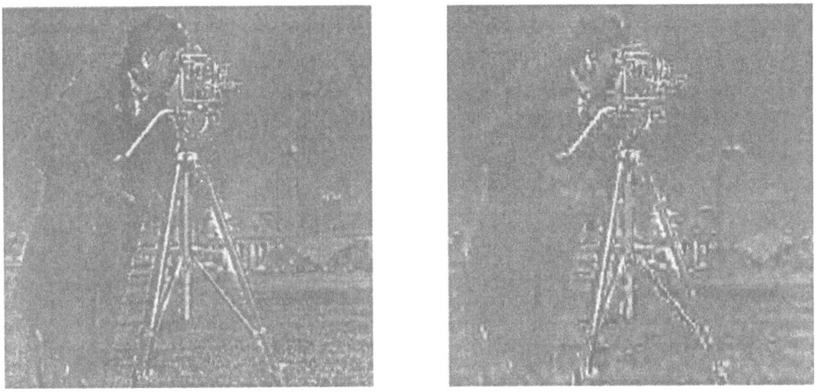

Fig. 3.3. Texture: left, coding residue; right, subband coded texture at 0.15 bpp

4. Results and Conclusion

The compression achieved with the above strategy is equal to 24 (0.33 bpp) for the addition of the strong edge and fine-texture components of Figs. 3.2 and 3.3. The result is shown in the right image of Fig. 4.1. For comparison, the application of the JPEG standard [12] at the same bit-rate is shown in the left image. The subjective quality of the described technique is significantly better because of the good rendition of the strong edges. The PSNR value (25.5 dB) is also larger than for the JPEG reconstruction (24 dB). The artifacts produced by the block-based DCT coding -blockiness in smooth areas and ringing in the neighborhood of strong transitions- are not present in the result of the two-component coding scheme. However, a different kind of visual artifacts may be observed. A certain smoothing effect is visible in some parts of the image and there are some missing objects, for instance the neck of the shirt has been almost removed.

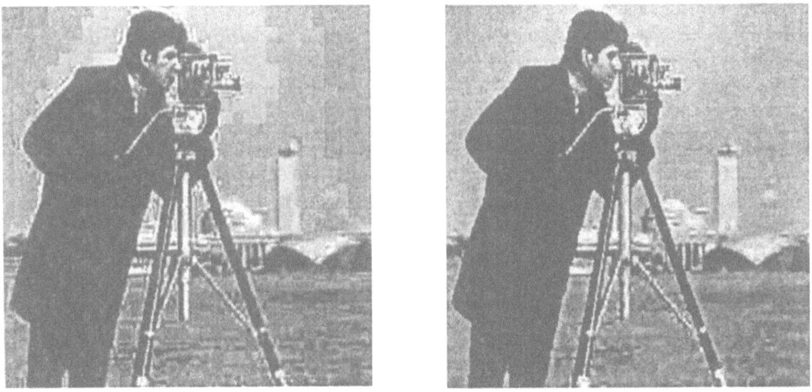

Fig. 4.1. Results: left, JPEG at 0.35 bpp; right, presented technique at 0.33 bpp

The separate coding of strong edges permits the adaptation of the coding scheme to the visual perception of the images, avoiding unnatural degradations produced by waveform coding techniques at high compression ratios. A number of different artifacts are introduced by this method at low bit-rates. It is hoped that such effects are more naturally perceived than those of waveform coding techniques by the subjective judgement of the observer.

Mathematical Morphology provides powerful operators to perform shape analysis. Morphological operators are very useful for the detection of edge features in 'perceptually motivated' Second Generation image coding applications, as has been shown in the present paper. The morphological operators involved in the coding of the primary component, i.e. the watershed and the morphological Laplacian, perform very efficiently compared to more conventional techniques for edge extraction, like the LGO operator, or the diffusion filters iteratively applied for the interpolation of smooth areas reported in previous works. The new morphological interpolation technique intended for scattered data interpolation described in section 2. has proven to be faster than linear diffusion techniques for the generation of the smooth component from edge features, with similar quality of the interpolation results.

References

1. S. Carlsson. Sketch based coding of grey level images. *EURASIP, Signal Processing*, 15(1):57–83, July 1988.
2. J. R. Casas and L. Torres. A feature-based subband coding scheme. To be presented at ICASSP'96, May 1996.
3. T. N. Cornsweet. *Visual Perception*. Academic Press, New York, 1970.
4. P. Grattoni and A. Guiducci. Contour coding for image description. *Pattern Recognition Letters*, 11:95–105, February 1990.
5. F. Marqués, J. Sauleda, and A. Gasull. Shape and location coding for contour images. In *Picture Coding Symposium*, pages 128.6.1–18.6.2, Lausanne, Switzerland, March 1993.
6. D. Marr. *Vision*. Freeman, New York, 1982.
7. F. Meyer and S. Beucher. Morphological segmentation. *Journal of Visual Communication and Image Representation*, 1(4):21–46, September 1990.
8. X. Ran and N. Farvardin. A perceptually motivated three-component image model. Part I: Description of the model. *IEEE Transactions on Image Processing*, 4(4):401–415, April 1995.
9. X. Ran and N. Farvardin. A perceptually motivated three-component image model. Part II: Application to image compression. *IEEE Transactions on Image Processing*, 4(4):430–447, April 1995.
10. J. A. Robinson. Image coding with ridge and valley primitives. *IEEE Transactions on Communications*, 43(6):2095–2102, June 1995.
11. P. Soille. Spatial distributions from contour lines: an efficient methodology based on distance transformations. *Journal of Visual Communication and Image Representation*, 2(2):138–150, June 1991.
12. G. K. Wallace. The JPEG still picture compression standard. *Communications of the ACM*, 34(4):30–44, April 1991.

Efficient Representations with Quantized Matching Pursuit

Vivek K Goyal[1], Martin Vetterli[1], and Nguyen T. Thao[2]

[1] Dept. of Electrical Engineering and Computer Science, Univ. of California, Berkeley, CA 94720 and Département d'Électricité, École Polytechnique Fédérale de Lausanne, CH-1015 Lausanne, Switzerland
[2] Dept. of Electrical and Electronic Engineering, Hong Kong Univ. of Science and Technology, Clear Water Bay, Kowloon, Hong Kong

Summary. We propose that a quantized matching pursuit representation be viewed as a set of linear inequality constraints on the input. It is shown that a standard (linear) reconstruction method can give estimates that do not satisfy the constraints on the input. This new framework leads to a *consistent* reconstruction algorithm which, in experiments for source coding of an $\mathbb{R}^N$-valued source, lessened MSE distortion by as much as a factor of seven.

1. Introduction

The matching pursuit algorithm—introduced to the signal processing community in the context of time-frequency analysis by Mallat and Zhang [1]—is an adaptive basis method for finding linear combinations that approximate a given signal vector. It has recently been successfully applied to low bit rate video compression [2, 3, 4]. In these applications, as in all compression applications, the coefficients must be quantized. We will see that when coefficients are quantized in matching pursuit, linear reconstructions may be inconsistent and hence suboptimal. The use of a consistent reconstruction method can in some cases dramatically lower reconstruction error.

2. Quantized Matching Pursuit

2.1 Unquantized Matching Pursuit Algorithm

Let $\mathcal{D} = \{\varphi_k\}_{k=1}^{M} \subset \mathbb{R}^N$ span $\mathbb{R}^N$. Also assume the φ_k's are distinct and normalized such that $||\varphi_k|| = 1$ for all k. $\mathcal{D}$ is called the *dictionary* of vectors. The matching pursuit algorithm, summarized in Table 2.1, finds a linear combination of elements of $\mathcal{D}$ to approximate $f \in \mathbb{R}^N$:

$$f = \sum_{i=0}^{n-1} \alpha_i \varphi_{k_i} + R_n f. \qquad (2.1)$$

The output of a matching pursuit expansion is not only the coefficients (α_0, $\alpha_1, \ldots$), but also the indices ($k_0, k_1, \ldots$). Papers by Mallat and his coworkers give a thorough treatment of the properties of matching pursuit [1, 5, 6].

Table 2.1. Matching Pursuit Algorithm

1. Let $i = 0$. Let $R_0 f = f$.
2. Let $\varphi_{k_i} = \text{argmax}_{\varphi \in \mathcal{D}} |\langle \varphi, R_i f \rangle|$ and $\alpha_i = \langle \varphi_{k_i}, R_i f \rangle$.
3. Let $R_{i+1} f = R_i f - \alpha_i \varphi_{k_i}$. Unless stopping criterion is satisfied, increment i and go to step 2.

2.2 Discussion

Coefficients are quantized in any digital computer implementation of matching pursuit. If the quantization is coarse, as it must be for moderate to low bit rate compression applications, the effects of quantization may be significant. Heretofore, little work has been done to understand the qualitative effects of coefficient quantization in matching pursuit.

Define *quantized matching pursuit* (QMP) to be matching pursuit in which the coefficients are quantized as $\hat{\alpha}_i = q_i(\alpha_i)$, where the q_i's are (scalar) quantization functions. We are assuming that the quantization of α_i occurs before the residual $R_{i+1} f$ is calculated, and that the quantized version is used in determining the residual so that quantization errors do not propagate to subsequent iterations.

2.3 Consistency

Let $Q : X \to Y$ be a quantization function. We say that $\hat{x} \in X$ is a *consistent estimate* of $x \in X$, or a *consistent reconstruction*, if $Q(\hat{x}) = Q(x)$ [7]. In words, we would say that an estimate is consistent if it has the same quantized version as the original. Consistency depends only on the deterministic properties of Q, and not on statistical properties of the X-valued source. Specializing to our application of QMP to an $\mathbb{R}^N$-valued source and assuming p iterations, $X = \mathbb{R}^N$ and $Y = \mathcal{D}^p \times q_1(\mathbb{R}) \times \ldots \times q_p(\mathbb{R})$.

Reconstructions from a quantized matching pursuit representation are generally computed by using the quantized coefficients in (2.1), giving

$$\hat{f} = \textstyle\sum_{i=0}^{p-1} \hat{\alpha}_i \varphi_{k_i}. \tag{2.2}$$

A shortcoming of this reconstruction is that it disregards the effects of quantization; hence it can produce inconsistent estimates. Simulations reported in [8, §3.3.3] show that the probability of (2.2) giving inconsistent estimates can be large. The dependence on $\mathcal{D}$ and the scalar quantizers is complicated.

A QMP representation can be viewed as a set of linear inequality constraints on the original signal. At the $(i+1)$-st iteration, the selection of k_i gives the constraints[1]

$$\left|\left\langle \varphi_{k_i}, f - \textstyle\sum_{\ell=0}^{i-1} \hat{\alpha}_\ell \varphi_{k_\ell} \right\rangle\right| \geq \left|\left\langle \varphi, f - \textstyle\sum_{\ell=0}^{i-1} \hat{\alpha}_\ell \varphi_{k_\ell} \right\rangle\right|, \quad \forall\, \varphi \in \mathcal{D}. \tag{2.3}$$

[1] Here and in (2.4), consider the summations to be empty for $i = 0$.

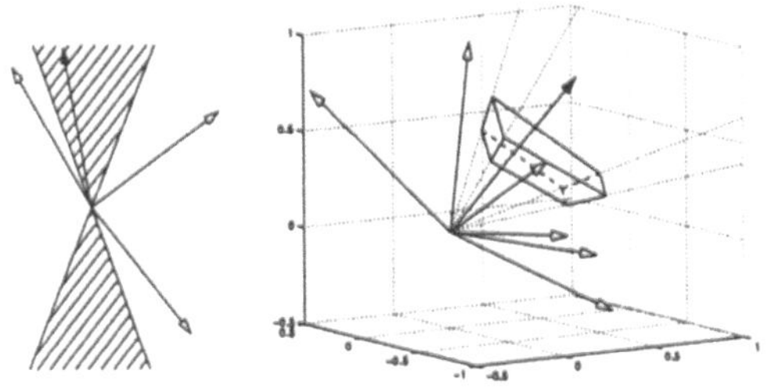

(a) (b)

Fig. 2.1. Illustrations of linear inequality constraints on the input that arise from a one iteration QMP representation. The vectors shown compose $\mathcal{D}$; φ_{k_0} has a solid arrowhead. (a) Constraint (2.3) in $\mathbb{R}^2$; (b) Constraints (2.3) & (2.4) in $\mathbb{R}^3$.

For each element of $\mathcal{D}\backslash\{\varphi_{k_0}\}$, (2.3) specifies a pair of linear half-space constraints on the original signal f with boundaries passing through $\sum_{\ell=0}^{i-1}\hat{\alpha}_\ell\varphi_{k_\ell}$. It can be shown [9] that these define two infinite convex polyhedral cones situated symmetrically with their apexes at $\sum_{\ell=0}^{i-1}\hat{\alpha}_\ell\varphi_{k_\ell}$. An example of such constraints in $\mathbb{R}^2$ for $i=0$ is shown in Fig. 2.1(a). If φ_{k_0} is the vector with the solid arrowhead (chosen from all of the marked vectors), the source vector must lie in the hatched area.

The value of $\hat{\alpha}_i$ gives the constraint

$$\left\langle\varphi_{k_i}, f-\sum_{\ell=0}^{i-1}\hat{\alpha}_\ell\varphi_{k_\ell}\right\rangle \in q_i^{-1}(\hat{\alpha}_i). \tag{2.4}$$

This specifies a pair of planes, perpendicular to φ_{k_i}, between which $f-\sum_{\ell=0}^{i-1}\hat{\alpha}_\ell\varphi_{k_\ell}$ must lie.[2] Constraints (2.3) and (2.4) are illustrated in Fig. 2.1(b) for $N=3$ and $i=0$. The vector with the solid arrowhead was chosen among all the marked dictionary vectors as φ_{k_0}. Then the quantization of α_0 implies that the source vector lies in the volume shown.

By being explicit about the constraints as above, we see that, except when zero is an interior point of $q_i^{-1}(\hat{\alpha}_i)$ for some i, the partition cells defined by QMP are convex.[3] (The partition cells are intersections of cells of the form shown in Fig. 2.1(b).) Therefore, consistent reconstructions can be computed through the method of alternating projections [10]. (Reconstruction using linear or quadratic programming is also possible [9].) One would normally start with an initial estimate given by (2.2). Given an estimate $\hat{f}$, the algorithm given in Table 2.2 performs the one "most needed" projection; namely, the first projection needed in enforcing (2.3)–(2.4). Among the possible projections in enforcing (2.3), the one corresponding to the largest deviation from consistency is performed. For notational convenience we assume uniform quantization with $q_i(\alpha_i)=m\Delta \iff \alpha_i\in\left[(m-\frac{1}{2})\Delta,\,(m+\frac{1}{2})\Delta\right)$; steps 5 and 6 could easily be adjusted for general quantizers.

[2] Technically, q_i must be such that $q_i^{-1}(\hat{\alpha}_i)$ is always convex, i.e. an interval of $\mathbb{R}$.

[3] "Hourglass" cells resulting from $0\in q_i^{-1}(\hat{\alpha}_i)$ do not pose problems in reconstruction [9].

Table 2.2. Algorithm for "Most Needed" Projection

1. Initialize c (counter of number of QMP steps $\hat{f}$ is consistent with) to zero.
2. Let $\bar{f} = \hat{f} - \sum_{i=0}^{c-1} \hat{\alpha}_i \mathcal{D}_{k_i}$ (summation is empty for $c = 0$).
3. Find $\varphi = \text{argmax}_{\varphi \in \mathcal{D}} |\langle \varphi, \bar{f} \rangle|$. If $\varphi = \varphi_{k_c}$, go to step 5; else go to step 4.
4. ($\hat{f}$ is not consistent with k_c.) Let $\tilde{\varphi}_{k_c} = \text{sgn}(\langle \varphi_{k_c}, \bar{f} \rangle) \varphi_{k_c}$, $\tilde{\varphi} = \text{sgn}(\langle \varphi, \bar{f} \rangle) \varphi$, and then $\hat{f} = \hat{f} - \langle \tilde{\varphi}_{k_c} - \tilde{\varphi}, \bar{f} \rangle (\tilde{\varphi}_{k_c} - \tilde{\varphi})$, a projection prescribed by (2.3). Terminate.
5. ($\hat{f}$ is consistent with k_c.) If $\langle \varphi_{k_c}, \bar{f} \rangle \in \left[\hat{\alpha}_c - \frac{\Delta}{2}, \hat{\alpha}_c + \frac{\Delta}{2}\right)$, go to step 7; else go to step 6.
6. ($\hat{f}$ is not consistent with $\hat{\alpha}_c$.) Let

$$\beta = \text{sgn}(\langle \varphi_{k_c}, \bar{f} \rangle - \hat{\alpha}_c) \cdot \min \left\{ \left| \langle \varphi_{k_c}, \bar{f} \rangle - (\hat{\alpha}_c + \tfrac{\Delta}{2}) \right|, \left| \langle \varphi_{k_c}, \bar{f} \rangle - (\hat{\alpha}_c - \tfrac{\Delta}{2}) \right| \right\}.$$

   Then let $\hat{f} = \hat{f} - \beta \varphi_{k_c}$, a projection prescribed by (2.4). Terminate.
7. ($\hat{f}$ is consistent with $\hat{\alpha}_c$.) Increment c. If $c = p$, terminate ($\hat{f}$ is consistent); else go to step 2.

2.4 An Example in $\mathbb{R}^2$

Consider quantization of an $\mathbb{R}^2$-valued source. Assume that two iterations will be performed with the four element dictionary

$$\mathcal{D} = \left\{ \begin{bmatrix} \cos \frac{(2k-1)\pi}{8} & \sin \frac{(2k-1)\pi}{8} \end{bmatrix}^T \right\}_{k=1}^{4}.$$

Even if the distribution of the source is known, it is difficult to find analytical expressions for optimal quantizers.[4] Since we wish to use fixed, untrained quantizers, we will use uniform quantizers. It will generally be true that $\varphi_{k_0} \perp \varphi_{k_1}$, so it is sensible to use equal quantization step sizes for α_0 and α_1.

The partitions generated by matching pursuit are very intricate. In Fig. 2.2, the heavy lines show the partitioning of the first quadrant when the quantizer reconstruction points are $\{m\Delta\}_{m\in\mathbb{Z}}$ and decision points are $\{(m+\frac{1}{2})\Delta\}_{m\in\mathbb{Z}}$ for some quantization stepsize Δ.[5] In this partition, most of the cells are squares, but there are also some smaller cells. The fraction of cells that are not square goes to zero as $\Delta \to 0$.

This quantization of $\mathbb{R}^2$ gives concrete examples of the inconsistency resulting from using (2.2). The linear reconstruction points are indicated in Fig. 2.2 by o's. The light line segments connect these to the corresponding optimal reconstruction points. Such a line segment crossing a cell boundary indicates a case of (2.2) giving an inconsistent estimate.

---

[4] The issue of optimal quantizer design is considered for the case of a source with a uniform distribution on $[-1, 1]^2$ in [8, §3.3.2].

[5] The partition is somewhat different when the quantizer has different decision points, *e.g.* $\{(m+\frac{1}{2})\Delta\}_{m\in\mathbb{Z}}$ [8, §3.3.2]. The qualitative conclusions are unchanged.

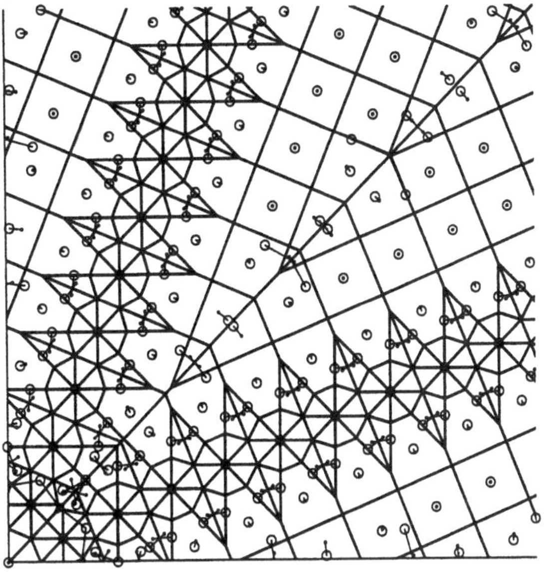

Fig. 2.2. Partitioning of the first quadrant of $\mathbb{R}^2$ by matching pursuit with a four element dictionary is given by the heavy lines. Linear reconstruction points are marked by o's and connected to Optimal reconstruction points by light line segments. (Optimality is with respect to a uniform source distribution.)

3. Source Coding using Quantized Matching Pursuit

This section explores the efficacy of using QMP for lossy compression of vectors in $\mathbb{R}^N$. Most lossy compression can be viewed as compressing vectors in $\mathbb{R}^N$, although the source distribution will depend on the application. The application may also give coding constraints (such as requiring a fixed bit rate), and may suggest a relevant distortion metric. Here we will measure rate by entropy, thus implicitly allowing variable bit rates, and measure distortion by MSE. Experimental results will be given for autoregressive sources, but distributional knowledge will not be used in the design.

3.1 Experimental Results

With no distributional assumptions, we expect the best performance with a dictionary that is "evenly spaced" on the unit sphere or a hemisphere. Thus dictionaries formed from sets of maximally spaced points on the unit sphere [11] were used. For simplicity, the inner product quantization is uniform. It is unlikely that any other fixed quantization would do better over a large class of source distributions. Furthermore, the quantization stepsize Δ is constant across iterations. This is consistent with equal weighting of error in each direction. Experimental results with other types of dictionaries were presented in [8, 12], but there consistent reconstructions were not computed.

The experiments involve quantization of a zero mean Gaussian AR source with correlation coefficient $\rho = 0$ or $\rho = 0.9$. Source vectors are generated by forming blocks of N samples. Rate is measured by summing the (scalar)

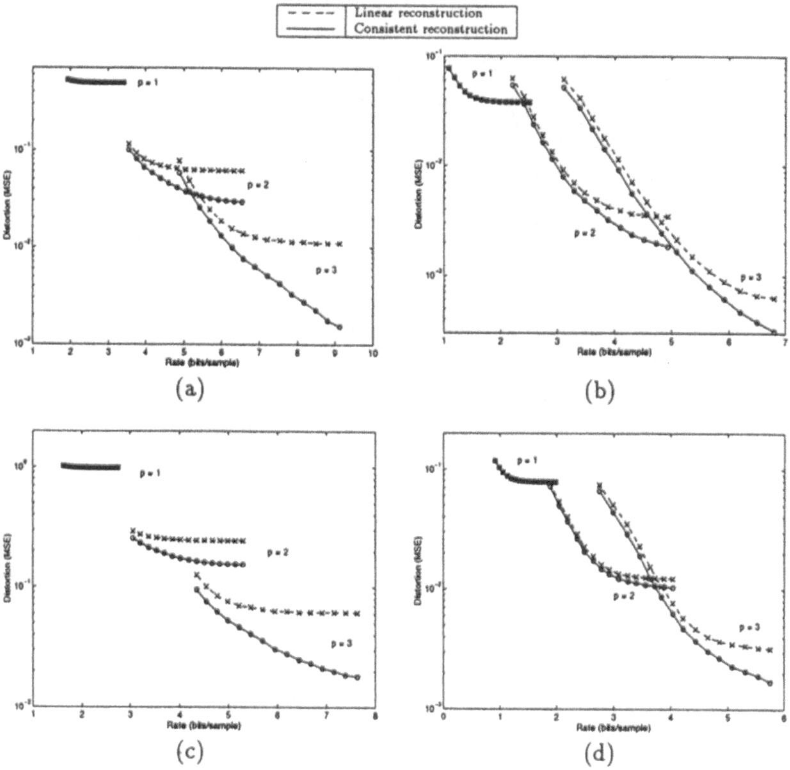

Fig. 3.1. Simulation results: (a) $N = 3$, $M = 7$, $\rho = 0$; (b) $N = 3$, $M = 7$, $\rho = 0.9$; (c) $N = 4$, $M = 11$, $\rho = 0$; (d) $N = 4$, $M = 11$, $\rho = 0.9$.

sample entropies of $k_0, k_1, \ldots, k_{p-1}$ and $\hat{\alpha}_0, \hat{\alpha}_1, \ldots, \hat{\alpha}_{p-1}$, where p is the number of iterations of the algorithm.

Fig. 3.1 gives simulation results obtained with $(N, M) = (3, 7)$ and $(4, 11)$. The $\times$'s, which are connected by dashed lines, are $D(R)$ points resulting from using linear reconstruction. The o's, connected by solid lines, are $D(R)$ points obtained with consistent reconstruction. Traversing each curve from left to right corresponds to varying Δ from $10^{-0.2}$ to $10^{-1.6}$. Since consistency is not an issue for a single-iteration expansion, the curves coincide for $p = 1$. For both $\rho = 0$ and $\rho = 0.9$, consistent reconstruction gives significantly lower distortion, although the difference is greater for $\rho = 0$.

3.2 A Few Possible Extensions

The experiments of the previous subsection are the tip of the iceberg in terms of the possible design choices. In this subsection, a few possible variations are presented along with plausibility arguments for their application.

In dictionary design, issues of interest include not only rate-distortion (R-D) performance, but also storage requirements, complexity of inner product computation, and complexity of largest inner product search. A distortion only (as opposed to R-D) approach was given in [6, Ch. 8]. There is no *a priori* reason to use the same dictionary at every iteration. Given a p iteration estimate, the entropy of k_p becomes a limiting factor in adding the results of an additional iteration. To reduce this entropy, it might be useful to use coarser dictionaries as the iterations proceed. Another possibility is to adapt the dictionary by augmenting it with samples from the source.

Instead of having a fixed number of iterations, it may be useful to use a stopping criterion based on the energy of the residue. This would create a guaranteed upper bound on the error and might have a favorable R-D impact.

The experimental results that have been presented are based on having separate entropy codes for each k_i and each $\hat{\alpha}_i$. Significant reduction in rate, at the cost of increased coding complexity, can be achieved by coding the indices as a vector [12]. Other coding possibilities exist, including discarding the indices that correspond to zero quantized coefficients. Using non-uniform quantizers, for example exploiting the fact that the probability density of α_0 will be small near zero, may also have a favorable impact on R-D performance.

References

1. S. G. Mallat and Z. Zhang, "Matching pursuits with time-frequency dictionaries", *IEEE Trans. Signal Proc.*, vol. 41, pp. 3397–3415, Dec. 1993.
2. R. Neff, A. Zakhor, and M. Vetterli, "Very low bit rate video coding using matching pursuits", *in Proc. SPIE VCIP*, vol. 2308, pp. 47–60, 1994.
3. M. Vetterli and T. Kalker, "Matching pursuit for compression and application to motion compensated video coding", *in Proc. IEEE ICIP*, 1994.
4. T. Kalker and M. Vetterli, "Projection methods in motion estimation and compensation", *in Proc. IS & T/SPIE*, San Jose, CA, 1995.
5. Z. Zhang, *Matching Pursuit*, PhD thesis, New York Univ. (NYU), 1993.
6. G. Davis, *Adaptive Nonlinear Approximations*, PhD thesis, NYU, 1994.
7. N. T. Thao and M. Vetterli, "Reduction of the MSE in R-times oversampled A/D conversion from $O(1/R)$ to $O(1/R^2)$", *IEEE Trans. Signal Proc.*, vol. 42, pp. 200–203, Jan. 1994.
8. V. K Goyal, "Quantized overcomplete expansions: Analysis, synthesis and algorithms", Technical Report M95/57, UC-Berkeley/ERL, July 1995, (URL: `http://www-wavelet.eecs.berkeley.edu/~vkgoyal/erl-95-57.html`).
9. V. K Goyal, M. Vetterli, and N. T. Thao, "Quantized overcomplete expansions in $\mathbb{R}^n$: Analysis, synthesis and algorithms", submitted to *IEEE Trans. Info. Theory*, Feb. 1996.
10. D. C. Youla, "Mathematical theory of image restoration by the method of convex projections", in H. Stark, editor, *Image Recovery: Theory and Application.* Academic Press, 1987.
11. R. H. Hardin, N. J. A. Sloane, and W. D. Smith, "Library of best ways known to us to pack n points on sphere so that minimum separation is maximized", URL: `ftp://netlib.att.com/netlib/att/math/sloane/packings/`.
12. V. K Goyal, M. Vetterli, and N. T. Thao, "Quantization of overcomplete expansions", *in Proc. IEEE Data Compression Conf.*, pp. 13–22, 1995.

Applications

Non Uniform Multiresolution Method for Optical Flow Computation

Isaac COHEN, Isabelle HERLIN

AIR Project,
INRIA, Rocquencourt
B.P. 105, 78153 Le Chesnay CEDEX, France.
Email Isaac.Cohen@inria.fr, Isabelle.Herlin@inria.fr

Summary. In this paper we propose a non uniform multiresolution method defining a new approach for coarse to fine grid generation. It allows to locally increase the resolution of the grid according to the studied problem. Each added node refines the grid in a region of interest and increases the numerical accuracy of the solution in this region. We make use of such a method for solving the optical flow equation with a non quadratic regularization scheme allowing the computation of optical flow field while preserving its discontinuities. This new scheme is used for processing oceanographic and atmospheric image sequences.

1. Introduction

The increasing number of satellites dedicated to environmental monitoring allows to characterize natural phenomena with different physical measures. Furthermore, the regular spatial and temporal sampling allows to characterize short range evolution of atmospheric and oceanographic phenomena by processing an image sequence. For example, Sea Surface Temperature (SST), altimetry and ocean color measurements can be used simultaneously or separately for studying sea surface streams.

In this paper we define a complete framework for processing large image sequences for a global monitoring of short range oceanographic and atmospheric processes. This framework is based on a non uniform multiresolution method defining a new approach for coarse to fine grid generation. It allows to locally increase the resolution of the grid according to the studied problem. Each added node refines the grid in a region of interest and increases the numerical accuracy of the solution in this region. We make use of such a method for solving the optical flow equation with a non quadratic regularization scheme. This scheme allows the computation of optical flow field while preserving its discontinuities.

2. Non Uniform Multigrids

Computing an optical flow field over an image sequence using a classical approach leads to the solution of a large set of linear equations in case of a quadratic regularizer or to an iterative solution for a non quadratic one. In both approaches space discretization (i.e. image tessellation) is an important issue since it defines the accuracy of the solution and the numerical complexity of the algorithm. In this section we propose a selective multi-resolution approach. This method defines a new approach for coarse to fine grid generation. It allows to increment locally the resolution of the grid according to the studied problem. The advantage of such a method is its lowest numerical complexity and its higher accuracy. Each added node refines the grid in a region of interest and increases the numerical accuracy of the solved problem in this region.

2.1 Grid subdivision scheme

In this section we describe the non uniform multiresolution scheme and the coarse to fine grid generation. The grids are generated by a recursive subdivision of the triangles. Each grid will be used as a tessellation of the domain (*i.e.* the image) for solving the minimization problem. This minimization is performed through the solution of the associated Euler-Lagrange equation. It is generally a Partial Differential Equation (PDE) solved through a Finite Element Method (FEM). This method is generic and can be used for most minimization problems but it constrains the type of admissible tessellation. The tessellation must fulfill the *conform triangulation* requirement of the FEM scheme [2], *i.e.: any face of any n-simplex* $\mathcal{T}_1$ *in the triangulation is either a subset of the boundary of the domain, or a face of another n-simplex* $\mathcal{T}_2$ *in the triangulation.*

This requirement limits the type of n-simplex and the subdivision scheme that can be used for an automatic non-uniform cell subdivision. Indeed, different mesh refinement techniques were proposed in computer vision [11, 12] but most of them do not fulfill this requirement.

Hierarchical triangular decomposition methods are differentiated on the basis of whether the decomposition is into three (*ternary*) or four (*quaternary*) parts. Ternary decompositions are formed by taking an internal point of one of the triangles $\mathcal{T}$ and joining it to the vertices of $\mathcal{T}$. Quaternary decompositions are formed by joining three points, each one on a different side of a given triangle. In case of a ternary decomposition, the described surface is usually continuous at every level. However, the triangles are often thin and elongated and furthermore the equiangularity is not satisfied. In the case of a quaternary decomposition, each triangle can be adjacent to a number of triangles on each of its sides and the resulted surface is not continuous unless all triangles are uniformly splitted [10].

In the following, we present our subdivision method (illustrated in fig. 2.1) based on triangular cells. These cells are well adapted for domain triangulation and allow to derive the simple recursive subdivision scheme: For each triangle $\mathcal{T}_1$ to be subdivided:

Step 1 Find the adjacent triangle $\mathcal{T}_2$ sharing the largest edge of $\mathcal{T}_1$ (if this edge belongs to the domain boundary then subdivide $\mathcal{T}_2$ into two triangles),

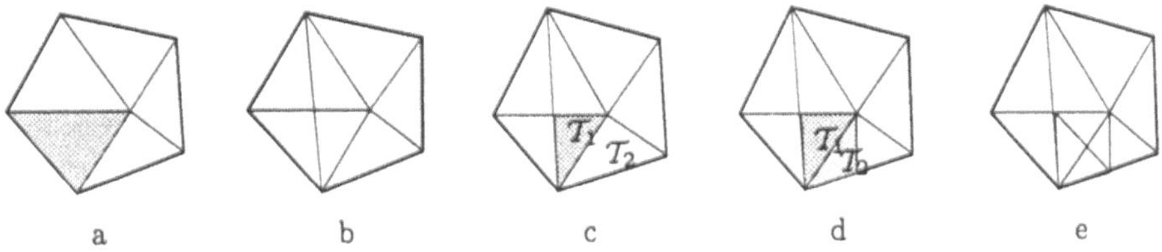

Fig. 2.1. Illustration of the recursive subdivision scheme used to construct the different grid levels of the non uniform multigrid method.

Step 2 If the shared edge is the largest edge of $\mathcal{T}_2$ then subdivide into four triangles the block $\mathcal{T}_1 \bigcup \mathcal{T}_2$. Otherwise process the triangle $\mathcal{T}_2$ (*i.e.* goto step 1).

2.2 A Multigrid Scheme Adapted to Motion Computation

Given a triangulation $\mathcal{T} = \bigcup_i \mathcal{T}_i$ of the image domain, we have to construct a multiresolution grid such that the grid resolution increases only near moving structures. With such a scheme, the coarse to fine pyramid is built optimally since we refine the grid resolution only in the vicinity of interesting regions. These regions are characterized as regions where a motion is detected. In the following we will prove that motion detection can be calculated with the norm of the estimated motion field along the direction of the image gradient.

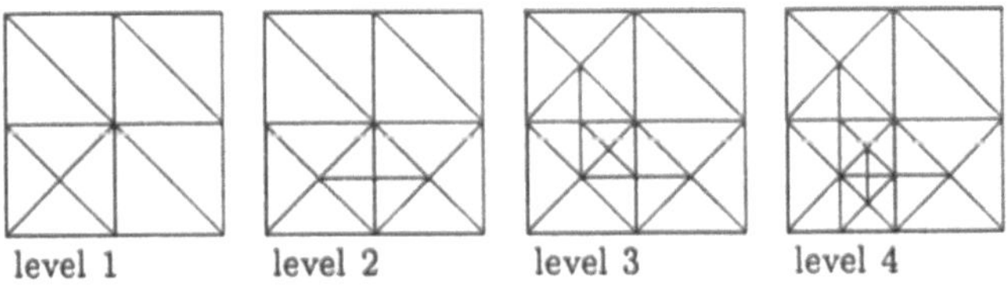

Fig. 2.2. An illustration of the non-uniform subdivision scheme at four different levels.

Let I represent the image brightness and $\overrightarrow{\nabla I}$ its gradient vector field, then the optical flow equation can be rewritten as:

$$\frac{dI}{dt} = \frac{\partial I}{\partial t} + \overrightarrow{\nabla I}.\overrightarrow{w} = I_x u + I_y v + I_t = 0 \tag{2.1}$$

where $\overrightarrow{w} = (u, v)$ is the optical flow. This equation yields [13]: $\frac{dI}{dt} = \frac{\partial I}{\partial t} + ||\overrightarrow{\nabla I}||w_\perp$, where $w_\perp$ is the norm of the component $\overrightarrow{w_\perp}$ of the motion field

$\overrightarrow{w}$ along the direction of $\overrightarrow{\nabla I}$. If the flow constraint equation is satisfied (*i.e.* $\frac{dI}{dt} = 0$) and $\|\overrightarrow{\nabla I}\| \neq 0$, we obtain:

$$\overrightarrow{w_\perp} = -\frac{\partial I/\partial t}{\|\overrightarrow{\nabla I}\|} \frac{\overrightarrow{\nabla I}}{\|\overrightarrow{\nabla I}\|} \tag{2.2}$$

Although $w_\perp$ does not always characterize image motion due to the aperture problem, it allows to locate moving points. Indeed, $w_\perp$ is high near moving points and becomes null near stationary points.

The definition of $w_\perp$ gives the theoretical proof of the motion measure D, defined by Irani *et al* [6] and used by several authors [7, 11]:

$$D(x, y, t) = \frac{\sum_{(x_i, y_i) \in W} |I(x_i, y_i, t+1) - I(x_i, y_i, t)| \, |\nabla I(x_i, y_i, t)|}{\sum_{(x_i, y_i) \in W} |\nabla I(x_i, y_i, t)|^2 + C} \tag{2.3}$$

where W is a small neighborhood and C a constant used to avoid numerical instability. This motion measure, defined as *residual motion*, is a particular form of $w_\perp$ where the numerator and the denominator are summed over a small neighborhood.

The subdivision scheme is based on a split strategy. We start with a coarse tessellation of the image and split each cell $\mathcal{T}$ of the triangulation according to the norm $w_\perp$ summed all over the cell: $w_\perp^{\mathcal{T}}$. A cell is subdivided while $w_\perp^{\mathcal{T}}$ is greater than a given threshold and while its area is greater than another one. Figure 2.2 illustrates a four levels subdivision scheme. This coarse to fine grid can also be used for other problems in computer vision: for example, one can use $|\nabla I|^2$ for a grid generation focusing on edge structures.

3. Non Quadratic Optical Flow Computation

The differential techniques used for computing the optical flow are based on the image flow constraint equation (Eq. 2.1). This equation, based on the assumption that the image irradiance remains constant, relates the temporal and spatial changes of the image irradiance $I(x, y, t)$ at a point (x, y) to the velocity (u, v) at that same point [5]. Equation (2.1) is not sufficient for computing the image velocity (u, v) at each point since the velocity components are constrained by only one equation; this is the aperture problem. Therefore, most of the techniques use a regularity constraint that restrains the space of admissible solutions of equation (2.1) ([1] and references therein). This regularity constraint is generally quadratic and therefore enforces the optical flow field to be continuous and smooth. But, true discontinuities can occur in the optical flow and they are generally located on the boundary between two surfaces representing two objects with different movements. This type of discontinuity occurs for example on temperature front in SST images, and cloud boundary in atmospheric images. Recovering this discontinuity is necessary for further analysis of oceanographic and atmospheric images.

In this section we present a method for computing optical flow based on a non-quadratic regularization technique. This method constrains the space of admissible solution of Eq. (2.1) to the space of functions with bounded variation, i.e.: $BV_1 = \left\{ f = (f_1, f_2) \text{ such that } \int_\Omega |\nabla f_1| + |\nabla f_2| \, dxdy < +\infty \right\}$. This allows to preserve sharp signals as well as discontinuities in the space of L^1 functions. The optical flow problem can be stated as the minimization of the functional:

$$\int_\Omega \sqrt{u_x^2 + u_y^2} + \sqrt{v_x^2 + v_y^2} + (I_x u + I_y v + I_t)^2 \, dxdy, \tag{3.1}$$

where u_x and u_y (resp. v_x and v_y) represent the partial derivatives of u (resp. v) with respect to x and y. The solution of the minimization problem (3.1) leads to a set of nonlinear PDE. These equations are solved through the associated evolution equations, or equivalently, the gradient descent method [4, 8]. This time-dependent approach means that we solve the evolution equation:

$$\begin{cases} \frac{\partial u}{\partial t} + \mathcal{D}u + (uI_x^2 + vI_xI_y + I_xI_t) = 0 \\ \frac{\partial v}{\partial t} + \mathcal{D}v + (uI_xI_y + vI_y^2 + I_yI_t) = 0 \\ +\text{Boundary conditions:} \nabla u_{|\Gamma} = \nabla v_{|\Gamma} = 0 \\ +\text{Initial estimation } u(0, x, y) = u_0(x, y). \end{cases} \tag{3.2}$$

where $\mathcal{D}$ is the nonlinear operator defined by:

$$\mathcal{D}f = -\frac{\partial}{\partial x}\left(\frac{f_x}{\sqrt{f_x^2 + f_y^2}}\right) - \frac{\partial}{\partial y}\left(\frac{f_y}{\sqrt{f_x^2 + f_y^2}}\right) \tag{3.3}$$

and a stationary solution of Eq. (3.2) characterizes also a minimum of Eq. (3.1). This evolution equation can be solved with a difference or a finite element method. Rudin *et al* [9] proposed a finite difference method for image deblurring, by using similar equations solved over a rectangular tessellation of the image. In this paper we make use of a finite element method allowing the use of arbitrary tessellations of the image domain by taking into account image motion in order to reduce the numerical complexity of the algorithm and to increase its accuracy near moving structures. This is performed with the non uniform multiresolution method described in section 2. The use of a FEM gives a natural way for sampling the solution over the different grid levels by making use of the analytical representation of the solution.

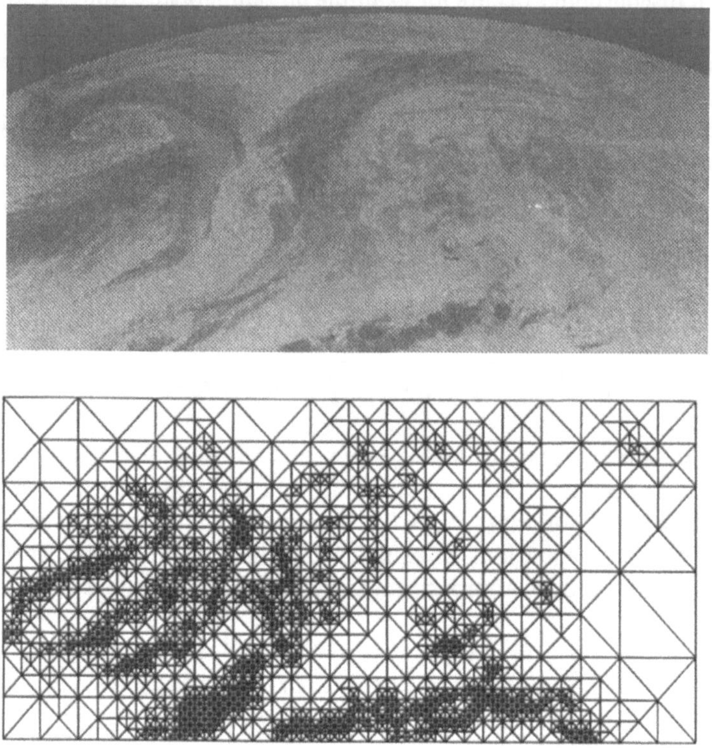

Fig. 4.1. A frame of the Meteosat infrared image sequence and the associated grid at the finest resolution.

4. Experimental Results

Our main objective is to derive atmospheric circulation from an image sequence of infrared measurements. In this section, we considere a Meteosat image sequence in order to compute clouds motion. For this purpose, we generate the non-uniform multiresolution grids according to the method described in section 2.2. Figure 4.1 displays a frame of the image sequence and the grid obtained at the finest resolution. This grid is used to solve Eq. (3.2) with a finite element method. The displacement field is displayed in Fig. 4.2.

5. Conclusion

In this paper we address some problems encountered in processing very large environmental image sequences representing the evolution of a physical phenomenon. The paper presents a non uniform multiresolution scheme allowing an efficient coarse to fine grid generation for optical flow computation. This method allowing an important reduction of the algorithmic complexity while having a higher accuracy near moving structures.

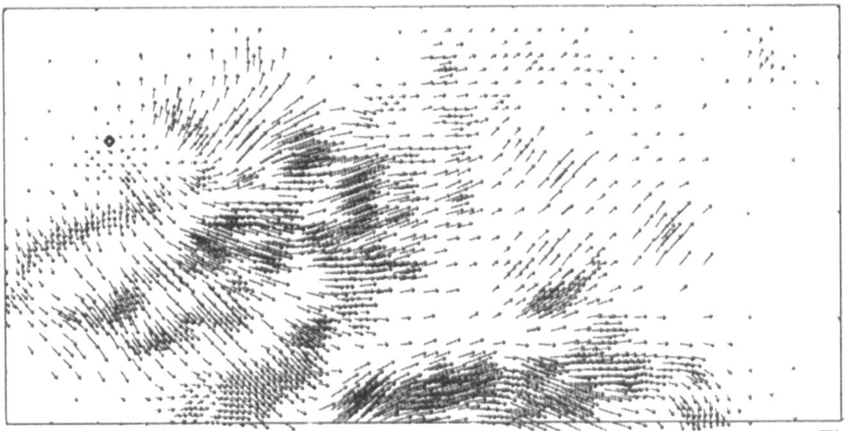

Fig. 4.2. A plot of the computed optical flow characterizing clouds motion. The black quadrangle represents the detected vortex [3].

We are currently studying the comparison between the computer vision approach and the classical method used by oceanographic and atmospheric researchers which deal with more elaborated physical models.

Acknowledgement. We thank M. Desbois and A. Szantai from the LMD for providing the atmospheric image sequence.

References

1. J.L. Barron, D.J. Fleet, and S.S. Beauchemin. Performance of optical flow techniques. *International Journal of Computer Vision*, 12(1):43–77, February 1994.
2. P. G. Ciarlet. *The finite element methods for elliptic problems.* NORTH-HOLLAND, Amsterdam, 1987.
3. I. Cohen and I. Herlin. A motion computation and interpretation framework for oceanographic satellite images. In *IEEE, Computer Vision Symposium*, pages 13–18, Florida, November 1995.
4. R. Glowinski. *Numerical Methods for Nonlinear Variational Problems.* Springer-Verlag, New-York, 1984. Springer Series in Computational Physics.
5. B.K.P. Horn and G. Schunck. Determining optical flow. *Artificial Intelligence*, 17:185–203, 1981.
6. M. Irani, B. Rousso, and S. Peleg. Detecting and tracking multiple moving objects using temporal integration. In *Proceedings of the Second European Conference on Computer Vision 1992*, pages 282–287, May 1992.
7. J.R. Muller, P. Anandan, and J.R. Bergen. Adaptive-complexity registration of images. In *IEEE Proceedings of Computer Vision and Pattern Recognition*, pages 953–957, 1994.

9. L.I. Rudin, S. Osher, and E. Fatemi. Nonlinear total variation based noise removal algorithms. In *Ecoles CEA - EDF - INRIA; Problèmes Non Linéaires Appliqués: Modélisation Mathématique pour le traitement d'images*, pages 149–179, March 1992.
10. H. Samet. *The Design and Analysis of Spatial Data Structures.* Addison-Wesley, 1989.
11. R. Szeliski and H.Y. Shum. Motion estimation with quadtree splines. Technical report, DEC Cambridge Research Lab, March 1995.
12. M. Vasilescu and D. Terzopoulos. Adaptive meshes and shells: Irregular triangulation, discontinuities, and hierarchical subdivision. In *IEEE Proceedings of Computer Vision and Pattern Recognition*, pages 829–832, June 1992.
13. A. Verri and T. Poggio. Motion field and optical flow: Qualitative properties. *IEEE Transactions on Pattern Analysis and Machine Intelligence*, 11(5):490–498, May 1989.

Appariement de Trajectoires Issues d'un Système de Vision Stéréoscopique en Mouvement de Convergence

Samira AIT KACI AZZOU[1] Slimane LARAB[1] Juan Lopez CORONADO[2]

1 Laboratoire Images
Institut d'informatique USTHB
BP 32, EL Alia Alger, Algérie
Tel : 51-55-75 poste(645)
Fax : (01)-50-64-19

2 Department of systems and automatic engineering
Universidad de Valladolid, Paseo del cauce S/N
47011 Valladolid, Espagne
Tél : (983) 42 33 58

Résumé. Cet article décrit une méthode de résolution du problème d'appariement dans un système de vision stéréoscopique non anthromorphe muni de dex caméras en mouvement de convergence. Généralement, on effectue, à chaque prise de vue, l'appariement des images stéréoscopiques et le calcul de profondeurs. Une ;ise en correspondance entre les cartes de profondeurs successives est donc nécessaire. Pour éviter cette dernière étape, nous proposons une approche qui consiste à apparier les trajectoires de primitives images générées par des caméras en rotation. Nous montrons que l'appariement des trajectoires peut être réalisé sans connaissance préalable du modèle géométrique du système de vision.

Mots-clés : Mouvement de convergence - Vision stéréoscopique - Appariement

Summary. This paper describes the matching problem solving method in "not-anthropomorphic" stereoscopic vision system with two cameras in convergence movement. in general, ateach picture raking, the stereoscopic image correspondence and the depths calculus are done. Matching the successive depth maps is then necessary. To avoid this last step, we suggest an approach that consists on matching features trajectories generated by cameras in rotation. Is shown that, matching trajectories can be done without preliminary knowledge of the system vision geometrical model.

Key Words : Convergence Movement - Stereoscopic Vision - Matching

INTRODUCTION

La principale différence entre les algorithmes de vision stéréo et la perception visuelle humaine vient du fait que la vision humaine est essentiellement dynamique; nos globes oculaires sont perpétuellement en mouvement ainsi que notre tête[FAU88].

De même, pour un système informatique de vision, le mouvement de convergence des capteurs peut améliorer la perception de l'environnement. Ainsi, utiliser des caméras mobiles en vision stéréoscopique permettra de produire un ensemble d'images prises en des endroits différents, ce qui fournit une carte de profondeurs plus complète.

C'est pourquoi, de nombreux chercheurs se sont intéressés ces dernières années à la notion du mouvement. Mais, la majorité d'entre eux se sont penché sur le problème d'estimation du mouvement et de la structure de l'environnement en utilisant soit l'approche du flot optique[BOU88], [AN88], [LJ89], [SUN92], ou bien l'approche basée sur la mise en correspondance des primitives [TH84], [FLT87], [AN88],[FM93].

La notion du mouvement n'a pas été utilisée intensivement pour résoudre le problème de l'appariement stéréoscopique de primitives images dans le cas où les positions des caméras sont inconnues. Il a par contre été abordé dans le domaine de la stéréo où ces positions sont supposées connues[GRI85], [OK85].

Un système de vision stéréoscopique dont les caméras sont en mouvement fournit un couple d'images à chaque prise de vue, et donc une séquence d'images est obtenue par chaque caméra.

Le traitement de ces couples d'images pour la reconstruction 3D de la scène, en exploitant l'approche de Marr[MAR82] nécessite les étapes suivantes :

- mise en correspondance de chaque couple d'images,

- calcul de la carte de profondeurs pour chaque prise de vue,

- mise en correspondance des primitives 3D dans les cartes de profondeurs successives[AN88].

Les difficultés de cette approche réside dans le traitement des ambiguïtés pour chaque couple et dans l'appariement des cartes de profondeurs.

Pour pallier à ces inconvénients, nous proposons une démarche qui consiste à :

- élaborer les trajectoires des primitives images,

- réaliser la mise en correspondance de ces trajectoires,

- déduire la carte de profondeurs.

L'élaboration des trajectoires de primitives est un thème déjà abordé : mise en correspondance des primitives dans une séquence d'images [DF90], [XR88]; [MEY93], c'est pourquoi, notre intérêt c'est porté sur la résolution du problème de mise en correspondance des trajectoires images.

Dans cet article, nous proposons une méthode de résolution de ce problème pour un système de vision stéréoscopique non anthropomorphe dynamique. Trois algorithmes sont proposés, dont le dernier "Apparie_Mouv0" permet de réaliser l'appariement des trajectoires sans connaissance préalable du modèle géométrique des caméras et du mouvement.

II. LES ASPECTS THEORIQUE

II.1. DESCRIPTION DU SYSTEME

La modelisation géométrique d'un système de vision stéréoscopique non anthropomorphe, équipé de deux cameras en mouvement de rotation, se présente comme indiqué par la figure Fig.1

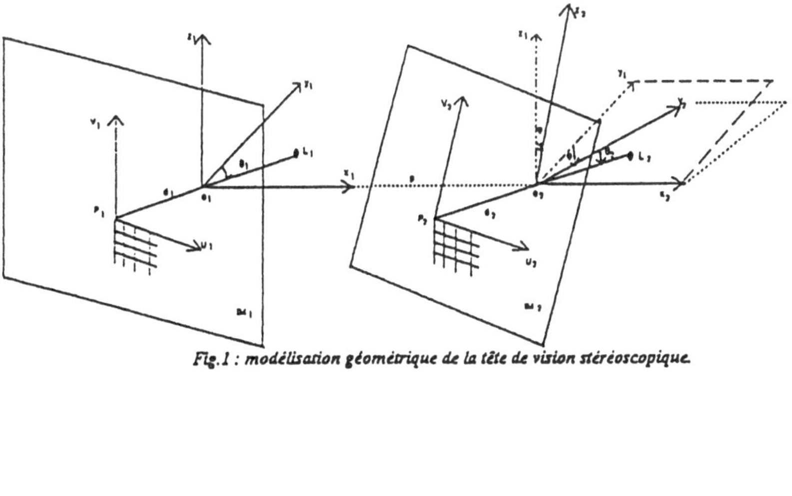

Fig.1 : modélisation géométrique de la tête de vision stéréoscopique.

Nous supposons que le centre de rotation o_1 (resp. o_2) de la caméra1 (resp. caméra2) est situé sur l'axe optique P_1L_1(resp. P_2L_2). Cette hypothèse n'est pas une contrainte; car dans ce qui suit, nous pourrons constater que les algorithmes développés sont indépendants des positions des centres de rotation o_1 et o_2.

Notons : d_1(resp. d_2) la distance o_1P_1 (resp. o_2P_2).

θ_1 (resp. θ'_2) est l'angle de rotation de la caméra1 (resp. caméra2) autour de l'axe o_1z_1 (resp. o_2z_2).

φ : angle d'inclinaison de la caméra2 par rapport à la caméra1.

p : est la distance entre les points o_1 et o_2

II.2. LES PRIMITIVES A APPARIER

Tout point M(x,y,z) de l'espace tridimensionnel se projette en un point $M^1{}_k(u_k{}^1,v_k{}^1)$ (resp. $M^2{}_k(u_k{}^2, v_k{}^2)$)sur IM1 (resp. IM2) à la k^{eme} rotation de la caméra1 (resp. caméra2).

La séquence de points $M^1{}_k$ (resp. $M^2{}_k$) obtenue après k rotations de la caméra1 (resp. caméra2) constitue la trajectoire $t^1{}_k$ (resp. $t^2{}_k$) du point M^1 (resp. M^2).

Les coordonnées des points images $u_k{}^i$ et $v_k{}^i$ (i=1,2) dépendent des paramètres intrinsèques de la caméra telles que la distance focale et la taille du pixel. Considérer la trajectoire des points images $M_k{}^i$ $(u_k{}^i,v_k{}^i)$ (i=1..2) comme primitive à apparier, nous impose le calibrage des caméras. C'est pourquoi, il serait plus intéressant de travailler avec la trajectoire du rapport $R_k{}^i=u_k{}^i/v_k{}^i$ (i=1..2) qui nous permet de s'affranchir de certains de ces paramètres (voir Fig.2).

Les R^k (i:1,2) s'écrivent pour un point objets M de la manière suivante :

$$R_1^k(M) = a_1.\cos(\theta_1 + k.dt) + b_1.\sin(\theta_1 + k.dt) \quad \text{avec} \quad \begin{cases} a_1 = x/z \\ b_1 = -y/z \end{cases}$$

$$R_2^k(M) = a_2.\cos(\theta_2 + k.dt) + b_2.\sin(\theta_2 + k.dt) \quad \text{avec} \quad \begin{cases} a_2 = \dfrac{a_1 - p/z}{b_1.\sin(\varphi) + .\cos(\varphi)} \\ b_2 = \dfrac{b_1.\cos(\varphi) - \sin(\varphi)}{b_1.\sin(\varphi) + .\cos(\varphi)} \end{cases}$$

et dt est le pas de rotation des caméras.

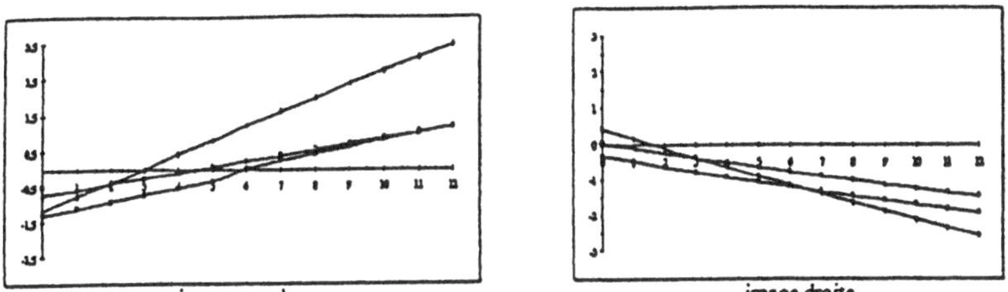

image gauche   image droite

fig.2- Trajectoir R=f(dt)

II.3. STRATEGIE D'APPARIEMENT

II.3.1. HYPOTHESES

Nous supposons que :

- Les caméras sont identiques.
- L'appariement de primitives dans une séquence d'images a été réalisé

Notons :

$T^1=\{t_1(M_i)/i<=n\}$ où $t_1(M_i)$ est la trajectoire du point M_i obtenue après rotation de la caméra1 et n le nombre de points dans IM_1

$T^2=\{t_2(M_j)/j<=m\}$ où $t_2(M_j)$ est la trajectoire du point M_j obtenue après rotation de la caméra2 et m le nombre de points dans IM_2

- Les points d'impact des axes optiques sont déduits à l'issue de l'étape d'élaboration des trajectoires

II.3.2. PRINCIPE DE BASE

Cette méthode suppose que θ_1 et θ_2 sont connus ainsi que le pas de rotation dt.

Pour cela, considérons les fonctions $R_1^k(M)$ et $R_2^k(M)$ pour un point objet M(x,y,z).

Connaissant les trajectoires sur IM1 et IM2, les paramètres a_1,b_1, a_2 et b_2 se calculent de la façon suivante :

$$a_1 = \frac{R_1^i(M).\sin(\theta_1+(i+1).dt) - R_1^{i+1}(M).\sin(\theta_1+i.dt)}{\sin(dt)}$$

$$b_1 = \frac{R_1^{i+1}(M).\cos(\theta_1+i.dt) - R_1^i(M).\cos(\theta_1+(i+1).dt)}{\sin(dt)}$$

$$a_2 = \frac{R_2^j(M).\sin(\theta_2+(j+1).dt) - R_2^{j+1}(M).\sin(\theta_2+j.dt)}{\sin(dt)}$$

$$b_2 = \frac{R_2^{j+1}(M).\cos(\theta_2+j.dt) - R_2^j(M).\cos(\theta_2+(j+1).dt)}{\sin(dt)}$$

L'expression de tg(φ) peut être déduite comme suit

$$tg(\varphi) = \frac{b_1 - b_2}{b_1 . b_2 + 1}$$

Nous remarquons que l'exploitation de tout couple de trajectoires en correspondance fournit la valeur

$t\varphi = \dfrac{b_1 - b_2}{b_1 \cdot b_2 + 1}$ égale à $tg(\varphi)$. En conséquence, nous pouvons énoncer la proposition suivante :

PROPOSITION 2

Une condition nécessaire pour qu'une trajectoire $t_1(M_i)$ générée sur IM_1 soit homologue à $t_2(Mj)$ générée sur IM_2 est que tφ calculée à partir de ces trajectoires doit être égale à $tg(\varphi)$.

Quelle est la valeur de tφ engendrée à partir d'un couple ($t_1(M_i)$,$t_2(M_j)$) non en correspondance?

Soit tφ la valeur engendrée par ($t_1(M_i)$,$t_2(M_j)$) et à pour expression :

$$t\varphi = \frac{b_{1i} - \dfrac{b_{1j} - tg(\varphi)}{b_{1j} \cdot tg(\varphi) + 1}}{b_{1i} \cdot \dfrac{b_{1j} - tg(\varphi)}{b_{1j} \cdot tg(\varphi) + 1} + 1} = \frac{tg(\varphi) \cdot (b_{1i} \cdot b_{1j} + 1) + (b_{1i} - b_{1j})}{-tg(\varphi) \cdot (b_{1i} - b_{1j}) + (b_{1i} \cdot b_{1j} + 1)} \neq tg(\varphi) \text{ sauf si } \quad b_{1i} = b_{1j}$$

Par conséquent, cette proposition ne peut etre considérée comme une condition suffisante d'appariement correct. Elle peut être, néanmoins, exploitée pour aboutir à un algorithme de mise en correspondance. La solution que nous proposons possède une démarche analogue à celle en vigueur dans la transformée de Hough[HOU62] : il s'agit d'engendrer toutes les possibilités pour $tg(\varphi)$ - notée - tφ et de retenir celle qui recueille le maximum de suffrages. Cette proposition constitue le principe de base de l'algorithme d'appariement nommé "Apparie_Mouv2"; le "2" fait référence aux deux paramètres supposés connus : le pas de rotation dt et l'angle initial θ_1(resp. θ_2) pour la caméra1 (resp. caméra2).

II.3.3. ALGORITHME : APPARIE_MOUV2

DEBUT
- Pour toute trajectoire $t_1(M_i)$ de T^1

Faire

    - Calculer les paramètres a_1 et b_1
    - Pour toute trajectoire $t_2(M_i)$de T^2

    Faire

        - Calculer les paramètres a_2 et b_2
        - Evaluer tφ pour le couple ($t_1(M_i)$,$t_2(M_j)$) : $t\varphi = \dfrac{b_1 - b_2}{b_1 \cdot b_2 + 1}$
        - $H_{t\varphi}(t_1(M_i),t_2(M_i))=1$

    Fait

Fait
- Rechercher la valeur de tφ ayant engendré le plus grand score;
- $H_{t\varphi} \leftarrow H_{\overline{t\varphi}}$ où $\overline{t\varphi}$ vérifie Card $\left[H_{\overline{t\varphi}}\right] = \text{Max Card}_{t\varphi} \left[H_{t\varphi}\right]$

FIN

II.3.3.1. CONSISTANCE DE L'ALGORITHME

"Apparie_Mouv2" peut-il produire une fonction $H_{t\varphi}$ où $t\varphi \neq tg(\varphi)$ et $Card(H_{t\varphi})$ atteint le nombre acceptable de trajectoires en correspondance?

Sachant que $tg(\varphi)$ est une valeur particulière parmi toutes les valeurs possibles de $t\varphi$, la probabilité pour qu'une valeur de $t\varphi \neq tg(\varphi)$ recueille un nombre d'appariement >1 ne relève que du hasard des nombres. Cette probabilité diminue et tend vers zéro pour un nombre d'appariement proche du score réalisé par $tg(\varphi)$.

ETUDE DES DIFFERENTS CAS

Soient (t_i^1, t_i^2) et (t_j^1, t_j^2) les trajectoires des points images Mi(xi,yi,zi) et Mj(xj,yj,zj) engendrées par la caméra1(resp. caméra2).

a) : Erreur de correspondance croisée

Supposons que l'algorithme réalise les correspondance suivantes (t_i^1, t_j^2) et (t_j^1, t_i^2) en générant deux valeurs identiques de $t\varphi$. L'expression suivante est alors déduite.

$$2.c1.d1.\left(tg(\varphi)^2+1\right)=0 \Rightarrow c1.d1=0 \text{ avec } c1 = y_i.y_j + z_i.z_j \text{ et } d1 = y_i.z_j - z_i.y_j$$

Ainsi, (t_i^1, t_j^2) et (t_j^1, t_i^2) est possible si et seulement si : c1=0 ou d1=0.

La disposition de Mi, Mj qui permet de produire ce cas d'ambiguïté est illustré par la figure Fig. 3.

La valeur de $t\varphi$ calculée à partir de cette correspondance est égale à $tg(\varphi)$ si d1=0 et à $(-cotg(\varphi))$ si c1=0.

b) cas général d'appariement erroné

Soient t_k^2, t_l^2 les trajectoires des points Mk(xk,yk,zk) et Ml(xl,yl,zl) générées par la caméra2. Peut-on aboutir aux couples de correspondance suivants : (t_i^1, t_k^2) et (t_j^1, t_l^2)?

En procédant comme dans le premier cas, nous obtenons :

$$c2.d3 - c3.d2 = 0 \text{ avec } \begin{cases} c2 = y_i.y_k + z_i.z_k \\ d2 = y_i.z_k - z_i.y_k \end{cases} \text{ et } \begin{cases} c3 = y_l.y_j + z_l.z_j \\ d3 = y_l.z_j - z_l.y_j \end{cases}$$

Ce cas de correspondance ne peut être réalisé que si Ml vérifie $zl=\gamma.yl$ où $\gamma=(yj+\alpha.zj)/(\alpha.yj-zj)$ avec$\alpha=c2/d2$.

c) cas d'ambiguïté

Peut-on aboutir aux couples de correspondance suivants : (t_i^1,t_i^2) et (t_i^1,t_j^2)?

Une telle hypothèse conduit à : $b_{1i}^2.[b_{2j}-b_{2i}]=b_{2i}-b_{2j} \Rightarrow b_{1i}^2=-1$. Ceci indique que ce cas d'ambiguïté est impossible sauf si $b_{2i}=b_{2j}$, (voir cas a)

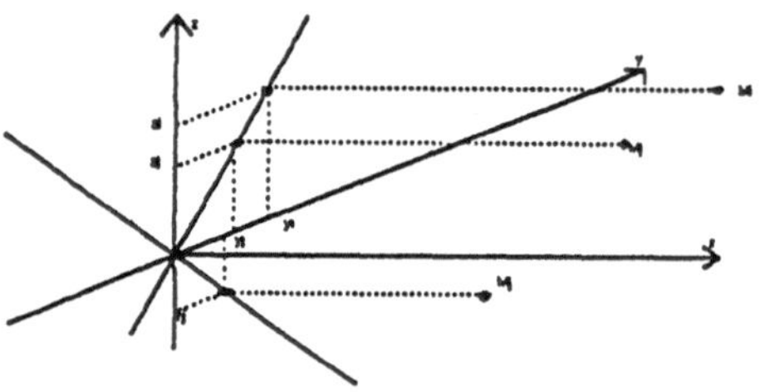

Fig. 3 Cas d'ambiguïtés

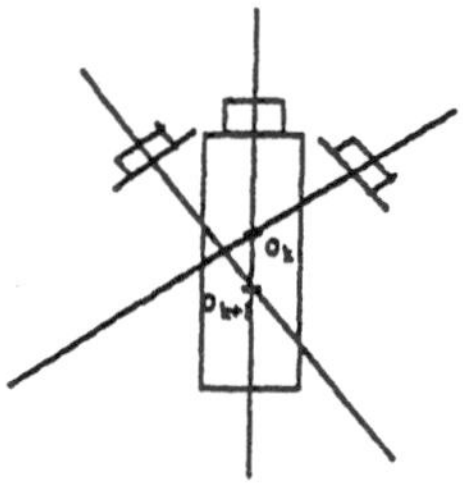

Fig. 4 Rotation de la caméras

Ces ambiguïtés peuvent être levées en introduisant des contraintes supplémentaires d'unicité, d'ordre et de voisinage.

REMARQUE

Dans le cas pratique, le centre de rotation n'est pas situé sur l'axe optique (voir Fig. 4) où O_k (resp. O_{k+1}) est le centre de rotation après k (resp. k+1) rotations. Les calcules effectués dans l'algorithme restent valables; car $tg(\varphi)$ est en fonction de b_1 et b_2 qui sont indépendants de la distance inter-centres de rotation, et se calcul à partir de deux positions d'un point image sur sa trajectoire.

II.3.4.UN AUTRE ALGORITHME D'APPARIEMENT DE TRAJECTOIRES "APPARIE_MOUV1"

II.3.4.1. PRINCIPE DE BASE

L'algorithme précédent Apparie_Mouv2 suppose que la position initiale des caméras est connue. Ceci est donc un inconvénient du moment que notre but est d'éviter la calibration de ces dernières. L'algorithme Apparie_Mouv1 nous permet d'omettre cette contrainte.

Les paramètres b_1, b_2 peuvent s'écrire de la façon suivante :

$$b_1 = C_1.\sin(\theta_1)+ C_2.\cos(\theta_1) \text{ où : } C_1 = R_1^0 \text{ et } C_2 = \frac{R_1^i - C_1.\cos(i.dt)}{\sin(dt)}$$

$$b_2=K_1.\sin(\theta_2)+K_2.\cos(\theta_2) \text{ où : } K_1 = R_2^0 \text{ et } K_2 = \frac{R_2^j - K_1.\cos(j.dt)}{\sin(dt)}$$

L'injection des nouvelles expressions de b_1 et b_2 dans la formulation suivante : $b_2 = \frac{b_1.-tg(\varphi)}{b_1.tg(\varphi)+1}$

nous conduit à l'équation ci-dessous :

$$r_1.X_1+r_2.X_2+r_3.X_3+r_4.X_4+r_5.X_5+r_6.X_6+r_7.X_7+r_8.X_8=1$$

Les σi (i=1..S) sont évalués et ils sont en fonction de $R_0{}^1$, $R_1{}^1$, $R_0{}^2$, $R_1{}^2$ et du pas de rotation dt par contre les paramètres Xi (i=1..S) sont en fonction de θ_1, θ_2 et φ.

Le calcul des paramètres Xi (i : 1..S), nécessite huit trajectoires de points, obtenues par chacune des cameras, ou quatre trajectoires de segments de droites.

tg(φ) s'écrit : $tg(\varphi)^2 = \frac{X_5}{X_1.X_7}$

Etant donnee que ce parametre est constants, nous pouvons énoncer la proposition suivante:

Proposition 3

Une condition nécessaire pour qu'un ensemble de trajectoires $t_1(M_i)$, (i : 1..S) genere par la caméra 1 soit homologue à l'ensemble $t_2(M_j)$ (j : 1..8) généré par la caméra 2 est que, **la valeur de tφ calculée en utilisant ces trajectoires soit égale à tg(φ).**

Quelle est la valeur de tφ engendrée à partir d'un couple de huit trajectoires ($t_1(M_i)$,$t_2(M_j)$) (i,j=1..8) non en correspondance?

La valeur de tφ calculée en utilisant des trajectoires non en correspondance est une valeur quelconque qui ne dépend que des coordonnées des points considérés. Cette proposition peut être alors exploitée pour aboutir à un algorithme d'appariement. La solution que nous proposons possède une démarche analogue à celle en vigueur dans la transformée de Hough[HOU62] : il s'agit d'engendrer toutes les possibilités pour tg(φ) - notée - tφ et de retenir celle qui recueille le maximum de suffrages.

II.3.4.2. DESCRIPTION DE L'ALGORITHME APPARIE_MOUV1

```
DEBUT
-Pour tout t1(Mi) de T1
faire
        - Pour tout G1={t1(Mk), k:1,8}
        faire
                -Pour tout t2(Mj) de T2
                faire
                        - Pour tout G2={t2(ML)) L:1,8}
                        faire
                                - Calculer Xn (n : 1,8)
                                - tφ² = X5 / (X1.X7)
                                - Htφ(G1, G2)=1
                        Fait
                Fait
        Fait
Fait
- Rechercher la valeur de tφ ayant engendré le plus grand score;
- Htφ ← Htφ̄ où tφ̄ vérifie Card [Htφ̄] = Max_tφ Card [Htφ]
FIN.
```

Remarque :

G^1(resp. G^2) est l'ensemble des trajectoires obtenues à partir de 4 segments voisins sur IM_1(resp. IM_2).

II.3.4.3. CONSISTANCE DE L'ALGORITHME

L'algorithme établit la correspondance entre les trajectoires de segments, tel que chaque groupe de 4 segments G^1 et G^2 effectivement en correspondance engendre une même valeur de $t\varphi = tg(\varphi)$. Est-il possible qu'une valeur de $t\varphi$ différente de $tg(\varphi)$ soit engendrée par un nombre important de groupe de segment G^1 et G^2 sachant que les segments dans G^1 ne sont pas les homologues de ceux dans G^2. Partant de cette hypothèse, la valeur de $t\varphi$ calculée à partir du système formé par G^1 et G^2 est une valeur quelconque qui dépend des coordonnées des extrémités des segments considérées. Ainsi, si un autre couple de quadruplé (G^1,G^2) en fausse correspondance génère la même solution $t\varphi$ différente de $tg(\varphi)$, ceci ne relève que du "hasard des nombres".

II.3.5. ALGORITHME D'APPARIEMENT DE TRAJECTOIRE "APPARIE_MOUV0"

II.3.5.1 PRINCIPE DE BASE

L'algorithme précédent Apparie_Mouv1 suppose que le pas de rotation des caméras est connu. L'algorithme Apparie_Mouv0 nous permet d'omettre cette contrainte.

Sachant que

$$C1 = R_0^1,\ C2 = \frac{R_1^1 - C1.\cos(dt1)}{\sin(dt1)},\ K1 = R_0^2 \text{ et } K2 = \frac{R_1^2 - K1.\cos(dt2)}{\sin(dt2)}$$

où dt1 (resp. dt2) constitue le pas de rotation de la caméra1 (resp. caméra2).

En procèdent de la même façon que pour Apparie_Mouv1, nous obtenons l'expression suivante : .

$$\boxed{\sigma 1.Z1 + \sigma 2.Z2 + \sigma 3.Z3 + \sigma 4.Z4 + \sigma 5.Z5 + \sigma 6.Z6 + \sigma 7.Z7 + \sigma 8.Z8 = 1}$$

Les σi (i=1..8) sont évalués et ils sont en fonction de $R_0{}^1$, $R_1{}^1$, $R_0{}^2$ et $R_1{}^2$ par contre les paramètres Zi (i=1..8) sont en fonction de θ_1, θ_2, φ., dt1 et dt2.

L'expression de tg(φ) est : $\boxed{tg(\varphi)^2 = \frac{Z5}{Z1.Z7}}$

Comme pour Apparie_Mouv1, la ***proposition*** 2 constitue le principe de base de l'algorithme "Apparie_Mouv0".

II.3.5.2. DESCRIPTION DE L'ALGORITHME APPARIE_MOUV0

Début

-Pour tout $t_1(M_i)$ de T^1
faire
- Pour tout $G^1=\{t_1(M_k), k=1,8\}$
Faire
-Pour tout $t_2(M_j)$ de T^2
Faire
- Pour tout $G^2=\{t_2(M_L)\ L=1,8\}$
Faire
- Calculer Zn (n = 1..8)
- $t\varphi^2 = \frac{Z_6}{Z_1.Z_8}$
- $H_{t\varphi}(G^1, G^2)=1$
Fait
Fait
Fait
Fait
- Rechercher la valeur de tφ ayant engendré le plus grand score;

- $H_{t\varphi} \leftarrow H_{\overline{t\varphi}}$ où $\overline{t\varphi}$ vérifie $\text{Card}\left[H_{\overline{t\varphi}}\right] = \text{Max}_{t\varphi}\ \text{Card}\left[H_{t\varphi}\right]$

FIN.

<u>Remarque :</u>

G^1(resp. G^2) est l'ensemble des trajectoires obtenues à partir de 4 segments voisins sur IM_1(resp. IM_2).

II.4. COMPLEXITE TEMPORELLE DES ALGORITHMES

a) Algorithme Apparie_Mouv2

Nous mesurons cette complexité par le nombre de trajectoires traitées.

Soit n le nombre de points dans IM_1 et IM_2, la complexité de "Apparie_Mouv2" = $Card(IM_1).Card(IM_2)=n^2 =O(n^2)$

b) Algorithmes Apparie_Mouv1 et Apparie_Mouv0

Huit trajectoires de points ou 4 trajectoires de segments ou encore 2 trajectoires de quadrilatères sont sélectionnées dans IM_1.

Soit n le nombre de points dans IM_1 et IM_2, la complexité de "Apparie_Mouv1" = $Card(IM_1).Card(IM_2)=n^2.(n-1)^2 =O(n^4)$

Cette complexité est réduite en introduisant la contrainte de voisinage, et devient :

Complexité = $Card(IM_1).Card(IM_2)= n.(Nbvoisins).n.(Nbvoisins)=n^2.(Nbvoisins)^2 = O(n^2)$

où Nbvoisins est le nombre de voisins 2D possibles pour un quadrilatère.

III. RESULTATS D'EXPERIMENTATION

Les trois algorithmes développés ont été expérimenté sur des données générées aléatoirement par ordinateur.

Pour des données non bruitées, chacun des trois algorithmes, a donné les résultats escomptés.

III.1. APPLICATION DES ALGORITHMES AUX DONNEES BRUITEES

Les données sont générées aléatoirement par ordinateur et ensuite entachées d'un bruit δ comme déjà expliqué pour l'algorithme de suivi.

III.1.1. ALGORITHME APPARIE_MOUV2

Les resultats de l'expérimentation montre à travers le tableau 1, que pour un seuil de tolérance$|\varphi max - \varphi min|$ de l'ordre de 1°, l'algorithme Apparie_Mouv2 donne des résultats satisfaisants.

Nous pouvons constater que ni la variation de la valeur de l'angle d'inclinaison φ (voir tableau 1), ni celle de l'angle de rotation dt (voir Tableau 2), n'influe sur le taux de mise en correspondance. Cependant, la variation des positions initiales des caméras joue un rôle important dans la robustesse de l'algorithme. Nous avons remarqué que pour des caméras en mouvement de convergence, plus le champ de vision des deux caméras est grand plus l'algorithme est robuste. En effet, le tableau 1 montre que pour un champ de vision de 30°, l'algorithme fournit des résultats satisfaisant pour un bruit de l'ordre de 1 pixel alors que dans le tableau 2 où le champ de vision est de 90°, l'algorithme est robuste pour un bruit d'au plus 3 pixels.

tg=30°, td=60°, dt=3°						
	φ=5°			φ=3°		
Bruit δ (pixel)	[φmin,φmax]	nombre de correspondances correctes	nombre de correspondances erronées	[φmin,φmax]	nombre de correspondances correctes	nombre de correspondances erronées
0.1	[4 87-5 16]	10	0	[2.91-3 12]	10	0
0 3	[4 43-5 43]	10	2	[2.60-3.51	14	4
0.5	[4 90-4 99]	7	0	[2.94-2.99]	6	0
1	[4 66-4 92]	7	1	[2.86-3 01]	5	1

Tableau 1- Résultats d'expérimentation des algorithmes Apparie_Mouv2 avec variation de φ

tg=0°, td=90°, φ=5°						
	dt=5°			dt=3°		
Bruit δ (pixel)	[φmin,φmax]	nombre de correspondances correctes	nombre de correspondances erronées	[φmin,φmax]	nombre de correspondances correctes	nombre de correspondances erronées
0 1	[4 99-5 00]	10	0	[4 98-5 00]	10	0
0 5	[4 94-5 01]	10	0	[4 94-5.00]	10	0
1	[4 92-5 00]	10	2	[4 63-5 00]	11	1
3	[4 56-5 00]	10	2	[4 70-5.04]	9	1
4	[4 47-5 16]	8	5	[4 50-4 96]	9	0

Tableau 2- Résultats d'expérimentation des algorithmes Apparie_Mouv2 avec variation de dt

III.1.2. APPLICATION D' APPARIE_MOUV1 ET APPARIE_MOUV0

L'application de ces deux algorithmes aux mêmes données fournissent le même resultat. L'expérimentation montre à travers le tableau 3 et le tableau 4, que pour un seuil de tolérance $|\varphi\,max - \varphi\,min|$ de 5 degrés, ces deux algorithmes donnent des résultats satisfaisant pour un bruit de l'ordre de 0.1 pixels.

θg=30°, θd=60°, dt=3°						
	φ=5°			φ=3°		
Bruit δ (pixel)	[φmin,φmax]	nombre de correspondances correctes	nombre de correspondances erronées	[φmin,φmax]	nombre de correspondances correctes	nombre de correspondances erronées
0 001	[4 60-6 81]	24	0	[2.53-3 05]	24	0
0 05	[6 03-10 96]	24	0	[3 89-4 17]	16	8
0 1	21 64°	8	0	14 55°	8	0

Tableau 3- Résultats d'expérimentation des algorithmes Apparie_Mouv1 et Apparie_Mouv0 avec variation de j

$\theta g=0°, \theta d=90°, \varphi=5°$						
	dt=5°			dt=3°		
Bruit δ (pixel)	[φmin,φmax]	nombre de correspondances correctes	nombre de correspondances erronées	[φmin,φmax]	nombre de correspondances correctes	nombre de correspondances erronées
0.001	[4.23-7.35]	24	0	[4.72-7.52]	24	0
0.05	[5.10-8.08]	16	0	[16.21-20.01]	16	0
0.1	[14.24-17.15]	16	0	[8.56-13.73]	16	0

Tableau 4 Résultats d'expérimentation des algorithmes Apparie_Mouv1 et Apparir_Mouv0 avec variation de dt

CONCLUSION

Dans cette article nous avons apporté une contribution pour l'utilisation du mouvement de convergence des caméras dans le processus de vision stéréoscopique non anthropomorphe.

Nous avons présenté trois algorithmes d'appariement. Le premier suppose que les positions initiales des caméras sont connues; cette algorithme est un précurseur pour les autres, car il nous a permis de démontrer que l'appariement des trajectoires est possible dans un système de vision stéréoscopique non anthropomorphe. A travers l'algorithme Apparie_Mouv1, nous pouvons dire que nous avons pu réaliser un appariement de trajectoires sans connaissance préalable des modèles géométriques des cameras. Vient par la suite l'algorithme Apparie_Mouv0 qui nous montre que les trajectoires peuvent être exploitées pour la mise en correspondance de primitive sans connaissance du mouvement des caméras ni de leurs modèles géométriques.

BIBLIOGRAPHIE

[ANS88] : J.K.Aggarwal and N.Nandhakumar
On the computation of motion from sequences of images : a review
Proceedings of IEEE vol 76, N°8, August 1988.

[BOU88] : P.Bouthemy
Modeles et méthodes pou l'analyse du mouvement dans une séquence d'images,
TIPI, avril 1988.

[DF90] : R.Deriche and O.Faugeras
Tracking line segments
Image and vision computing, vol 8, N°4, Nov 1990.

[FAU88] : O.D.Faugeras
Quelques pas vers la vision artificielle en trois dimensions
T.S.I.-technique et science informatiques, vol.7, n°6, 1988.

[FM93] : O.Faugeras and S.Maybank
Motion from point matches : Multiplicity of solutions
International Journal of computer Vision, 4, 225-245, 1993

[FLT87] : O.Faugeras, F.Lustman, G.Toscani
Calcul du mouvement et de la structure à partir de points et de droites.
Rapport de Recherche Num 661, INRIA Mai 1987.

[GRI85] : W.E.L.Grimson
Computational experiments with a feature based stereo algorithm
IEEE Transaction Of Pattern Analysis and Machine Intelligenc Vol PAMI 7, Num 1, Janvier 1985.
[HOU62] : Hough
Méthod and means for recognizing complex patterns.
U.S.Patent 3069654, 18 decembre 1962.
[LAR93]: S.Larabi
Mise en correspondance d'une séquence d'images prise par une caméra en rotation
Rapport interne, Institut d'informatique USTHB, juin 1993.
[LJS9] : S.P.Lou and R.C.Jain
Motion detection in spatio-temporel space
CVGIP 45, 224-250, 1989.
[MAR82] : D.Marr
Vision : A computational investigation into humain representation and processing of visual information.
Freemann and Company San-Francisco, 1982
[MEY93] : F.Meyer
Suivi de régions et analyse des trajectoires dans une séquence d'images.
Thèse pour le titre de Docteur de l'université de Rennes I, Université de Rennes 1993
[OK85] : Y.Ohta et t.Kanade
Stereo by intera and inter-scanline search using a dynamic programming
IEEE Transaction Of Pattern Analysis and Machine Intelligenc Vol PAMI 7, Num 2, Mars 1985.
[[TH84] : G.Toscani
Systemes de calibration et perception du mouvement en vision artificielle
These de Docteur en science, université Paris-sud, 1987
[SUN92] : V.Sundareswaran
A fast methode to estimate sensor translation
Computer Vision-ECCV 92, secons European conference on computer vision,
Santa Margarita Ligure, Italy, may 19_22, 1992, Proceedings
[XR88] : M.XIE et P.RIVES
Un algorithme efficace pour la mise en correspondance des segments 2D dans une sequence d'images.
Rapport de recherche Num 929, INRIA, Nov 88

Analyse Microlocale et Tomographie Géométrique

Jean-Clarence NOSMAS

Laboratoire J.A. DIEUDONNE, URA 168,
Univ. de NICE-SOPHIA ANTIPOLIS, France
E-mail nosmas@math.unice.fr

Résumé. Nous montrons, dans ce travail comment, en utilisant des techniques d'Analyse Microlocale, il est possible de localiser les discontinuités d'une fonction f, connaissant les points où sa transformée de RADON n'est pas régulière.

Les résultats obtenus servent à justifier des algorithmes "géométriques" de reconstruction en Tomographie médicale 2D ou 3D que nous mettons au point actuellement.

Mots-Clefs : transformation de RADON, transformation de RADON atténuée, Tomographie, variété Lagrangienne, distribution lagrangienne, opérateur intégral de Fourier, support singulier, spectre singulier.

1. Introduction

Nous nous intéressons aux problèmes de reconstruction en Tomographie en vue d'applications à la Tomographie Médicale.

Rappelons que "reconstruire", dans ce contexte, c'est calculer f connaissant sa transformée de Radon Rf (que nous appellerons "sinogramme"). Cette opération d'inversion de R, quand on l'effectue numériquement, est coûteuse en temps de calcul et en place mémoire; il est donc intéressant de ne l'effectuer que quand c'est nécessaire ou bien d'imaginer des algorithmes de reconstruction plus efficaces.

Le but de ce qui suit est de montrer comment obtenir une information pertinente sur f directement à partir de Rf, ie sans reconstruction préalable. Cette information peut soit être utilisé pour un "diagnostic", soit être exploitée pour reconstruire f rapidement moyennant une hypothèse à priori sur la structure de f (constante ou affine "par morceaux" par exemple)

Les articles de J.P. THIRION sur la "Tomographie Géométrique" [7] et de V.P. PALAMODOV [5] sont à l'origine de ce travail.

Dans [7] on admet implicitement le résultat suivant de Tomographie 2D:

Si f est singulière sur une courbe C, en un sens que nous préciserons plus loin, Rf est singulière sur la courbe duale (les "points" de la courbe duale, sont les droites tangentes à la courbe C) et vice-versa.

En d'autres termes, il est théoriquement possible de localiser les singularités de f directement à partir de son sinogramme.

Ce résultat est en fait un cas particulier d'un énoncé plus général qui s'applique également en Tomographie 3D. Nous le démontrons à l'aide de techniques d'Analyse Microlocale, techniques "bien connues" des spécialistes (et d'eux seuls ?...) qui font appel aux notions de support, support singulier, spectre singulier ("wave front" en anglais),variétés et distributions Lagrangiennes, Opérateurs Intégraux de Fourier (OIF ou "Fourier Integral Operator" en anglais).

Ces techniques sont bien adaptées car les différentes transformations de RADON, la transformation de RADON "atténuée" et certaines transformations de RADON "généralisées" sont des OIF "classiques".

Notre but étant de valider des méthodes numériques plus géométriques de reconstruction ou de diagnostic, pour aller plus loin, il faut être capable de "lire" la structure de f sur son sinogramme. Suivant le modèle de fonction choisi, à priori, pour f, on saura ainsi quel type de singularité il faudra "detecter" sur le sinogramme.

Le cadre théorique de la "Transmission", introduit par A. HIRSCHOWITZ et A. PIRIOU [4] semble bien adapté à l'étude de ce problème. Il devrait permettre une approche systématique dans tous les cas d'intérêt pratique. Les résultats que nous avons obtenus dans cette direction font l'objet d'un travail en préparation et les algorithmes numériques qui en découlent sont en cours de test.

Le plan que nous avons adopté est le suivant:

Nous rappelons brièvement, dans le paragraphe 2, des résultats géométriques "classiques", implicites ou épars dans la littérature, concernant les Transformations de RADON usuelles. Nous renvoyons à [2],[3],[1],[6],[8] (liste non exhaustive...).

Dans paragraphe 3, nous explicitons, sur des exemples "génériques", quelques conséquences des résultats du paragraphe précédent.

2. Propriétés géométriques des transformations de RADON

Etant donnés les entiers n et k ($k < n$),on note $G(k,n)$ la Grassmanienne des k-plans (sous-espaces affines de dimension k) de $\mathbb{R}^n$.

Rappelons quelques propriétés de $G(k,n)$:

* C' est une variété algébrique lisse, réelle de dimension (k+1)(n-k); en effet, si l est un (n-k)-plan vectoriel, l'ouvert des k-plans transverses à l s'identifie aux applications affines de $\mathbb{R}^k$ dans $\mathbb{R}^{n-k}$;

** $G(k,n)$ est équipée d'une mesure positive dg qu'on peut décrire de la manière suivante :

tout k-plan p est l'image par une rotation r de $\mathbb{R}^n$ (ie un élément de $SO(n)$) d'un k-plan parallèle à $\mathbb{R}^k \times \{0\}$; ce dernier est caractérisé par son intersection m avec $\{0\} \times \mathbb{R}^{n-k}$;

On peut donc paramètrer $G(k,n)$ par $SO(n) \times \mathbb{R}^{n-k}$: $p = (r,m)$.

La mesure dg est définie par :

$$\int_{G(k,n)} u(p)dg := \int_{SO(n)\times \mathbb{R}^{n-k}} u(r,m)\, drdm$$

où dr désigne la mesure de Haar de $SO(n)$

Définition 2.1. *Etant donnés un entier k, une fonction f (resp u), C^∞ à support compact sur $\mathbb{R}^n$ (resp $G(k,n)$), , la transformée de RADON de f est la fonction sur G(k,n) définie par:*

$$R_k(f)(p) := \int_p f d\mu(p)$$

où $d\mu(p)$ désigne la mesure de Lebesgue sur p;

La transformée de RADON duale de u est la fonction sur $\mathbb{R}^n$ définie par:

$$R_k^* u(x) := \int_{SO(n)} u(p_x) dr$$

où p_x est le k-plan $x + r(\mathbb{R}^k \times \{0\})$

Remarque 2.1. *quand $k = 1$, on parle de transformation "en rayons X".*

Pour interpréter plus géométriquement cette définition, on introduit la sous-variété Z de $G(k,n) \times \mathbb{R}^n$

$$Z := \{(p,x) \in G(k,n) \times \mathbb{R}^n /\ x \text{ appartient à } p\ \}$$

et on note π (resp ω) la projection de Z sur $\mathbb{R}^n$ (resp sur $G(k,n)$).

Lemme 2.2. *π et ω sont des fibrations; en outre, π est propre .*

Proposition 2.1.

$$R_k(f) = \omega_*(\pi^* f)$$

$$R_k^*(u) = \pi_*(\omega^* u)$$

où π^* désigne l'opérateur "image réciproque" par π ($\pi^* f := f \circ \pi$) et ω_* l'opérateur "image directe" par ω, c'est à dire l'opérateur d'intégration sur les fibres de ω

Etant donnée une distribution T sur une variété X, on note supp sing(T) le support singulier de T ie l'ensemble des points de X au voisinage desquels T ne concide pas avec une fonction C^∞.

Si T est la distribution associée à une fonction "affine par morceaux", le support singulier de T est constitué par le "bord" des "morceaux" ie les points de discontinuité de la fonction ou de sa dérivée (au sens des distributions) si elle est continue.

Pour préciser la nature de l' "irrégularité" au voisinage d'un point du support singulier de T, on introduit la notion de Spectre Singulier de T (SS(T) ou "wave front set"):

brièvement, c'est un sous-ensemble fermé de T^*X, le fibré cotangent de X, qui se projette sur le support singulier (cf [1] pour un exposé détaillé élémentaire).

Par exemple, si C est une courbe (resp une surface) lisse dans $\mathbb{R}^2$ (resp $\mathbb{R}^3$) d'équation $\phi = 0$ et si $\{x \in \mathbb{R}^2 (resp \mathbb{R}^3) / \ \phi(x) \leq 0\}$ est compact, la distribution T associée à la fonction caractéristique de ce compact est une distribution *Lagrangienne classique* (cf plus loin) et

$$supp\ sing(T) = C \text{ et } SS(T) = N_0^* C.$$

La mesure μ portée par la courbe C $\mu := \phi^* \delta$ est un autr e exemple significatif de distribution lagrangienne dont le spectre singulier est égal à $N_0^* C$.

$N_0^* C$ est un "modèle" de sous-variété Lagrangienne; on sait associer à une telle sous variété Λ des classes de distributions, les *Distributions Lagrangiennes* , dont le spectre singulier est inclus dans Λ.

Les *Opérateurs Intégraux de Fourier* sont des opérateurs dont le noyau-distribution est une distribution lagrangienne.

La proposition 2.1 montre que le noyau-distribution Δ_k de R_k est la mesure "portée" par Z définie par:

$$< \Delta_k, u \otimes v >:= \int_Z u(p)v(x) dg \otimes dx_Z.$$

Comme conséquence nous avons la

Proposition 2.2. *R_k est un Opérateur Intégral de Fourier (OIF) elliptique associé à $N_0^* Z$, le fibré conormal de Z dans $G(k,n) \times \mathbb{R}^n$; cet opérateur se prolonge donc en un opérateur continu de $C_0^{-\infty}(\mathbb{R}^n)$, l'espace des distributions à support compact dans $\mathbb{R}^n$, dans $C^{-\infty}(G(k,n))$,l'espace des distributions sur G(k,n).*

En d'autres termes, nous avons :

$$supp(\Delta_k) = supp\ sing(\Delta_k) = Z \text{ et } SS(\Delta_k) = N_0^* Z$$

Il est important de savoir que le spectre singulier d'une distribution se "comporte bien" sous l'action des OIFs en ce sens que nous avons l'inclusion-CLEF suivante :

$$SS(R_k f) \subset (N_0^* Z)' \circ SS(f) \ (*)$$

où $\circ$ désigne la composition des relations et

$$(N_0^* Z)' = \{(p, q\,; x, \xi) / (p, q\,; x, -\xi) \in N_0^* Z\}$$

Pour certaines classes de distributions, in particulier celles qui nous intéressent ie les distributions Lagrangiennes classiques, cette inclusion est une égalité.

L'inclusion (*) contient l'essentiel de l'information géométrique pertinente; nous l'expliciterons en partie dans le paragraphe suivant.

Remarque 2.3. *On a pour* R_k^* *des énoncés "symétriques" avec* Z *remplacé par* $Z_s := \{(x,p) \in \mathbb{R}^n \times G(k,n) \;\; x$ *appartient à* $p\}$

En outre, $R_k^* R_k$ *est un opérateur de convolution sur* $\mathbb{R}^n$ *inversible (un opérateur pseudo-différentiel elliptique)*

Expérimentalement, l'architecture des scanners fait qu'on obtient toujours la restriction du sinogramme à des sous-variétés de $G(k,n)$, par exemple à la sous-variété des k-plans parallèles à une direction donnée ou à celle des k-plans passant par un point en géométrie conique.

la remarque suivante montre qu'il est possible d'extraire une information partielle "projection" par "projection".

Remarque 2.4. *Soit* S *une sous-variété de* $G(k,n)$ *et* i_S *l'injection de* S *dans* $G(k,n)$*, l'opérateur* i_S^* *de restriction à* S *est un OIF associé au fibré conormal* N_S^* *de* $\{ \; (i_S(x), x) \in S \times G(k,n) \; \}$ *dans* $S \times G(k,n)$ *.*

On a donc l'inclusion

$$SS(i_S^* u) \subset (N_S^*)' \circ SS(u)$$

3. Application à la Tomographie 2D ou 3D

Examinons le cas le plus simple où k = 1 et n = 2; dans ce cas, $G(1,2)$ est naturellement isomorphe à $\mathbb{R}^2$.

Théorème 3.1. *Soit* C *une courbe algébrique lisse et* f *une distribution à support compact dont le spectre singulier est contenu dans* N_0^*C *le fibré conormal à* C*.*

Si C^* *désigne la courbe duale de* C*, alors nous avons l'inclusion*

$$supp\; sing(Rf) \subset C^*$$

Remarque 3.2. *L'inclusion* $SS(f) \subset N_0^*C$ *implique (on projette sur* $\mathbb{R}^2$*),* $supp\; sing(f) \subset C$*.*

Il n'est pas nécessaire que C soit régulière pour pouvoir conclure; on peut remplacer cette hypothèse par une hypothèse de régularité sur N_0^*C.

Si C est régulière, N_0^*C est une sous-variété lagrangienne de $T_0^*\mathbb{R}^2$; nous avons alors la

Proposition 3.1. *Soit* Λ *une sous variété lagrangienne de* $T_0^*\mathbb{R}^2$*; l'ensemble* Σ *défini par :*

$$\Sigma := (N_0^* Z)' \circ \Lambda$$

est une sous-variété lagrangienne de $T_0^*G(1,2)$ *et*

$$\Lambda = (N_0^* Z_s)' \circ \Sigma$$

Remarque 3.3. *Si Λ est algébrique, Σ est algébrique et, si C (resp C^*) désigne la projection de Λ (resp Σ) sur $\mathbb{R}^2$ (resp $G(1,2)$), C et C^* sont des courbes algébriques (éventuellement singulières) duales l'une de l'autre.*

Dans cette situation, nous avons le

Théorème 3.4. *Si f est une distribution lagrangienne associée à Λ, Rf est une distribution lagrangienne associée à Σ (et vice versa) et, si $supp\ sing(Rf) = C^*$, alors $supp\ sing(f) = C$*

Que Λ soit algébrique nous permet d'utiliser des résultats de Géométrie Algébrique classique, en particulier de préciser la topologie du support singulier de f, connaissant celle de Rf (existence de cusps, points doubles etc..).

Remarque 3.5. *Les résultats précédents se généralisent donc au cas k=2 et n =3.*

Si Λ est simplement lisse, nous avons pour tout k et n le

Théorème 3.6. *Soit Λ comme dans la proposition 3.1, f est une distribution lagrangienne associée à Λ et C^* le support singulier de Rf .*

Si C désigne l'enveloppe des k-plans de C^* (on suppose qu'une telle enveloppe existe) alors

$$supp\ sing(f) \subset C$$

Il est possible de faire une étude similaire pour les transformations de Radon atténuées en supposant la fonction d'atténuation régulière(la géométrie est la même) , mais en pratique, les modèles choisis pour cette fonction sont des modèles discontinus ou peu réguliers. Il est donc nécessaire pour tenir compte de cette "contrainte expérimentale" de travailler avec des OIFs "à symboles singuliers" étudiés par R.B. MELROSE et G.A. UHLMANN, ou bien, pour un modèle particulier de fonction d'atténuation (constante par morceaux par exemple) de faire une étude directe.

References

1. Chazarain-Piriou : *Introduction à la théorie des Equations aux Dérivées Partielles linéaires,* Gauthier-Villars 1981.
2. Guillemin-Sternberg : *Geometric Asymptotics,* Math. Surveys and Monographs, Amer. Math. Soc. 1977.
3. Helgason :*Groups and geometric analysis,* Academic Press 1984.
4. Hirschowitz-Piriou : *Propriétés de Transmission pour les Distributions Intégrales de Fourier,* Comm in P.D.E.4(2),113-217 1979.
5. Palamodov : *Some singular problems in Tomography,* Translations of math. monographs, vol. 81 p. 123 -140. Amer. Math. Soc. 1990.
6. Quinto : *The dependance of the generalised Radon Transform on defining measures,* Trans. Amer. Math. Soc. 257, 1980, p. 331-346.

7. Thirion : *Segmentation of tomographic data without Segmentation* IEEE transactions of Medical Imaging 1992.
8. Rouvière : *Transformations de RADON*, Cours de DEA, Université de Nice-Sophia Antipolis 1994-1995.

An Efficient Wavelet-based Facial Edge Detection Approach in Model-based Coding*

Jian-Feng Liu John Chung-Mong Lee

Department of Computer Science
Hong Kong University of Science and Technology, Hong Kong

Abstract. Model-based coding is a new data compression scheme for very low bit rate image transmission. It consists of image analysis, parameter transmission, and image synthesis, in which edge detection based facial feature estimation is probably one of the most essential process. As many of the existing edge detection methods are not able to produce a satisfying and efficient result for this specific coding application, we are proposing herewith a new wavelet -based facial edge detection approach. Our approach applies the coarse-to-fine strategy using discrete wavelet transform for distinct facial edge feature detection, and extract the local modulus maxima of a continuous wavelet transform as the edge candidates by choosing a suitable continuous wavelet. In terms of speed and quality of the output when used for model-based coding, experiments show, our method outperforms many of the existing edge detection approaches, such as Prewitt's method, Canny's method, Marr-Hildreth's method, etc. It makes full use of the multiresolution properties of wavelet transform and can be used in conjunction with other methods for extremely low bit rate transmission of videophone pictures.

Key Words: Wavelet, Edge detection, Model-based coding, Multiresolution

1 Introduction

Model-based coding is a new emerging technique for data compression of image sequences based on the "a priori" information of the scene content. The term "model based" refers to the use of models of the objects in the scene for the reconstruction of the input images at the receiving end. The method offers the advantage of extremely low bit rate because only the necessary parameters to animate the model are transmitted.

The principle idea of a model-based coding scheme is shown in Fig. 1. The object of a video sequence is analysed using computer vision or image processing techniques to yield parameter information about the shape, location and motion. This parameter information is employed to synthesize a model of the objects by using computer-graphics methods. Tracking techniques are used to make the model

*. This research was supported by Hong Kong Telecom Institute of Information Technology under grant HKTIIT93/94.EG02

mimic the movements of the object it represents. The parameters obtained from image analysis are then coded and transmitted to the receiver which will reconstruct the model.

The success of a model-based coding scheme depends on the accurate parameterisation of the original objects in the scene (image analysis) and the ability to reconstruct sufficiently realistic images at the receiver (image synthesis). Besides tremendous work in the area of image synthesis, a more difficult task is the detection, location and description of the parameters of the objects in this image, where facial edge detection is most essential. In fact, to estimate facial features, we have to match a synthetic model with the result of edge detection applied on the original picture. In our experiments, the CANDIDE face model is used. In order to estimate distinct facial features, we leave on the model only those parts that correspond to facial features.

As for facial edge detection, by now, much effort has contributed to the detection of image edges, and some of the approaches can indeed yield a satisfying output. However, for the detection of facial image edges, there exist some special problems that need to be coped with:

a. The algorithm has to be simple and fast;

b. The illumination of the faces in videophone pictures is usually inhomogeneous which may result in some unwanted edges that are hard to delete.

c. Only the obvious facial features such as the edges of eyes, nose, mouth, ears, etc. need to be detected.

The approaches in reference [1]-[8] either are much time consuming or can not produce a satisfying facial edge detection result because no effective step is taken to eliminate the effect of illumination without deteriorating detection quality. Therefore, in the following sections, we propose an efficient facial edge detection approach that can overcome the above problems. Our proposal is based on the multiresolution representation of discrete wavelets and the multiscale edge detection of continuous wavelets.

The following sections are organized as follows. Section 2, which is the most important part of the paper, addresses the theory of wavelets and the detailed facial edge detection method using wavelets; Section 3 briefly addresses the corresponding facial feature matching and estimation method which obtains the parameters for transmission; Section 4 then shows some experimental results using the proposed facial edge detection approach in comparison with other edge detection methods, and finally section 5 concludes the paper.

2 Wavelet Based Facial Edge Detection

In this section, first, we'll discuss the theory of multiresolution representation using discrete wavelets; then we'll introduce the continuous wavelet transform which is used as a powerful edge detection tool; finally, we'll present the detailed facial edge detection method using both discrete and continuous wavelets.

2.1 Multiresolution Representation Using Discrete Wavelets and the Continuous Wavelets as Edge Detector

Recently, wavelet has attracted many people's attention. Wavelet transform has become well known as useful tools for various signal processing applications [11] - [14]. As an important application, a multiresolution representation of discrete wavelets provides a simple hierarchical frame work for interpreting and processing the image information. In some sense, the information of the image at a coarse resolution can not only provide the "context" of the image, but also eliminate some unwanted details such as noise or irregular intensity, whereas the finer details of the image correspond to the particular "modalities". It is therefore natural to first analyse the image information at a coarse resolution and then increase the resolution. This is called a coarse-to-fine processing strategy. At a coarse resolution, the image information is characterized by very few samples, and most noisy details such as those caused by non-uniform illumination are mostly eliminated. Hence, the coarse information processing can be performed quickly and satisfactorily. The finer details are characterized by more samples, but the prior information, derived from the context, constrains the computations and thus speed up the process. With a coarse-to- fine strategy, we can detect not only the necessary facial features in spite of inhomogeneous illumination, but also process the minimum amount of details which are sufficient to perform a detection task.

To perform the multiresolution representation of facial images, we apply the following fast discrete wavelet transform algorithm based on Mallat's[13].

Suppose $\{V_j\}$ is a given multiresolution analysis, Φ is the corresponding scaling function, $\Phi_{J,K} = \sqrt{2^J}\,\Phi\left(2^J x - k\right)$, $f \in V_J$ (J is a known integer) is a random signal, $C_{J,K}$ is the discrete coefficients, then we have:

$$f(x) = \sum_{k \in Z} C_{J,k} \cdot \Phi_{J,k}(x) \qquad (1)$$

The Fourier transform of Φ is characterized by:

$$\hat{\Phi}(\omega) = \prod_{p=1}^{\infty} H\left(e^{-i2^{-p}\omega}\right) \qquad (2)$$

where $H\left(e^{-i\omega}\right)$ is the transfer function of a discrete filter that satisfies:

$$\left|H\left(e^{-i\omega}\right)\right|^2 + \left|H\left(-e^{-i\omega}\right)\right|^2 = 1 \qquad (3)$$

The discrete filters $H = \left(h_n\right)$ $n \in Z$ whose transfer function satisfies (3) are called Quadrature Mirror Filters. Let $C_{0,k}$ represent the intensity of a discrete image, them after some derivation, we have:

$$C_{J,k} = C_{J+1,k} \bullet (H, H) \qquad (J=-1....-N) \qquad (4)$$

where $C_{J,k}$ is the approximation of a signal at a resolution, $C_{0,k}$ is the operation of convolving each row of C with H, and down sampling each row at the rate of 2:1, then convolving each column of the above result with H, and down sampling each column at the same rate. By applying equation (4) recursively, we can get the information at different resolution.

Now, let's address the next problem of edge detection. Presently. there exists a difficult problem for edge detection, i.e., a trade-off between localization and detection. On the one hand, a small edge mask (fine resolution) is sensitive to noise and sometimes detects redundant details, but it has a high localization precision. On the other hand, a large edge mask (coarse resolution) is robust to noise, but its localization precision is low. To extract the necessary edge information with high localization precision, we can make use of the multiscale property of wavelet transform.

As can be seen in reference [8], for a particular class of continuous wavelet, the image edges correspond to the local maxima of the wavelet transform. A maxima detection of the wavelet transform is equivalent to a classical edge detection. Here, to adapt to the human vision behaviour, we choose the first order derivative of Gaussian function as the continuous wavelet to detect facial edges:

$$\Psi(x) = 1/\left(\sqrt{2\pi} \cdot \sigma^3\right) \cdot \exp\left(\left(-x^2\right)/\left(2 \cdot \sigma^2\right)\right)$$

Since multiresolution representation has given us the multiresolution images of different image size (by downsampling). Therefore, through convolving these images of different sizes with the same continuous wavelet, we can acquire the equivalent multiscale edge detection results and this multiscale information enables us not only to denoise efficiently, but also to locate edges accurately.

2.2 Facial Edge Detection

The basic wavelet-based facial edge detection procedure is as follows:

Firstly, generate the multiresolution representation of facial images using discrete wavelet according to equation (4);

Secondly, at a certain coarse resolution, detect facial edges by extracting the local modulus maxima of the continuous wavelet transform. Choose a threshold properly (it's talked about later again in Section 4), so that only the distinct facial features are detected;

Thirdly, go to the next finer resolution. For each edge candidate detected at the previous coarser resolution, construct a wavelet transform mask over its corresponding area at the current resolution, compute and extract the edges with the same approach illustrated in Step 2.

Finally, if the current resolution is not the finest one (corresponding to original image), go back to step 3, otherwise, output the finer edge detection results and terminate.

Since many noisy details such as those caused by non-uniform illumination are eliminated at the coarse resolution, and the wavelet is first convolved with a coarse resolution representation (equivalent to a large edge mask) that can denoise efficiently, therefore, directed by the above processing result, the final output produced from the finest resolution representation not only has a high localization precision, but also can detect as many details as possible of only distinct facial features.

3 Facial Feature Matching and Estimation

After facial features are detected, we can apply the chamfer matching method to match facial features. By placing the model into different positions on the distance map produced from detected facial edges and calculating the average of the corresponding values, a measure of the goodness of fit is obtained which is called "edge distance". The position which minimizes the edge distance corresponds to the best fit of the model.

Having found the position of the minimum edge distance, a new search should be performed concerning the scale factor and the rotation angle which reduces the minimum distance even further.

Finally, we initiate an individual search for the left and right eye, the nose and the mouth within a small area of 10 pixels to refine the best matching position. Because the facial features have been clearly extracted in the edge detection stage, a very good estimation of the mouth, the nose and the eyes can usually be obtained.

4 Experimental Results

We implemented the proposed approach on the standard Miss America sequence, Clare sequence, and Lena image.

In the experiment, we choose Harr wavelet for the purpose of multiresolution representation, and we choose $\sigma = 1$ for the Gaussian wavelet. To adapt to different images, we adopt an adaptive thresholding method with the threshold changes proportional to the average of the wavelet transform maxima of the whole facial image. Through experimenting, we decided that only two neighbouring resolution levels are needed to eliminate the effect of non-uniform illumination while still keeping the obvious facial features to be detected.

Fig. 2 and 3 show our facial edge detection results of one frame of Miss America sequence and Clare sequence, and the detection result of Lena Image whose face is in a partial side view direction. Fig. 3 also shows some facial edge detection results using the conventional Prewitt's approach, mathematical morphology, Canny's approach and Mar-Hildreth's approach.

From Fig. 2 and Fig. 3, we can see although the faces of Miss America, Clare and Lena are non-uniformly illuminated, after applying our method, no unwanted edges appear on their faces while all the necessary distinct facial features remain. In contrast, other methods mentioned above either produce noisy edges because of irregular brightness, or some distinct edge features are missed by choosing a large threshold to remove those noisy edges caused by nonuniform illumination. Moreover, our edge detection process is able to complete only in several seconds on the 80486 personal computer, which is about two times or more faster than other methods. The above merits of our approach, as we have found, are much favourable for further model-based coding.

5 Conclusion

This paper has proposed an efficient edge detection approach for facial feature estimation based on continuous wavelets and multiresolution representation of images through discrete wavelets. Since this method is not only very fast and simple, but also much more effective than other methods, it represents a new foundation for further facial feature estimation and hence helps to achieve extremely low bit rate model-based coding. In fact, aside from facial edge detection in model-based coding, the proposed method is also suitable for those cases where the edge detection process needs not only to be fast, but also to be able to detect mainly distinct edge features. Of course, to make the proposed approach more widely applicable, the choice of an appropriate discrete wavelet is critical. The issue of how to select a discrete wavelet so that its representation can not only denoise efficiently but also retain the necessary detail, is a subject that we would like to look into more detail in the future

6 References.

[1] Prewitt, J.M.S, "Object Enhancement and Extraction. In Picture Processing and Psychopictorics," B.S. Lipkin and A. Rosenfeld (Eds.), New York: Academic Press,1970.

[2] Haralick, R.M., stermberg, S.R., and Zhuang, X., "Image Analysis Using Mathematical Morphology,". IEEE Trans. PAMI, Vol. 9, pp. 532-550, 1987.

[3] Canny, J, "A Computational Approach to Edge Detection," IEEE Trans. PAMI, Vol.8, pp. 679-698, 1986.

[4] D.Marr and E.Hildreth, "Theory of Edge Detection," Proc. R. Soc.Lond. B, Vol. 207, pp.187-217, 1980.

[5] Roberts, L.G, "Machine Perception of Three-Dimensional Solids," In Optical

and Electro-Optical Information Processing, J.P. Tippett et al. (Eds.), Cambridge, MA: MIT Press, 1965.
[6] Atkinson, J, Campbell F.W, "The Effect of Phase on the Perception of Compound Gratings. Vision Res," Vol. 14, pp. 159-162, 1974.
[7] Voorhees, H. and T. Poggio, "Detecting Textons and Texture Boundaries in Natural Images," Proc. First. Internat. Conf. on Computer Vision, London, England, pp. 250-258, June 1987.
[8] Liu Jian Feng, Qi Fei Hu, "A Wavelet Transform Approach to Image Edge Detection and Denoising," ICNNSP'95, to be held in Nan Jing.
[9] Liu Jian Feng, Qi Fei Hu, "A Wavelet Transform Approach to Image Spacial Filtering," Journal of Shanghai Jiao Tong University, Vol. 6, 1995 (to be published).
[10] Pearson, D. E., "Developments in Model-Based Video Coding," Proceedings of the IEEE. Vol.83, No. 6, pp. 892-906, 1995.
[11] M.Vetterli and C.Herley, "Wavelets and Filter Banks: Theory and Design," IEEE Trans. SP, Vol. 40, pp. 2207-2232, Sept. 1992.
[12] S.G.Mallat, "A Theory for Multiresolution Signal Decomposition: The Wavelet Transform," IEEE Trans. PAMI, Vol.11, No.7, pp. 674-693, 1989.
[13] Barrow, H.G., Tenenbaum, J.M., Bolles, R.C., and Wolf,H.C., "Parametric Correspondence and Chamfer Matching: Two New Techniques for Image Matching," Proc. 5th Int. Joint Conf. on Artificial Intelligence, Cambridge, DSA, pp. 659-663, 1977.

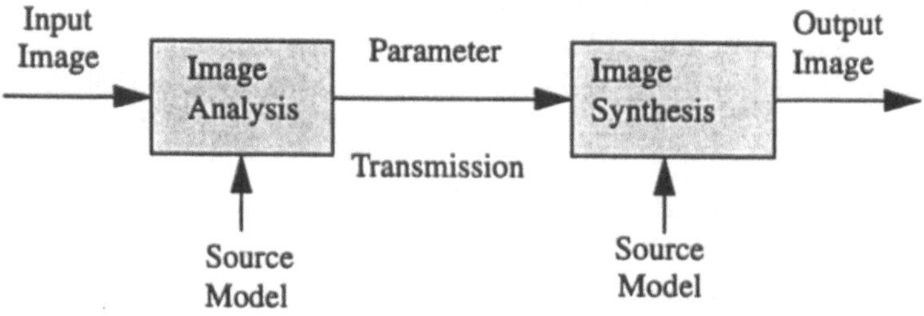

Fig. 1 Block diagram of model-based image coding systems.

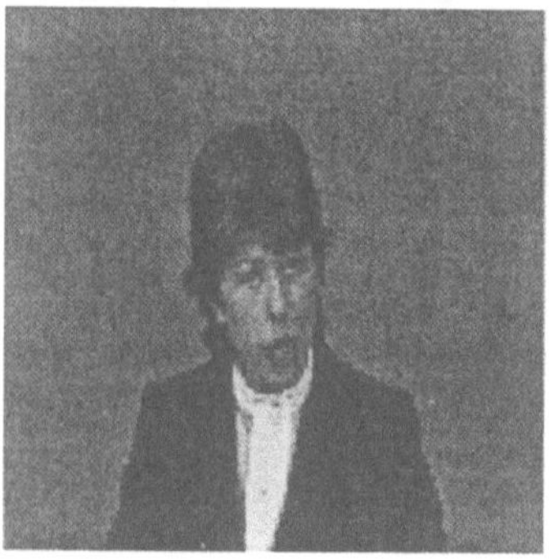

(a)

(b)

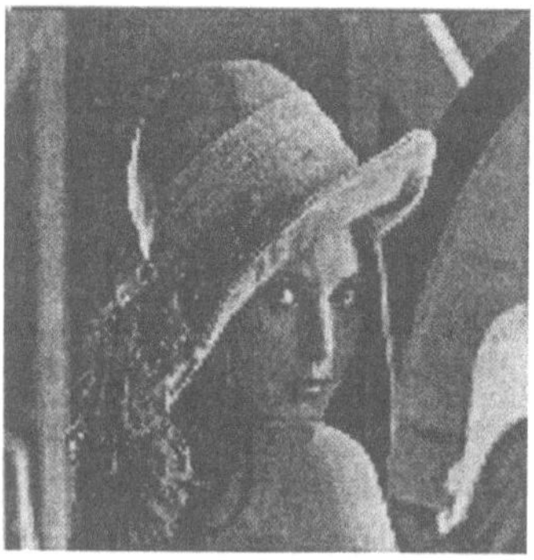

(c)

Fig. 2 Original image. (a) The original Clare image (at resolution 1); (b) The original Miss America image (at resolution 1); (c) The original Lena image (at resolution 1).

(a)

(b)

(c)

(d)

(e)

(f)

(m)

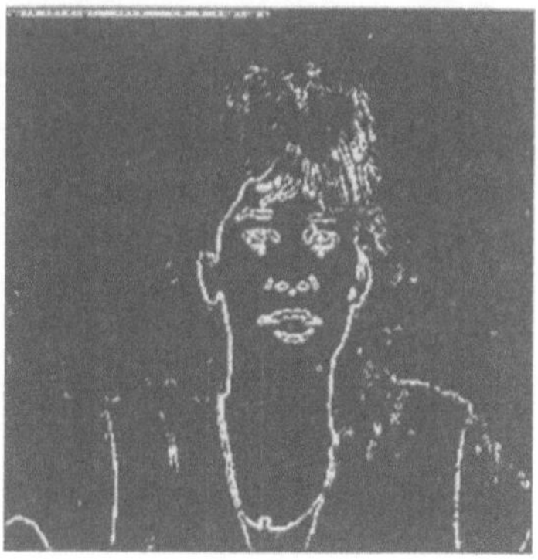

(n)

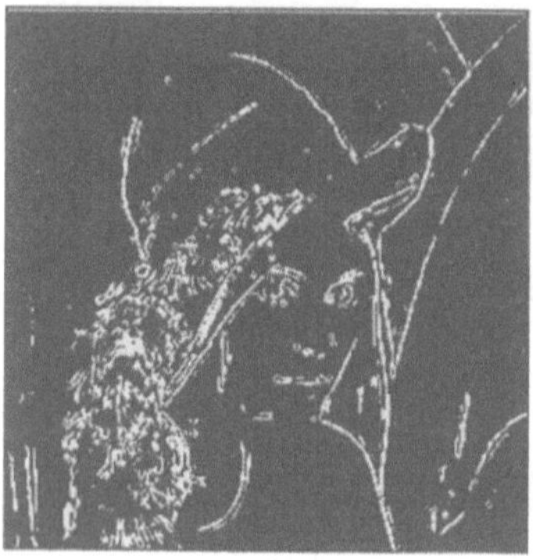

(o)

Fig. 3 (a), (b), (c) The corresponding facial edge detection results using the proposed approach; (d), (e), (f) The corresponding facial edge detection results using Prewitt's approach; (g), (h), (i) The corresponding facial edge detection results using mathematical morphology; (j), (k), (l) The corresponding facial edge detection using Canny's approach; (m), (n), (o) The corresponding facial edge detection using Marr-Hildreth's approach.

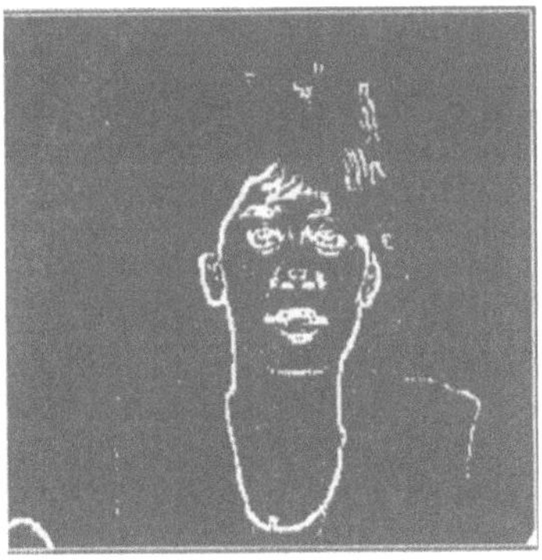

(g) (h)

(i) (j)

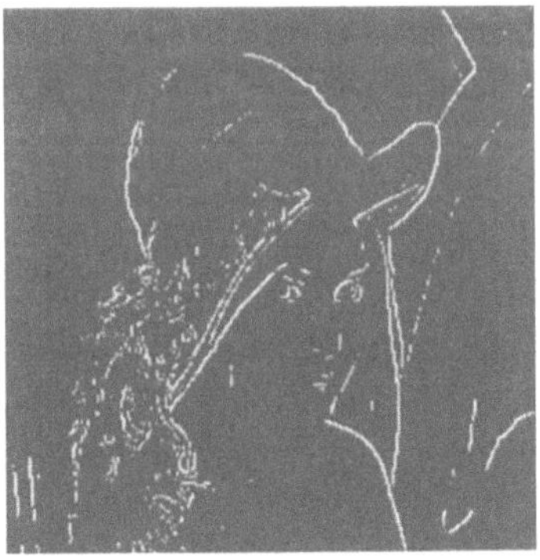

(k) (l)

A Kanizsa programme

V. Caselles[1], B. Coll[1], and J.M. Morel[2]

[1] Dpt. of Mathematics, Univ. Illes Balears. Ctra. Valldemossa km 7.5, Palma de Mallorca. Spain, dmivca0@ps.uib.es, dmitcv0@ps.uib.es
[2] CEREMADE, Univ. Paris-Dauphine 75775 Paris Cedex 16, France, morel@paris9.dauphine.fr

Summary. Based on the phenomenological description of Gaetano Kanizsa, we discuss the physical generation process of images as a combination of basic operations: occlusions, transparencies and constrast changes. These operations generate the essential singularities, which we call junctions. We deduce a mathematical and computational model to detect the "atoms" of the image: level lines joining T- or X-junctions. Then we propose the adequate modification of morphological filtering algorithms so that they smooth the "atoms" without altering the junctions. Finally, we give some experiments on real and synthetic images.
Key words: Kanizsa model, occlusions, transparencies, junctions, basic operations, level lines, morphological filtering, topographic maps.

1. Introduction and main results.

In this work, see also [2], we describe the physical generation process of images as a combination of occlusions, transparencies and contrast changes, in contrast with the clasical approach where the edges are frequently considered as the basic objects, [5]. We shall, following the psychologist and gestaltist Gaetano Kanizsa, [3], [4], define two basic operations for image generation: occlusion and transparency, according to which visual perception tends to remain stable with respect to these operations by detecting several kind of singularities which we call junctions. We intend to show that only "pieces of level lines of the image joining junctions" are the "atoms" of visual and artificial perception, and the main reason why level lines appear central is that they contain all of the image information invariant with respect to contrast changes.

The algorithm: Experimental Kanizsa programme

We call basic objects a class of mathematical objects, simpler to handle than the whole image but into which any image can be decomposed and from which it can be reconstructed. The main consequence in our analysis is that basic objects are all junctions of level lines and parts of level lines joining them, where junctions are in general the points of the image plane such that two level lines, with different levels, meet.

Before give the algorithm to decompose the image into the basic objects, we shall define the main junctions or singularities we consider.

Definiton of the main singularities. A *T-junction* at a point x where two level lines meet at point x and the half level lines (or branches) starting from x are four in number; two coincide and two take opposite directions. In the transparency phenomenon, the apparent crossing, which we call *X-junction*, consists in the meeting at a point x of two level lines, which locally create four angular regions.

The following junction detection algorithm, parameter-free except for two fixed thresholds eliminating quantization effects in space and grey level, will allow detect the basic perceptual-physical events discussed above in digital images.

Junction Detection Algorithm

• Fix an area threshold n and a grey level threshold b (in practice, $n = 40$ and $b = 2$).

• At every point x where two level lines meet: define $\lambda_0 < \mu_0$ the minimum and maximum value of u in the four neighboring pixels of x. We denote by L_λ the connected component of x in the set $\{y, u(y) \leq \lambda\}$ and by M_μ the connected component of x in the set $\{y, u(y) \geq \mu\}$. Let $\lambda_1 = min\{\lambda | \lambda \geq \lambda_0 | area(L_\lambda) > n\}$. Let $\mu_1 = max\{\mu | \lambda_1 \leq \mu \leq \mu_0 | area(M_\mu) > n\}$. If λ_1 and μ_1 have been found, if $\mu_1 - \lambda_1 \geq 2b$ and if the set $A = \{y, \mu_1 - b \geq u(y) \geq \lambda_1 + b\}$ has a connected component containing x with area larger than n, then retain x as a valid junction. If, in addition, the set A has a second connected component containing x (meeting the first one only at x) with area larger than n, then retain x as a possible X-junction.

After this segmentation of the image we propose the following adequate modification of morphological filtering algorithms in order to smooth the "atoms" without altering the junctions. Based on the Affine Morphological Scale Space (AMSS) model, proposed in [1], where the scale space can be modelized by a diffusive partial differential equation, we consider an approximation of this model, by fixing a neighborhood of the junction points (see [2]). The interpretation of the (AMSS) model is given in [1] and corresponds to an evolution equation for curves, proposed independently in [6] called Affine Scale Space. In order to discretize our considered model, we have adapted an very easy and most invariant numerical scheme (due to Luis Alvarez and Frederic Guichard) for (AMSS).

References

1. Alvarez L., Guichard F., Lions P.L. and Morel J.M. (1993): Axioms and fundamental equations of image processing, Report 9216, 1992 CEREMADE. Univ. Paris-Dauphine. Arch. for Rat. Mech. **16**, IX, pp. 200-257.
2. Caselles V., Coll B. and Morel J.M. (1995): A Kanizsa programme, Report CEREMADE. Univ. Paris-Dauphine.
3. Kanizsa G. (1979): Organization in Vision. N.Y. Praeger.
4. Kanizsa G. (1979): Grammatica del vedere, Il Mulino, Bologna.
5. Marr D. (1982): Vision, N. York, W.H. Freeman and Co.
6. Sapiro G. and Tannenbaum A. (1994): On affine plane curve evolution. Journal of Functional Analysis, **119**, 1, pp. 79-120.

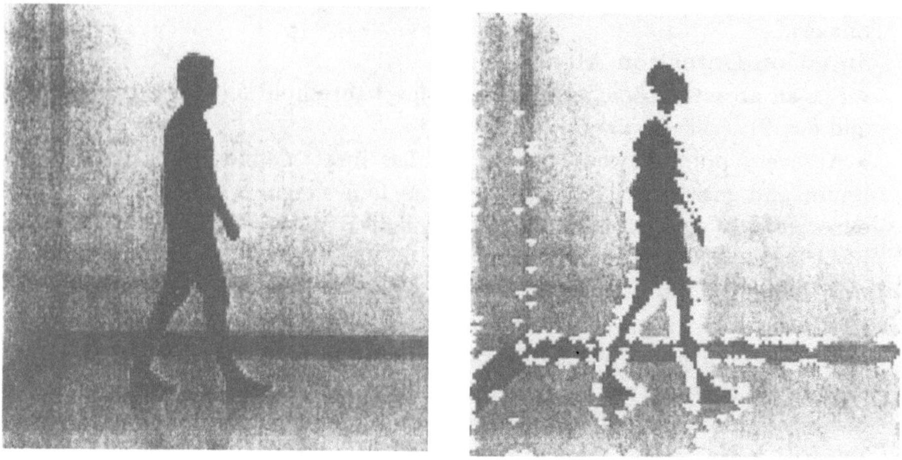

Experiment 1. Choice of the threshold for the Junction Detection Algorithm. On the left, the original image. On the right, the junctions detected after aplying the algorithm with threshold area n=40 and the grey level threshold b=2.

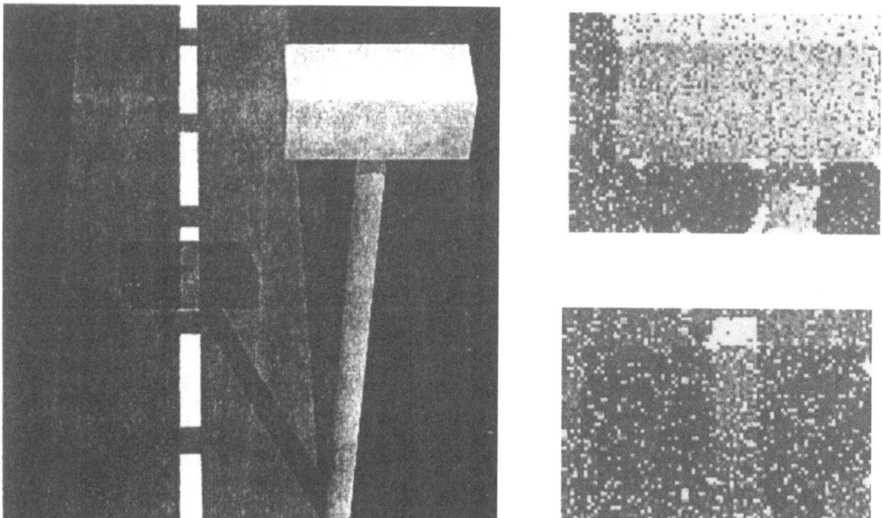

Experiment 2. Sensibility to noise. The big image is an original synthetic image. The other two are two details of the big one. The detail images have 40% of the pixels destroyed and replaced by an uniformly distributed random value between 0 and 255. In white, one can see the obtained T-junctions and X-junctions.

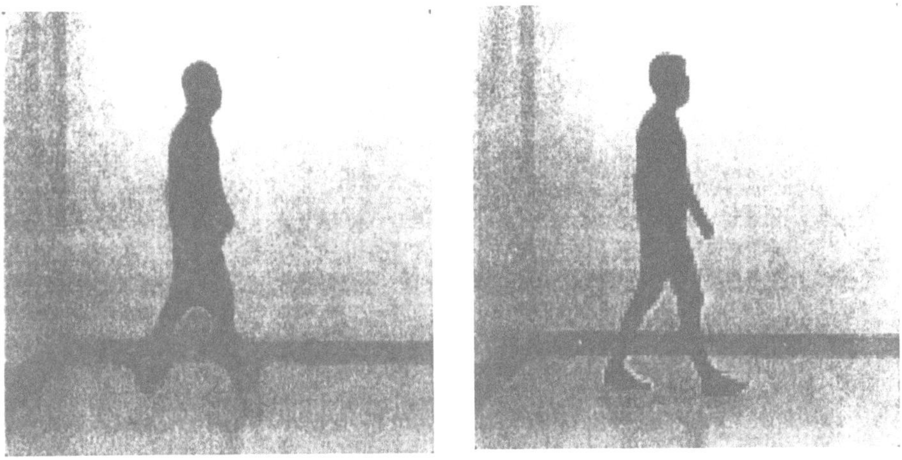

Experiment 3.Morphological filtering preserving Kanizsa singularities on a real image. On the left, we have applied a morphological filtering to original image of Experiment 1 at scale 30. On the right, application of the same morphological filtering, but fixing the junctions after applying the Junction Detection Algorithm, with an area threshold n=40 and contrast threshold b=2.

Experiment 4. Topographic maps. The morphological filtering applied by fixing the junctions computed from the Junction Detection Algorithm, simplifies the shape of level lines and permits to better visualize the whole topographic organization of level lines and junctions. On the left, the level lines of the original image alredy analysed in Experiment 1 displayed for all even levels. On the right, the level lines of the right image of Experiment 3, that is, after applying the morphological filtering by fixing the junctions, displayed for all even levels.

Lecture Notes in Control and Information Sciences

Edited by M. Thoma

1992–1996 Published Titles:

Vol. 186: Sreenath, N.
Systems Representation of Global Climate Change Models. Foundation for a Systems Science Approach.
288 pp. 1993 [3-540-19824-5]

Vol. 187: Morecki, A.; Bianchi, G.; Jaworeck, K. (Eds)
RoManSy 9: Proceedings of the Ninth CISM-IFToMM Symposium on Theory and Practice of Robots and Manipulators.
476 pp. 1993 [3-540-19834-2]

Vol. 188: Naidu, D. Subbaram
Aeroassisted Orbital Transfer: Guidance and Control Strategies
192 pp. 1993 [3-540-19819-9]

Vol. 189: Ilchmann, A.
Non-Identifier-Based High-Gain Adaptive Control
220 pp. 1993 [3-540-19845-8]

Vol. 190: Chatila, R.; Hirzinger, G. (Eds)
Experimental Robotics II: The 2nd International Symposium, Toulouse, France, June 25-27 1991
580 pp. 1993 [3-540-19851-2]

Vol. 191: Blondel, V.
Simultaneous Stabilization of Linear Systems
212 pp. 1993 [3-540-19862-8]

Vol. 192: Smith, R.S.; Dahleh, M. (Eds)
The Modeling of Uncertainty in Control Systems
412 pp. 1993 [3-540-19870-9]

Vol. 193: Zinober, A.S.I. (Ed.)
Variable Structure and Lyapunov Control
428 pp. 1993 [3-540-19869-5]

Vol. 194: Cao, Xi-Ren
Realization Probabilities: The Dynamics of Queuing Systems
336 pp. 1993 [3-540-19872-5]

Vol. 195: Liu, D.; Michel, A.N.
Dynamical Systems with Saturation Nonlinearities: Analysis and Design
212 pp. 1994 [3-540-19888-1]

Vol. 196: Battilotti, S.
Noninteracting Control with Stability for Nonlinear Systems
196 pp. 1994 [3-540-19891-1]

Vol. 197: Henry, J.; Yvon, J.P. (Eds)
System Modelling and Optimization
975 pp approx. 1994 [3-540-19893-8]

Vol. 198: Winter, H.; Nüßer, H.-G. (Eds)
Advanced Technologies for Air Traffic Flow Management
225 pp approx. 1994 [3-540-19895-4]

Vol. 199: Cohen, G.; Quadrat, J.-P. (Eds)
11th International Conference on Analysis and Optimization of Systems Discrete Event Systems: Sophia-Antipolis, June 15–16–17, 1994
648 pp. 1994 [3-540-19896-2]

Vol. 200: Yoshikawa, T.; Miyazaki, F. (Eds)
Experimental Robotics III: The 3rd International Symposium, Kyoto, Japan, October 28-30, 1993
624 pp. 1994 [3-540-19905-5]

Vol. 201: Kogan, J.
Robust Stability and Convexity
192 pp. 1994 [3-540-19919-5]

Vol. 202: Francis, B.A.; Tannenbaum, A.R. (Eds)
Feedback Control, Nonlinear Systems, and Complexity
288 pp. 1995 [3-540-19943-8]

Vol. 203: Popkov, Y.S.
Macrosystems Theory and its Applications: Equilibrium Models
344 pp. 1995 [3-540-19955-1]

Vol. 204: Takahashi, S.; Takahara, Y.
Logical Approach to Systems Theory
192 pp. 1995 [3-540-19956-X]

Vol. 205: Kotta, U.
Inversion Method in the Discrete-time Nonlinear Control Systems Synthesis Problems
168 pp. 1995 [3-540-19966-7]

Vol. 206: Aganovic, Z.;.Gajic, Z.
Linear Optimal Control of Bilinear Systems with Applications to Singular Perturbations and Weak Coupling
133 pp. 1995 [3-540-19976-4]

Vol. 207: Gabasov, R.; Kirillova, F.M.; Prischepova, S.V.
Optimal Feedback Control
224 pp. 1995 [3-540-19991-8]

Vol. 208: Khalil, H.K.; Chow, J.H.; Ioannou, P.A. (Eds)
Proceedings of Workshop on Advances in Control and its Applications
300 pp. 1995 [3-540-19993-4]

Vol. 209: Foias, C.; Özbay, H.; Tannenbaum, A.
Robust Control of Infinite Dimensional Systems: Frequency Domain Methods
230 pp. 1995 [3-540-19994-2]

Vol. 210: De Wilde, P.
Neural Network Models: An Analysis
164 pp. 1996 [3-540-19995-0]

Vol. 211: Gawronski, W.
Balanced Control of Flexible Structures
280 pp. 1996 [3-540-76017-2]

Vol. 212: Sanchez, A.
Formal Specification and Synthesis of Procedural Controllers for Process Systems
248 pp. 1996 [3-540-76021-0]

Vol. 213: Patra, A.; Rao, G.P.
General Hybrid Orthogonal Functions and their Applications in Systems and Control
144 pp. 1996 [3-540-76039-3]

Vol. 214: Yin, G.; Zhang, Q. (Eds)
Recent Advances in Control and Optimization of Manufacturing Systems
240 pp. 1996 [3-540-76055-5]

Vol. 215: Bonivento, C.; Marro, G.; Zanasi, R. (Eds)
Colloquium on Automatic Control
240 pp. 1996 [3-540-76060-1]

Vol. 216: Kulhavý, R.
Recursive Nonlinear Estimation: A Geometric Approach
244 pp. 1996 [3-540-76063-6]

Vol. 217: Garofalo, F.; Glielmo, L. (Eds)
Robust Control via Variable Structure and Lyapunov Techniques
336 pp. 1996 [3-540-76067-9]